Conceptual Design

Mogens Myrup Andreasen
Claus Thorp Hansen · Philip Cash

Conceptual Design

Interpretations, Mindset and Models

Springer

Mogens Myrup Andreasen
Department of Mechanical Engineering
Technical University of Denmark
Kongens Lyngby
Denmark

Philip Cash
Department of Management Engineering
Technical University of Denmark
Kongens Lyngby
Denmark

Claus Thorp Hansen
Department of Mechanical Engineering
Technical University of Denmark
Kongens Lyngby
Denmark

ISBN 978-3-319-19838-5 ISBN 978-3-319-19839-2 (eBook)
DOI 10.1007/978-3-319-19839-2

Library of Congress Control Number: 2015941342

Springer Cham Heidelberg New York Dordrecht London

Printed on acid-free paper

Springer International Publishing AG Switzerland is part of Springer Science+Business Media
(www.springer.com)

Foreword

Judging by the cover, this book is about conceptual design. And indeed if you are looking for a comprehensive system-level perspective of conceptualization as a driver for successful product development this is a book you should read. It presents conceptualisation methods used in design and product development and clearly underlines its importance in the context of the design process and well developed products.

But while you read this book you will find much more, a much deeper thinking and extensive study of design itself. This is a book about the essence of design, about creating the new and making of the better. It is about fostering changes through questioning the existent, the needs, and the methods we use, questioning the technology we create and use, and the future we want to shape. This book is a driver for thinking and rethinking about design, the nature of design and the importance of design in shaping our lives.

Presented thoughts and reasoning are built on direct evidence, on over 40 years of experience, thinking and questioning, systematic inquiry about design research and practice, methodology and models, processes and outcomes. This invaluable experience based on discussions with hundreds of PhD students and selecting thousands of design research papers, who I am honoured to say were in large numbers published in Design Society conferences, emanates from every chapter and thought.

In the end, conceptualisation is an essential part of product development. No matter whether one perceives it as the most important or just one of the steps in successful product development, it has to be executed with understanding beyond the just the product lines and profit margins. This book argues and presents a clear case for the need to understand the impacts of decisions through the products' life cycle. A well-reasoned framework of thinking with evidence from different domains brings the book closer to many readers from various fields.

Regardless of whether the reader is a student; researcher or experienced design practitioner the book offers a tuned and practical viewpoint, not only on conceptualisation, but also on designing with all the complexity involved.

The book will contribute to educate students in the multidisciplinary nature of design in a structured way. It will also serve as a reference for practitioners—working engineers, designers, and managers dealing with product development. It will have practical consequences in everyday work of many but the book will also be a reference point for all who think about design and search for unanswered questions or new answers about design and designing.

As a well-designed product the book is clear, memorable, relevant, and impactful and will become a central station on the way towards understanding design.

Zagreb, Croatia, 2015 Prof. Dorian Marjanović

Preface

This book's origins lie in design methodology, especially the work of Hubka and his articulation of design science and theory. In my youth, I had the privilege of working with Hubka and was one of the first 'believers' in the fundamentals of design science that also influence this book. Over my career, I have worked to build these ideas into the teaching and research at the Chair of Engineering Design at the Technical University of Denmark. First, established in 1952 it is now flatteringly known as home to the 'Copenhagen School' of design research.

Fundamentally, this book builds on and synthesizes many years of research and crystallizes these for the reader based on my, and my co-authors, extensive teaching experience. Throughout, the Design Society and its conferences have allowed us to establish a strong network from which we have found inspiration and insight into many aspects of this book, e.g. design behaviour, reasoning, understanding of value and mindset, socio-technical design, staging of teams, and many more. Fundamentally, we think that these dimensions are the prerequisites for using methods and executing successful, professional design work.

Throughout the writing process, my guiding star has been the question: "what to tell the students?" be they candidates, researchers, professionals, or anyone wishing to understand and work in design. The writing process has been influenced by many inspirations and has been forged in constructive conflict and discussion with a close group of valued colleagues. In particular, I have drawn much insight from the Summer School of Engineering Design Research, where I have spent many years in dialogue with PhD students and design scholars.

It is my core belief that effective design comes from professional, structured understanding, and skill, in the same way that a musician must understand and have insight into music. However, in both cases, the 'player' needs the instrument, the understanding, and practice in order to hone their skills. As in music, not every piece should be played in symphony but also in playful jam sessions where we realize our own creativity. In its nature design is to play, to imagine, and act on our imagination. I hope to capture these attributes in our interpretations, mindset, and models.

Over the last decades, design has grown as a research field, as a profession, and as a science with hundreds of new papers and books each year. In this book, I aim to bring this expanding field back to its fundamentals, explaining the totality of design in conjunction with its core concepts: conceptualization, synthesis, and reasoning. Thus I (and my co-authors) offer the reader a new, more cohesive world of thinking, concepts, models, and methods that equip them to tackle all manner of design challenges. All our experience and research leads us to believe this world is productive.

Kongens Lyngby, Denmark Mogens Myrup Andreasen
2015

Acknowledgments

Ken Wallace played a unique role in the crystallization of this book's core concepts, as mentor, and patient discussion partner, thank you Ken! Thanks to Tim McAloone and Anja Maier for giving valuable hints for the book's structure. Thanks to Mario Storga, Seppo Suistoranta, Christian Weber, Nathan Crilly, Boris Eisenbart, Lucienne Blessing, Jaap Daalhuisen, and to 'K&P' insiders Tom Howard, Martin Ravn, Jakob Parslov, Hans Peter Lomholt Bruun, Jakob Bejbro Andersen, and Jonas Torry-Smith for discussions and support. We would also like to extend a special thanks to master student Daniel Barreneche for his highly appreciated help in finalizing the manuscript.

Many of the people we quote have shown a unique graphical style, which we would like to preserve, although it might give the book an old-school flavour. It should be especially eye-catching for those who know our history.

Contents

Part II The Design Machinery

Chapter 1
Introduction: Conceptualization

Conceptualization is the process of creating something previously unknown or unseen: a new product. It is the core synthesis activity of design. In this book we do not try to spark wild ideas for new products, instead we aim to provide a process for, and deep understanding of, conceptualization. This takes us from exploration of human needs and opportunities to the goals, design processes, and reasoning required in creating good products.

To achieve this aim we explain the strengths and weaknesses of current design methods with respect to human ingenuity and cognition, and provide a deep insight into how designers use models and methods. We distil these insights into what we call mindset, articulated as models and heuristics throughout this book.

This chapter gives a foundation for this book, outlining our objectives and philosophy, as well as what we see as the book's role in practice, teaching, and research.

1.1 Argumentation

The **purpose** of this book is to empower the designer. Design is the key factor influencing products, services, systems, and activities in the modern world and is fundamentally linked to our welfare as a society. As such, any design should

M.M. Andreasen et al., *Conceptual Design*, DOI 10.1007/978-3-319-19839-2_1

at least satisfy the needs of the user. We address this by exploring the nature of design, enabling the reader to go beyond current textbooks' methods by creating a rich mindset for understanding and staging design.

This book is about the nature of conceptualization, written for those designers, educators, and researchers who want a deeper understanding of the nature and application of design.

Our objective is to support the reader in understanding and practicing design—whether as a designer, educator, or researcher. As such, we bring together the often-fragmented knowledge on design and enriching this with an understanding of fundamental design phenomena, theories, models, and concepts. Throughout, we highlight heuristics guiding how this knowledge can be used in practice, as well as explaining what this means for creating good products. We consider design activity from two perspectives as illustrated in Fig. 1.1 (Andreasen and Ahmed 2006).

Design in the broadest sense describes innovating or improving the means (either material or immaterial) for addressing human needs. Design is a social behaviour that is staged and celebrated in a community of practice. Central to all this is the **designer**, a denominator for the more or less central actors in the design activity, see Fig. 1.2. The prefix for these actors can be engineering, industrial, graphical, and so on—we do not distinguish as all designers build on the same fundamental principals.

Considering our broad view of the designer, you might ask, what products are we interested in? The professions mentioned above produce: engineered products, styled products, services, work spaces, complex systems, and design results related to industrial, public, and private sectors. Our aim is to explore the core of design and thus support all of these product types at a fundamental level. This also allows for many new, interdisciplinary, and non-engineering situations.

It is our experience that design methods only give a sparse insight into actual design, whilst giving the impression of rationality, which is not at all present.

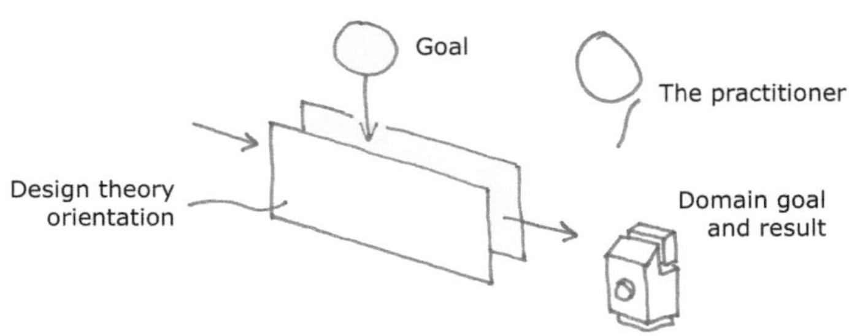

Fig. 1.1 The two dimensions of design activity: actual designing and reflective improvement

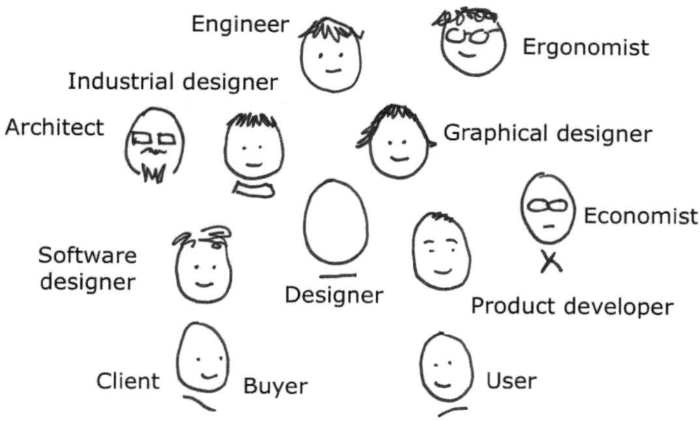

Fig. 1.2 This book's use of 'designer' relating to many types of actors

Therefore, we focus on providing a deeper understanding of design. This **philosophy** means we aim to supply the reader with the proper *mindset,* i.e. an internalized understanding of basic design phenomena, e.g. goal formulation, evaluation, coordination, sketching, etc. All of these phenomena build on conceptualization, which is at the core of the design effort. This mindset thus helps the designer create a more effective conceptualization strategy. We deal with this foundational understanding in **Part I** of this book.

> **Conceptualization strategy** ensures that the selected concept has a strong rationale, builds on the best available knowledge, and leads to successful business and need satisfaction.

Building on this core strategy, design is fundamentally teamwork. Designers act in self-organizing, self-propelling teams that interface with the organization and wider technical and professional contexts. We call this the 'design machinery' and discuss it in **Part II**. This deals with the staging of conceptualization.

Just as conceptualization cannot be separated from its staging, it can similarly not be treated without understanding the total design process and product lifecycle. Thus, our discussion of design merges the following dimensions:

- The **cognitive**, creative processes of the human mind, which generate ideas, understand value, and lead to decisions.
- The **evolutionary** process of gradual identification, clarification, concretization, and detailing of the design.
- The **expansion of scope** from ideas and concepts to composed, multidisciplinary product designs, leading to launch, lifecycle, and new business.

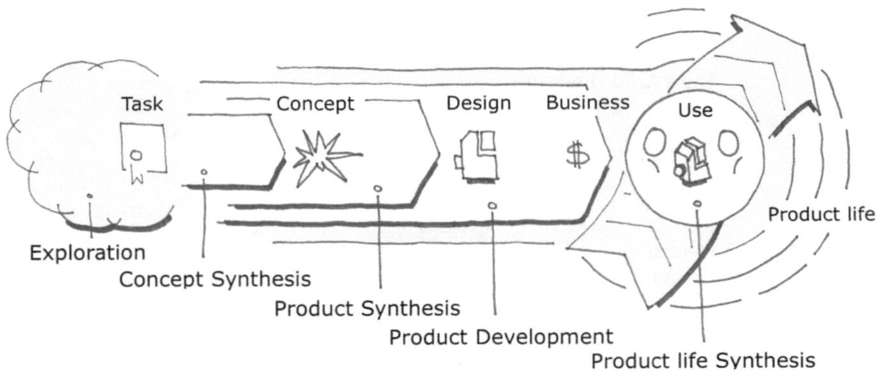

Fig. 1.3 Our **Encapsulation Design Model**: a framework describing the design process

These dimensions lead to the model of design we use as the guiding framework for this book, see Fig. 1.3. The backbone of this model is the sequence of design results: Task > Concept > Design > Business > Use. These are created via the sequence of activities: Exploration > Concept Synthesis > Product Synthesis > Product Development > Product Life Synthesis. This progression describes an expanded view of design that includes wider contextual activities and ends with the product lifecycle. In particular this last element is where the need is satisfied, new business is created, and side effects or environmental impacts are observed. We discuss these five stages in detail in **Part III** and call our framework the **Encapsulation Design Model**.

This model challenges the designer to see the design process as both a causal sequence of activities and, in the opposite direction, a layered means for judging the goodness of a product. In this way the designer must consider need satisfaction and product life, business value, design quality, and so on.

Design should be imagined backwards (starting from product life) and arranged forwards (starting from exploration).

This dual perspective is synthesized in the **Link Model** illustrated in Fig. 1.4. On one side the model shows the user perceiving a product through its use and need satisfaction; while on the other side it shows the designer imagining the user's need situation and value perception in order to identify design goals. This links the designer's reasoning from goal to product use and from product use to goal. We deal with this dual perspective in **Part IV**, where we talk about designing 'the good product'.

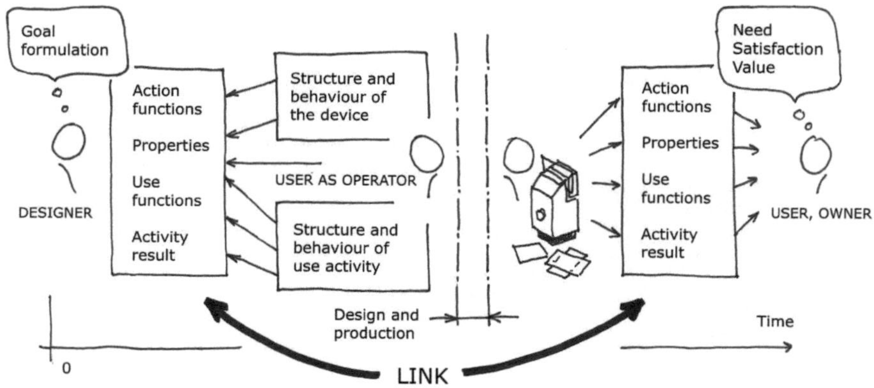

Fig. 1.4 The **Link Model** showing the user's value perception and the designer's reasoning

1.2 Practice, Research and Education

This book provides value for practitioners, students, and researchers in different ways, which we briefly outline here.

Design practice determines education and research aims, as well as providing the major channel through which design impacts society. Therefore, this book takes its point of departure from the scene of conceptualization and the articulations of concepts in society. When we talk about conceptualisation we refer to the phenomena encountered in practice. As such, practice provides the foundation for our insights and allows us to draw together our research and education experiences into a cohesive whole for the reader, as illustrated in Fig. 1.5.

With respect to **design research** we build on the founding idea of design methodology as "*the science of methods that are, or can be, applied in design*" (Roozenburg and Eekels 1996). Our early research on conceptualization developed from Hubka's Theory of Technical Systems (1973), a general theory of technical entities' structure and behaviour. Design research has matured since these early efforts, giving rise to today's repertoire of design theories, methods, and computer support. Hubka and Eder (1996) saw design science as "*a system of logically related knowledge which should contain and organize the complete knowledge about and for designing*". They distinguished between object knowledge, i.e. statements on the nature of technical artefacts, and process knowledge, i.e. statements on the design process. The limitations of this encyclopedic approach are broken by Horváth's (2004) structure of engineering design science. Here, he links design research and design practice into the whole of design science. However, at the heart of our understanding of design is that methods without consideration for the designer's situation or mindset are doomed to failure (Blessing and Andreasen 2005).

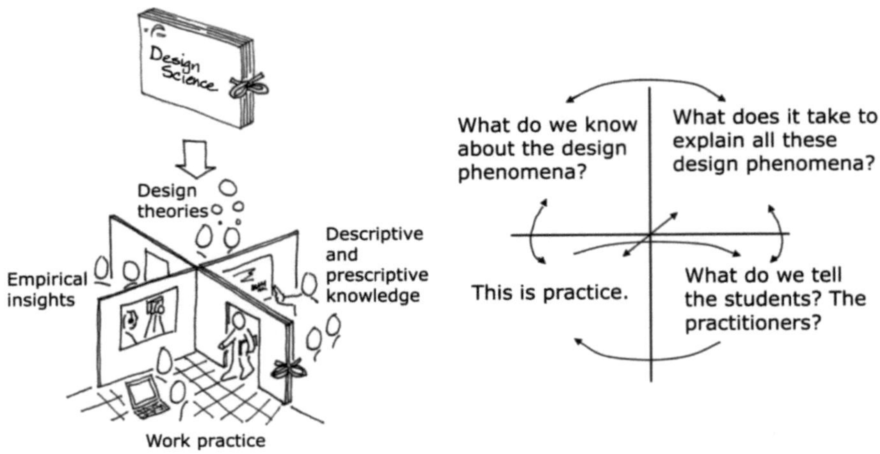

Fig. 1.5 Design science seen as four interrelated worlds (Andreasen 2008)

Natural science has, in all its forms, a common research paradigm studying nature's laws. Design research is about the designers' cognitive behaviour, the team, and organization, as well as understanding users, organizations, society, etc. Thus, design research today covers a much broader palette of research paradigms than its origins in engineering research's natural science approach. Design science comprises many overlapping and conflicting approaches and concepts. A reason for this apparent lack of precision compared to 'real sciences' is that many basic concepts are mental constructs, whose interpretation is necessarily embodied in researchers' models and theories. One example of such a theory is this book's Domain Theory. We highlight this theory because of its productivity and value in teaching, research, and practice (Andreasen 2011).

Finally, **design education** is an enormous challenge where we again see a split between 'natural science' and more holistic perspectives. We again take a synthesis approach. This book deals with both the technical insight necessary for mastering product synthesis (i.e. transforming human needs, ideas, and technologies into satisfying and valuable products) and the wider considerations of human ingenuity and staging design. Although design methodology provides the procedural support for design we must also consider how the designer bridges these stepping stones via their understanding, mindset, and creativity. This stepping stone idea of design is illustrated in Fig. 1.6.

Ultimately, our ideas for design education build on our experiences teaching the Design and Innovation program for Bachelor and Master's students at the Technical University of Denmark (McAloone et al. 2006), see the lead author 'working hard' in Fig. 1.7. This program integrates design thinking (e.g. creativity, product synthesis, innovation), social, and reflective competencies (e.g. sociotechnical systems, design research) and technical mastery.

Fig. 1.6 Methodological stepping stones bridged by the designers' creative mind

Fig. 1.7 *Left* Professors Boelskifte and Andreasen preparing the Design and Innovation program. *Right* Teaching situation from the program (*Courtesy* Torben Lenau)

1.3 This Book's Structure

This book contains four parts, summarized in Fig. 1.8. **Part I** sets **the scene of conceptualization**. This deals with the influences, conditions and possibilities affecting conceptualization, as well as the ecological, ethical, and legislative considerations in which we operate. The concept's role is explained and we explore how this can be developed or expanded to meet the requirements of the design situation.

Part II deals with **the design machinery** composed of: (1) the designers, their competences, and skills, (2) the staging of the design activity, and (3) creating products. Here Chap. 3 starts with design practice and explores how designers carry strategies, models, and methods into the design activity. Chapter 4 then discusses how design is staged, focusing on the frame in which a team is acting. Finally, Chap. 5 combines these elements to discuss creating products, as well as how to establish a team's design process. Here, we also start to explain the key factors influencing the creation of effective procedures and mindsets.

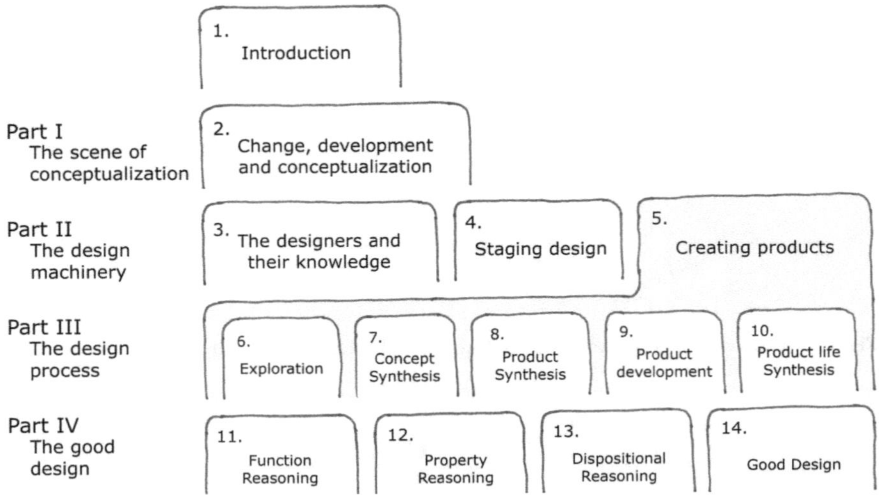

Fig. 1.8 The four parts and 14 chapters in this book

Part III deals with **the design process** based on our Encapsulation Design Model (Fig. 1.3). As such, the five chapters sequentially explain the steps from Exploration to Product Life Synthesis. **Exploration** describes the initiation and clarification of the concept from a number of perspectives. This feeds into **concept synthesis**, which is at the heart of this book, and describes the fundamental elements: creativity, systematic thinking, and visualization. Next, **product synthesis** explains how the product design is created by building on Domain Theory, which sees the product as a system of activities, organs, and parts. Finally, **product development** and **product life synthesis** examine the ideas of Integrated Product Development and linking the product design and its lifecycle explicitly at the conceptualization stage.

Finally, **Part IV** deals with **reasoning about good products**. Design is not about simply assembling a jigsaw but rather realizing the desired product behaviour. More technically: design determines the characteristics of a product, which realize the desired functions, properties, and value. The first three chapters in this part each examine one aspect of this logic, building on the Link Model: 'Function Reasoning', 'Property Reasoning' and 'Dispositional Reasoning'. Finally, Chap. 14 brings all of these together in order to discuss what makes 'good design', both in terms of a good product and good design practice.

Throughout the book we use many terms related to design and conceptualization that have specific meanings here and more general meanings in everyday language. As such, a summary of the terms used can be found in the Appendix.

References

Andreasen MM (2008) Consolidation of design research: symptoms, diagnosis, cures, actions?. Unpublished presentation at Design Society's board meeting, Eltville.

Andreasen MM (2011) 45 years with design methodology. J Eng Des 22(5):293–332. http://www.tandfonline.com

Andreasen MM, Ahmed S (2006) Thoughts on design research consolidation. Unpublished presentation at DS's advisory board meeting, Heraclion

Blessing L, Andreasen MM (2005) Teaching engineering design research. In: Clarkson J, Huhtala M (eds) Engineering design—theory and practice. In: A symposium in honor of Ken Wallace Engineering Design Centre, Cambridge, pp 32–41

Horváth I (2004) A treatise on order in engineering design science. Res Eng Des 15:155–181

Hubka V (1973) Theorie der Maschinensysteme (Theory of machine systems). Springer, Berlin

Hubka V, Eder WE (1996) Design science. Springer, Berlin

McAloone T, Andreasen MM, Boelskifte P (2006) A Scandinavian model of innovative product development. In: Krause F-L (ed) In: Procedings of 17th CIRP design conference 'the future of product development'. Springer, Berlin, pp 169–278

Roozenburg NFM, Eekels J (1996) Product design: fundamentals and methods. Wiley, Chichester

Part I
Conceptualization

Chapter 2
Change, Development, and Conceptualization: Setting the Scene

From a design perspective, our society is a result of the incremental development of numerous societal systems into a complex web over thousands of years. Societal and human demands are met by our efforts to develop new knowledge and technologies, and deploy these in products and systems. Our focus here is on this transition into products and systems via more or less industrialized design processes. In exploring this we take an 'onion peeling' approach to gradually dig down from the fundamental nature of developing societal systems to our final focus on conceptualization and design. We explore how the creation of influential, sustainable, and valuable products requires the designer to empathically and technically understand the wider societal context and consequences of their actions. Thus, we close this chapter by outlining the designer's role and challenges in this context, mirrored by what we see as the role of this book.

2.1 Describing Conceptualization: Peeling an Onion

In order to understand the role of conceptualization and the nature of concepts we must first take a broader look at our society and its development. In doing this we consider societal change, development, and innovation to better understand design, and thus conceptualizations', role. Thus, let us start peeling the onion (hopefully with fewer tears than in the real activity).

M.M. Andreasen et al., *Conceptual Design*, DOI 10.1007/978-3-319-19839-2_2

Our society is characterized by numerous interacting and interwoven systems that support our life, industry, and development. At the heart of these (from an engineer's perspective) are complex technological systems that form the basis for new product development, as well as the context in which these products will ultimately be used. The dominant driving force for change in these systems is human need, curiosity, and ingenuity. This is reflected in the competitive nature of human civilization, with individuals, groups, companies, and states all vying to better their peers in terms of, e.g. esteem, power, safety or health. From a design perspective we can reduce these drivers to the more neutral concepts of '**need**' and '**intention**'. We do not mention innovation here because it is the result of development, not a driver for it.

Human design activity is the 'machinery' through which products are developed. This normally takes place in companies and relates to the creation of tools, products, equipment, plants or complex systems, all embedded in the wider context of the market, society, and the environment (Hales and Gooch 2004) (see Chap. 9 for more discussion and illustration of this idea). Here, products can be one-of-a-kind, variant, series or mass-produced. Thus, the market forms the basic arena for design, with economy as the frame, technology as the fuel, and the perceived need for innovation (tied to the fundamental drivers noted above) as the driver.

Although this industrial setting might seem limited, the heart of all design efforts is the kernel 'concept'. This articulates the core idea, the response to the need and intention, the proposal for the product's realization, and the process of creating new business and need satisfaction. As such, when we deal with conceptualization we deal with the core of all design activities.

Example:
Creating a concept. Two designers were tasked with generating new concepts for an insulin injection device (an 'insulin pen') manufactured by Novo Nordisk. Novo Nordisk sells insulin and related products to diabetes patients worldwide. As part of the treatment schedule patients must regularly measure their blood sugar level and inject themselves with insulin.
As a first step the designers defined the current need, technological, and market situation to create an initial mission statement: "*The need for discretion in injection and the device's appearance, as well as the need for easy storage outside the home, are fundamental elements defining a product's position on the market. The key means for satisfying these needs are: minimal weight and size, and reduced time and complexity of injection. These should be integrated with user friendliness and safety*".
In navigating between the company's existing products, competitors' products, and the designers' own ideas, a concept emerged from a combination of these. This was called the 'Minimalistic device'. It had the following characteristics as is illustrated in Fig. 2.1:

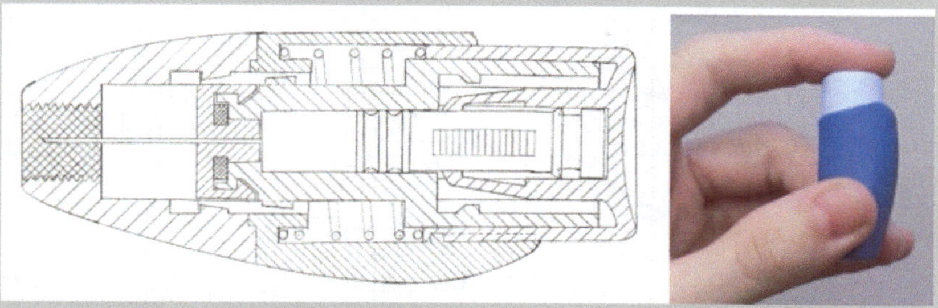

Fig. 2.1 *Left* Concept proposal for a 'minimalistic device'. *Right* Mock-up for showing the finger grip

- **Attributes**: a prefilled device with half the length of similar products, e.g. FlexPen®. The short length was achieved by detaching the dosing mechanism from the prefilled cartridge, which is now stored side-by-side.
- **Target group**: type 1 diabetic who needs rapid-acting insulin at his disposal.
- **Positioning**: small, low weight device that fits in a pocket and supports one-handed operation.

To allow themselves maximum freedom the designers separated their technical solution from the products' raison d'être, i.e. the *idea in* and the *idea with* the product. Here, the use activity is key to the patients and must be logical, simple, and safe.

- **The idea with the product**: discretion is satisfied by the small size of the device without sacrificing quality or safety in the operation and dosing.
- **The idea in the product**: the dosing mechanism is detached from the cartridge and dosing is achieved using the same mechanism as in the existing product.

Important questions for this concept (and in general) are: does it address the need, intention, and task, and is it tractable? In this case the designers felt the concept was satisfactory, although the assembly/disassembly required for use was not ideal. Other engineers at Novo Nordisk considered manufacturing, sales, and business perspectives to judge the concept. This resulted in mixed feedback and the concept was added to the company's idea bank but not progressed further (Courtesy of Jonas Mørkeberg Torry-Smith and Jacob Eiland).

Based on these considerations our onion peeling takes us through the following layers, illustrated in Fig. 2.2: **change and development in society** (Sect. 2.2), **knowledge and technologies** (Sect. 2.3), **industrial product development** (Sect. 2.4), and **conceptualization and design** (Sect. 2.5). Finally, we reach the core and explore the nature of conceptualization in Sect. 2.6.

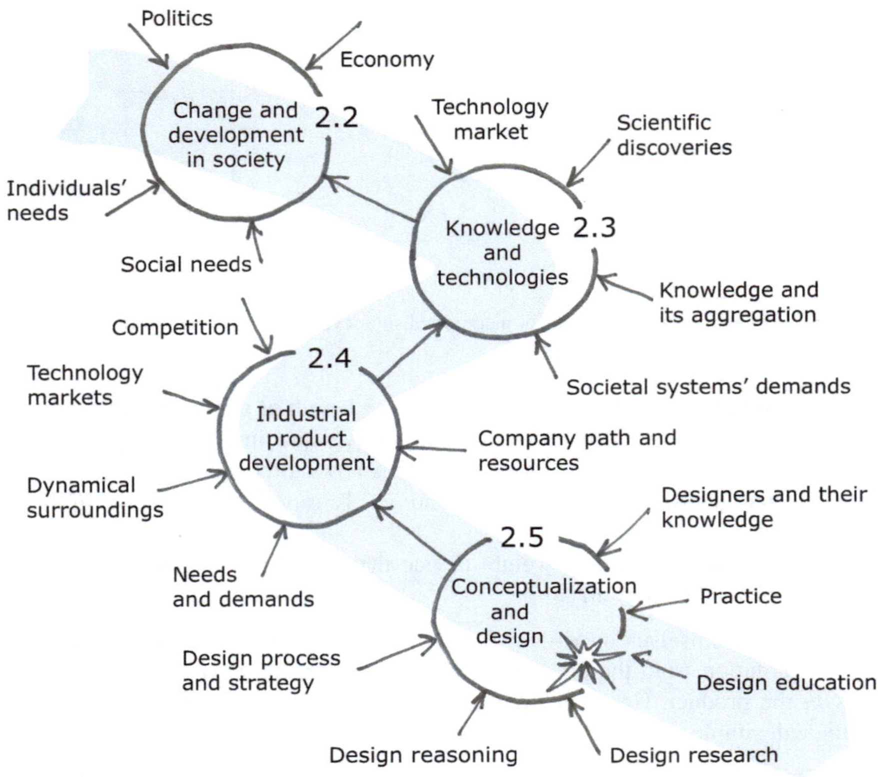

Fig. 2.2 From societal systems to conceptualisation and design in four layers

2.2 Change and Development in Society

In children's books we find exciting illustrations depicting the nature and complexity of our society's systems, e.g. Fig. 2.3. These show the 'anatomy' of our society with many interwoven systems supporting daily life and linking to larger systems such as trade, fabrication, finance, health care, education, transport, etc. Any new product that is conceptualized is based on activities and technologies belonging to these systems. Further, once the product is realized it will itself become part of these systems.

From this complex interaction between societal systems we can distil out a number factors relevant to design and conceptualization.

First consider **societal needs**. Typically societal systems are partially public, e.g. traffic signals, and partially private, e.g. cars. Public systems serve general societal needs while private elements serve individuals' needs. At the societal need level key drivers are societal necessity and industrial possibilities. **Individuals'**

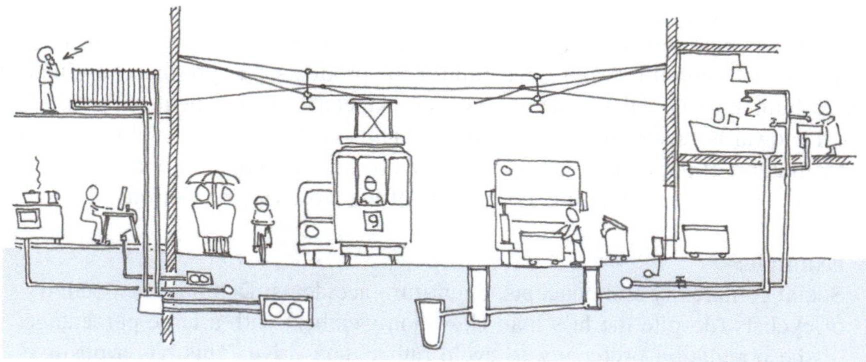

Fig. 2.3 A cross-section of a city street with communication, energy, and many other systems

needs, on the other hand, deal with how people live their lives in the context of the wider societal systems. This includes leisure, education, and all other personally important systems such as health, food, or media streaming services.

At a higher level still, all of these systems exist with respect to economic, environmental, and political structures. Here, **economy** defines how systems emerge and develop. In both large and small projects financing is a critical factor. Those involved in the development of new systems often experience economy as an 'iron ring' of limitations. In particular, designers are constrained by national economic considerations such as productivity, societal costs, and export value. As with economic factors the **environment** affects the scope of designers' activities. The pressure from a growing population, consumption of natural resources, and the damaging effects of human activity and need satisfaction are serious conditions that cannot be ignored. Further, as with all human endeavour **politics** affects how we realize our goals. Government and public institutions are involved in the creation and maintenance of public systems, and set the 'game rules' for societal and individual initiatives via legislation and regulation. Finally, consider **knowledge and technology**. Technologies are, on one hand, the machinery of economy, and on the other hand, the core of any societal system. Thus, the topic 'technology' forms the next layer of our onion.

Although these individual factors form something of a 'schoolmasterly' list interesting things happen when we consider the tensions between them.

Such a field of tension is *democracy, power, and knowledge*. The development of new systems and technologies is in the hands of companies deciding economy, politicians who formulate infrastructure and investments, and engineers and other carriers of technological knowledge. Ultimately, it is engineers and designers who develop and utilize technologies (Jørgensen 2008). However, it is by no means a given that these three entities will work together or are in line with what an educated citizenry might actually want or need.

Another example tension is *consumption, technology, and environment*. Since the 1960s, environmental considerations have invaded our consciousness. The explosion in technology and consumption of products has produced a situation where humanity is actively damaging our planetary environment and driving it rapidly towards unsustainable conditions. Here, technological development (and its facilitation of population growth, etc.) can be seen as a major contributor to this problem while at the same time posing one of the most viable solutions.

Example:
Social complexity and concepts. Right-turn accidents. Denmark is a country of cyclists (despite the less than Hawaiian weather) with a large percentage of the population preferring to cycle rather than drive. This has a positive impact on health and environment but comes at a price. The risk of being killed in an accident is five times higher for a cyclist than a car driver. In particular, the popular press has focused on accidents where right turning trucks cut-off and run over cyclists. There are about 50 such accidents per year and although the number is not large they have severe consequences for the cyclist, with about five cyclists killed annually.
Analyses of these right-turn accidents identified a variety of causes: wrongly adjusted truck mirrors, poor visibility, cyclists' behaviour, inappropriate road signs, and poor driving of the truck. The lack of a common cause means that it has been hard to effectively tackle this type of accident. However, because of their prominence in the press a private company arranged a competition looking for ideas to solving the problem. Over 1500 solutions were submitted, with the top three being: (1) to supply the truck with a light that activates with the right-turn signal and shows the truck's blind spot on the road, warning the cyclist, (2) to colour the cycle path where the cyclist and lorry's blind spot interfere, again warning (3) to automatically detect the cyclist's position in the danger area and alert the truck driver. These ideas were based on an ideal situation with responsible, considerate behaviour from all actors. However, in order to produce these solutions different actors must take responsibility and invest in them, e.g. in legislation for trucks or new training for cyclists. Key weaknesses are that not all trucks and not all crossings will be equipped with these concepts, and not all cyclists will understand them.
For the problem-solving, conceptualizing designer, balancing potential effects against investment is key. In particular, increasing the probability of positive effects and reducing negative effects. One example of a good concept, balancing these issues, is a Dutch proposal that builds on the idea of forcing the cyclist and truck into perpendicular positions for the right turn, as illustrated in Fig. 2.4.
The strength of this concept is that it clearly indicates its special design for right turns. Both cyclists and drivers have to be alert; neither have 'special rights'. However, the required investment is high and space-consuming, and has not yet been analysed for its efficacy in accident prevention. Further, to

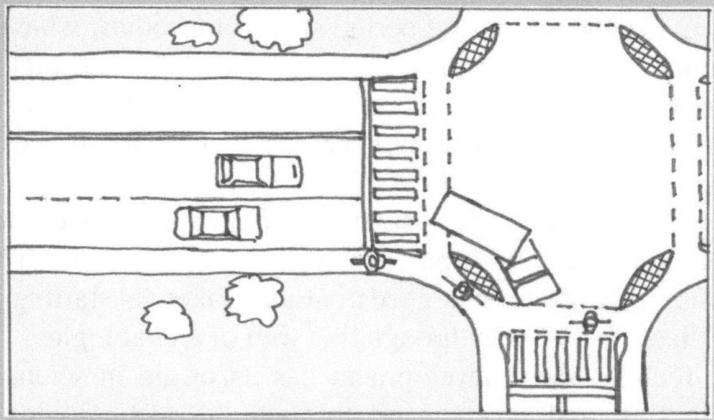

Fig. 2.4 A Dutch proposal for a lane layout to better position right turning cyclists and trucks

be a good solution, the stakeholders must accept it. In this context the proposal has clear advantages for cyclists and it is hard to imagine right-turn accidents with the new layout.

Just as in this case, real-world design problems are complex, with many actors and related systems. They are not clear: there is a need to do something but the diagnosis is weak, and the many actors have different value systems making it difficult to satisfy them all. Finally, for most actors it is easier if someone else solves the problem.

2.3 Knowledge and Technologies

Human knowledge is the result of cultural development and is composed of scientific discoveries, industrial advances and innovations, and human ingenuity and experience. Social and technological aspects are interwoven in a socio-technical totality, where a web of technologies orchestrates our daily lives. This makes us highly dependent on complex organizational, mechanical, and industrial systems. Underpinning all this are **state changes,** i.e. something changing from one state to another, often more valuable or practical. In Chap. 8 we discuss how state changes form the core of new product creation but it is sufficient here to acknowledge their importance.

We observe an endless number of natural changes: flowers grow, we get old, mountains are eroded, etc. In response to this, humans have used tools to intervene and bend nature to our will from our earliest history: catching fish, hunting animals, farming grain, building shelters and houses, using fire. Key to this human intervention is the use of tools or, more generally, **technology**.

Definition: A **technology** is a combination of material devices, procedural prescriptions, and intentions that are interwoven with humans' work and social activities, articulating and structuring humans' behaviour and life in society (after Jørgensen 2008).

When settlers moved west in the European colonization of North America they were dependent on that period's encyclopaedias, which for example explained the technology of soap making. Thus, our definition highlights the inseparability of the social and material realms, and points to how technologies define our lives. Ultimately, technology is executable; a device processes something, i.e. a state change happens (Arthur 2009).

Everywhere in the systems mentioned above we find technologies; agriculture, communication, transport, energy, health care, education, trade, and so on. Design aims to satisfy human needs but also takes its starting point in, is created by, and realizes its lifecycle through this web of technologies.

Technological development has its origin in science, is synthesized in design activities, and is managed through the technologies' deployment and service. However, it is also thought provoking to remember that many inventions are created without scientific insight. An example is the steam engine, which sparked the creation of new scientific fields to explain and optimize it, including thermodynamics, solid mechanics, fluid mechanics, and the theory of mechanisms. Based on this we might ask: what are the factors influencing the creation of knowledge and technology?

- **Scientific discoveries**: these are widely believed to be the main driver for technology development but in many cases industrial and entrepreneurial contributions are just as important.
- **Industrial development**: competition and market demand drives development and especially incremental change in the industrial sector. Many developments here are simply seen as new products despite actually representing new technologies.
- **Technology markets**: technologies have value and thus there is a market, need, and competition. Technologies can be purchased, licensed, or developed based on demand.
- **Societal systems**: developing and maintain societal systems drives a huge demand for new technologies. A 'good' example is our capacity for producing innovative military technologies.

Further to these, knowledge and technology development themselves provide a catalyst for further development. "*We hope in technology to make our lives better, to get us out of predicaments, to provide the future we want for ourselves and our children*" writes Arthur (2009). The real challenge in the development of societal systems and technologies is not the act of creation but to manage the consequences of our actions without shying away from new possibilities through fear or ignorance. Here, there are few objective truths and the risk of paralysis and shortsighted, partial decisions is high. As such, we have an obligation to understand our design efforts and their potential consequences but also to challenge the current status quo and develop new things. Thus, knowledge and technology are both the core of, and the means for, new product development.

2.4 Industrial Product Development

Industrial product development includes all the activities from ideation to launch, takes into account the whole product lifecycle, and incorporates the material cycle from raw supply to recovery and reuse. Product development describes a totality, independent of the type of product being developed, e.g. new, incremental or platform based.

In our onion metaphor the product development layer deals with a company's efforts where the following factors are key:

- **Needs and demands**: companies operate in arenas determined by their customer offering (consumers, public institutions, other companies) and thus face competition. Even a supplier of sub systems has to understand and respect the ultimate goal if they are to compete. As such, recognizing and interpreting needs is the main driver for new development.
- **Company path and resources**: typically, a company is largely determined by its path, i.e. its past products and experiences, and choice of new strategic directions. Here strategy should balance ambition, resources, competition and financial limitations.
- **Technology market**: technical knowledge and mastery is the fuel for product development whether it is available internally or brought in from the technology market.
- **Competition**: despite the many 'me too' products, competition drives companies to reflect on and strive for better customer-oriented features and qualities, for customization, and for more dynamic responses to the changing market.
- **Surrounding's dynamic**: all product types react to technology development and market changes through organizational change and striving to be agile and reactive.
- **Conceptualization and design**: the core of product development. These are the predominant activities in launching competitive products and thus form the final layer of our onion.

Based on these we can define product development from two perspectives, the user and the company.

Definition: Product development (user perspective) is the creation and launch of products with new or different functions and/or properties, which offer new or added value to the customer/user.

Definition: Product development (company perspective) is the use of exploration, design, manufacture, and marketing/sales to launch new, product-based business utilizing the company's resources.

Although products satisfy human needs these are not ease to observe, articulate or quantify. For the observer they are mental constructs based on interpretation of unsatisfactory situations. When products are developed their role as need satisfier should be predictable. Thus, the designer must be aware of new needs and need interpretations, and be able to translate these into the product and the business (see Chap. 6 for more).

Above, we saw the role of technologies and their relation to societal, industrial, and user's needs. Here, we see an **evolution** of general technologies in all systems, e.g. the increasing range of cars or efficiency of mobile networks. Similarly any single product also evolves, with contributions from different companies and occasional breakthroughs by innovative products. This gives a performance **S-curve**, through characteristic phases from introduction to growth and finally to maturity, where new technologies oust the old.

Companies' development processes normally follow a path or historical route paved by past products and technologies, and directed by strategic plans for the future. This builds on the, often incremental, steps between product generations, with reuse forming a basic building block for product development. However, judging a path's viability and redirecting via strategy are not easy.

Example:
Coloplast's path. Coloplast A/S is the world's leading supplier of intimate healthcare products (Fig. 2.5). The idea for core product area, ostomy bags, originated from a nurse in 1954: to attach a bag to the patient with an adhesive ring. The development included both Coloplast, at that time a plastic foil producer, and health professionals. The company's product development in this area follows a characteristic incremental pattern, where knowledge from every new launch is followed up and leads to innovations in materials, adhesives, and use. At the same time, the company grows internationally to develop sales channels, healthcare networks, and professional quality assurance. The company today is characterized by systematic development of networks and communication with patients, with the motto 'to create solutions that enable people to do more'.

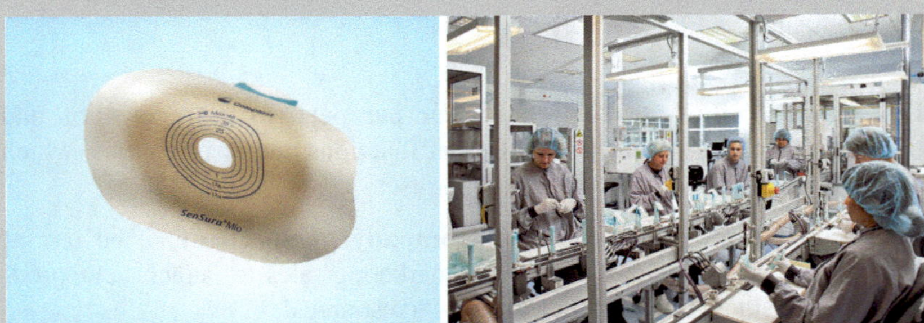

Fig. 2.5 Ostomy product from Coloplast A/S and its production facilities (courtesy Coloplast A/S)

For each project the designer must understand its link to these evolutionary patterns or risk being left behind. We must answer: what is the history and evolution of the relevant technologies? Where are the company's products and technologies on their S-curves? What is the company's path in terms of history and future strategy? And finally, how is the progression of the project related to the changing perceptions of need and market?

Although we introduce product development as a single, well-defined phenomenon it actually concerns the design process, strategic reasoning and designers' skills, independent of project size. Further, project size, complexity, and cost all influence the organizational setting in which these activities take place. In order to deal with this we discuss three general types of product development in this book: **new product design** where there is little prior domain knowledge; **incremental design** where company path and gradual improvement are dominant; and **platform based design** where product family and modularization are key.

2.5 Conceptualization and Design

Finally, we come to the core of our onion: *conceptualization and design*. This is dominated by a number of factors that we introduce here and explore throughout this book.

- **Design practice**: the designers' knowledge, ability to learn and approach to design work. These are applied in the context of the task and project, and integrated to form a team's community of practice.
- **Designers and their knowledge**: the designer is the carrier of knowledge, competences, and skills. Ultimately progress comes from the designers' cognitive activities.
- **Design process, methods, and staging**: practice is structured by the use of methods and shared design processes, in conjunction with the designers' own skills. These are brought together in the staging, i.e. the coordination of a team's knowledge, tools, and communication.
- **Design reasoning**: creating new products takes more than just planning and design knowledge. Design builds on fundamental reasoning about functions, properties, and the final product.
- **Design education and research**: design is not a science but is subject to scientific study from which key insight and direction can be drawn (indeed this forms the basis for this book).

This short sketch of design sets the scene for a general discussion of conceptualization before we explore these elements further in specific chapters. In particular, we come back to our onion metaphor to highlight that each layer influences and is influenced by the designers' work, summed up in Fig. 2.6. Conceptualization links to and influences each layer, while also reacting their influence in the formulation of goal and task. Going forward we need to understand what makes up a concept and conceptualization, and where this takes place.

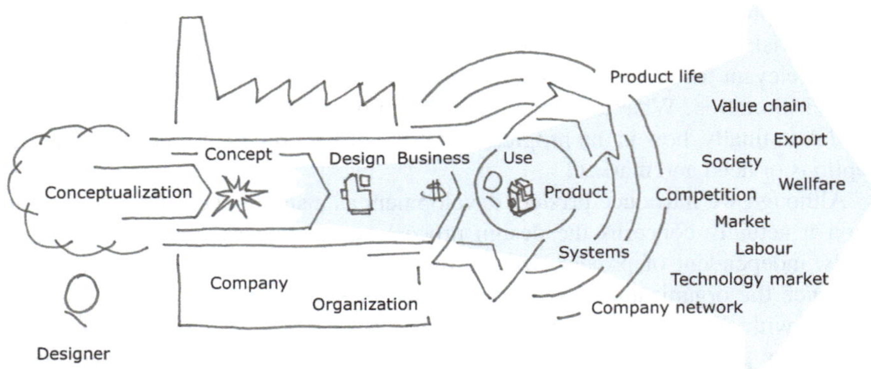

Fig. 2.6 Conceptualization and its influence/influenced by relationship with the wider world

2.6 Understanding 'The Conceptual'

A concept is an 'on the way' solution and thus partial and intermediate in its nature. Creating a concept is challenging because it must simultaneously answer the need and project goal, how the project can be realized, and combine multiple design entities. A concept mirrors the designer's understanding and interpretation of the design situation. This means clarity and precision in this understanding is critical because the concept is a point of no return.

2.6.1 The Need for 'The Conceptual'

Conceptualization describes the manipulation and combination of ideas. Concepts are at the core of a product, elucidating its utility and value. Thus, just as with a product, concepts have to be designed, drawn, and described so that others can understand and build on them.

Example:
Edison. Cartoonists' symbol for a creative idea is Edison's incandescent bulb. In reality Edison and his staff were the first product developers, using a systematic, professional approach to create not one, but hundreds of ideas and products. Fundamentally, Edison was a businessman, excelling at creating products from other peoples' ideas and inventions through experimentation and holistic thinking. One example was his experiments on adding sound to moving pictures. The concept was to record the sound on a steel wire and then (with difficulty) synchronize it with the pictures. Although rudimentary, efforts such as these paved the way for the incremental development of the modern film industry. Edison was so eager to test his idea that two staff members were asked 'to act'.

In everyday language ideas and innovation are almost overwhelmed with jargon and are imagined as almost mystical abilities in the popular consciousness. This mysticism and jargon typically drown out the majority of the actual work done in product development! As such, we focus on the real activities and design work that are responsible for producing new products.

The challenge for **conceptualization** is to create products by rethinking or imagining new situations and needs, and subsequently to translate these into new products with functions, utility, need satisfaction, excitement, and leading to a new level of sustainability.

Our society is flooded with new products, systems, and services, all begging for our attention (and money). These can range from seemingly identical products to true innovations, with many seemingly superfluous. Thus, we must consider the argument for creating a new product, does it have a compelling reason to exist or raison d'être? A precondition for this creative argument is conceptualization and professional design work.

2.6.2 What Is a Concept?

Design literature traditionally talks about ideas, principles and solutions: **ideas** are sudden initial thoughts, imaginations or visions; **principles** explain how things work; and **solutions** address problems with utility and value. None of these words is useful for describing the proposals created in during design. Hence we must add **concept** to our design vocabulary.

The design process is initiated by need perceptions, intentions, and ideas. During the process there is normally a strong desire to stop and check alternatives and answer the following:

- Are these design proposals good answers to the need, intention, and task?
- Do these proposals differentiate us from existing solutions or lead to really new products?
- Do these proposals represent the best possible use of available knowledge and technology?
- Are they tractable in the actual organization?

A design proposal at this stage, able to answer all of these questions, is what we call a **concept**. Thus, concept selection is a bottleneck in the development activity, where concepts are considered from two perspectives: do they contain the right answer, and can they be realized?

Definition: A **concept** is a design proposal that is detailed enough to justify if it is a good answer to the task and intention, and show a high probability of realization and success.

This definition is **preliminary**. We need a deeper understanding of concepts' nature, content and function before we can define them fully at the end of this chapter.

2.6.3 Concepts in the Literature

Most authors refer to a design entity called 'concept'. In the widely used book of Pahl and Beitz (2007) the design process is prescribed as four phases: clarification of the task, conceptual design, embodiment design and detailed design. Here the concept is the final output from the conceptual design phase. This phase is made up of establishing a function structure, searching for solutions and combining them into variants. Finally, the best variant is selected to give 'the concept'. In order to call something a concept in Pahl and Beitz's world it must describe:

- Is the task clear enough for us to create a design from this foundation?
- Do we need more information on the task or context?
- Can we reach our formulated goals within the economic frame?

Although the concept seems to concern clarification of insight, it is defined by the design's characteristics. Pahl and Beitz (2007) recommend that designers "*examine very carefully whether novel or more suitable paths may not be open to him. To that end he should have recourse to abstraction, which means ignoring what is particular or incidental and emphasizing what is general and essential*". The output from the concept phase is a final concept variant, where the solution principle and key embodiment features are determined. At an even more detailed level French (1985) sees *a scheme* as the output from of the concept phase. This describes solutions for the main functions as well as for cost, weight, and the main dimensions to be determined. Similarly, Roozenburg and Eekels (1996) state that in the concept phase "*broad solutions are to be generated and evaluated to provide for a suitable point of departure for embodiment design and detail design*". Here, selection of concepts is based on criteria related to use, visual impression, production, etc. Therefore, we can see that concepts must be more than just principles.

All of these authors focus on the concept as a foundation for embodiment and detailing, while our definition sees the concept as both the answer to need/intention and foundation for a tractable design and successful launch. We also see concepts described as 'complete' solutions in a preliminary state. This is in contrast to our more open interpretation where concepts can be partial and related to the reuse of known solutions.

2.6.4 Concept's Relativity and Meaning

Concepts contain 'differences that matter', i.e. they are superior to existing products in some way. A strong, innovative concept is one that sets a new agenda, reference or trend: '*since the appearance of this product we wouldn't dream of returning to the old ones*'. However, the majority of new products are variants or incremental improvements, and it is normal for new products to be based on previous or competitors' products. Here, the basic need satisfaction, use activity, principles, and functionality are known. In this context a product concept need not be described fully in all dimensions because the involved designers can fill in the gaps based on past products or experience, Fig. 2.7.

As we noted above the hype surrounding 'innovation' leads to the belief that a concept should describe a new principle, invention or idea based on a new physical phenomenon. This is seldom the case in practice and is not necessary for innovation. Many design scholars use the word 'principle' to articulate a design's basic (technical) mechanisms. For example, there exist many kinds of corkscrew, with differences in the pulling mechanism or in the choice of materials, handgrip, etc. As such, users' experience and perception of goodness goes far beyond the simple underlying principle.

In our preliminary definition we focused on the concept as the designed, specified, and modelled entity. However, the observer's interpretation and perception are just as important. These bring the concept into a context and add value to it, leading to the following heuristic.

> The concept is not just in the device but also in the **meaning** users give to the device and the context they use it in.

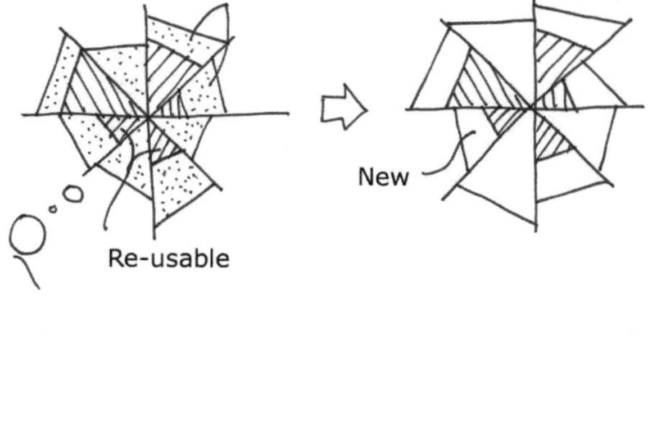

Fig. 2.7 A concept based on past products where the designer supplies the missing elements

Based on this we might ask: can we create a new concept just by bringing known products into a new context and establishing a new meaning? The answer is yes, although an established, successful product can lose its meaning and purpose when faced with a new or changed context.

2.6.5 Concepts' Composition

When a new product is launched several things can change. For example, new use activities are established that demand new systems and services, which lead to a new business strategy. When launched a product enters a unique lifecycle where it serves the user, creating several design entities: the product, new business, use activity, and product lifecycle—including several life activities (Hansen and Andreasen 2010). These are illustrated in Fig. 2.8 with respect to our Encapsulation Design Model. Each activity is an interaction between lifecycle systems, service, and actors. The use activity describes a cycle of activities in the hands of the user (see Chap. 10).

Central questions are: what belongs to the design project and what design entities need to be created? In the ideal world we would be able to manage every aspect of the task from company staging to resource allocation, sales, and distribution. Further, although the core result is the product and its use activity, we might also consider certain systems encountered during the lifecycle. The real goal of a project is to create a certain business result characterized by market share, customer identity, time span, or sales income. Thus, when companies strive for innovation they often innovate in multiple dimensions. For example, they might shift to a service business, change their product assortment accordingly, and create a new service delivery strategy.

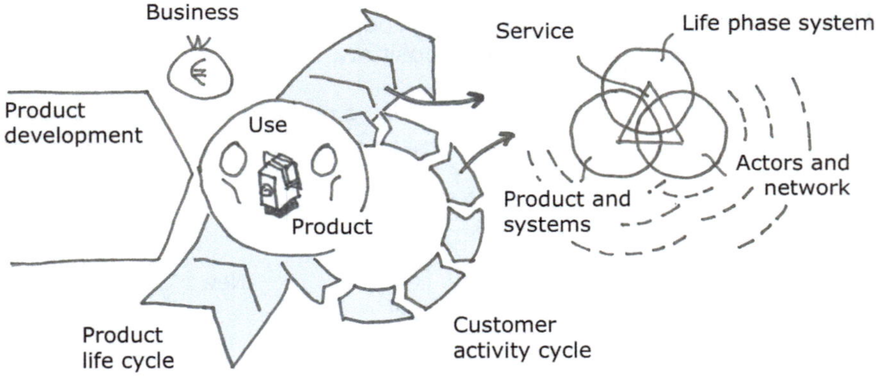

Fig. 2.8 Product development leading to various design entities

Innovation can point to a wide range of dimensions, not just in the product.

In Fig. 2.9 we illustrate a certain totality of design entity dimensions with respect to management, product planning, product development, and lifecycle concerns. Development projects can cover one or more of these dimensions. Although our focus is on new product development projects, in reality it might be relevant to create several developments in parallel with the product. Figure 2.9 shows an example plot for a project with rather substantial innovation activities. From this we can see that when we discuss what belongs to the design project we must consider this meta level, i.e. the level at which we get a complete picture of what needs to be innovated.

A product design activity is often followed by several other innovation activities; the designer needs to consider a **meta-level view** to identify and integrate these dimensions.

These development dimensions both influence the design activity and are influenced by the design/concept. We treat these alignment aspects in Chap. 13 on dispositional reasoning.

2.6.6 *Product Concept or Just Concept?*

Product development projects contain many different conceptualized entities (Olesen 1992). At the core is the **product concept**, created early in the design

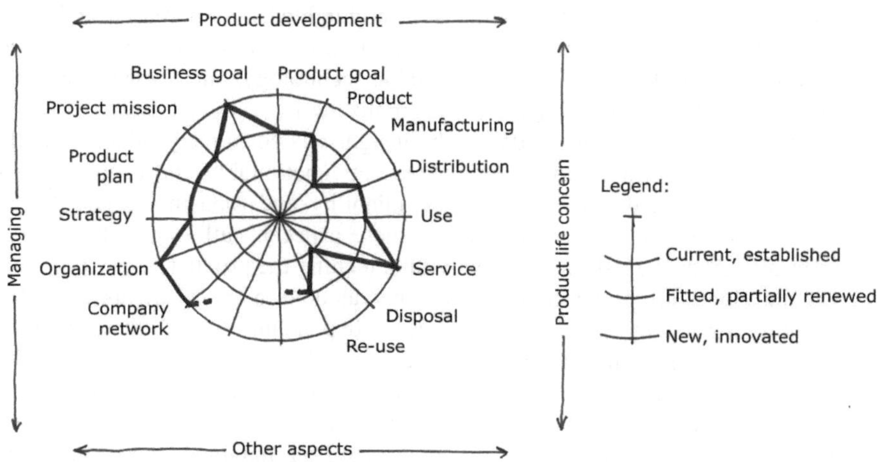

Fig. 2.9 Dimensions and levels of development related to industrial activity

activity and characterized by function, use, utility, and the actual design solution. The background and reason for the development project (and the product concept) comes from a **meta-level concept** that has been perceived or developed at a higher level in a company or organization, e.g. a new business idea concerning market, manufacture, and finance. The actual project may be one component of this meta-level concept. Finally, the product concept is normally composed of **partial concepts,** i.e. concepts that form part of the argumentation for the product and its realization, e.g. product structure, supply, production, distribution, sales, and use. The conceptual core can be linked to one or more partial concepts. This apparent expansion in the use of the word 'concept' should be understood broadly.

> A product concept's **argumentation** might rely on its own basic idea or on related partial concepts (dealing with, e.g. ideas for realization, distribution, and lifecycle).

When we discuss the flow and logic of the design process (Chaps. 5–10) we will further explore the product concept's composition.

> When creating a concept it is important to remember how it can be **composed,** i.e. combining both meta-level and partial concepts related to the product's composition.

2.6.7 Summing up Concepts

The idea of composing concepts might give the impression that they are simply jigsaws to be assembled. However, life is not so simple, their composition is dependant on many factors. One dimension highlighted by Andreasen and Hein (1987) is **clarification** of: intention and goal interpretation, need, user behaviour, application, competition and market, available technology, and knowledge. This clarification is related to the arena of conceptualization.

Another dimension of knowledge is the insight related to the product. Initially the product is based on assumptions concerning functionality and properties, its ease of production and assembly, and its reliability and robustness. The design activity should lead to factual insight into these aspects, outlining technological concerns as well as the risks associated with launching the product.

Although ideas are popularly thought to be valuable in and of themselves, this is not the case in reality. The value of an idea is dependant on its ability to deliver. As such, ideas need conceptualization to create value. This value grows as the product is synthesized and prototypes produce justified functionality and properties. A characteristic example is patents. Patents can describe solutions with no justification in the real world; instead they are built on imagined applicability and importance. Thus, once a patent has demonstrated its applicability in the hands of a producer its market value is increased.

The **arena of conceptualization** is the complex of evolving systems and technologies that serve our society. Our onion metaphor points to design as the core of innovation and development. In order to ensure vitality and value in the design activity we have to focus on the kernel: the concept. Here, the aim is to *ensure you are doing the right thing based on valid understanding of the need and the effects of the product*. **Conceptualization**, the nature of the product to be designed, and the nature of the need to be satisfied, points to the requirement for ideas and the combination of meta-level, relational, and sub-system concepts. Here, the aim is to *ensure you are making the best use of knowledge to create the best possible product*. Together, these point to a designer's dilemma: how to balance efforts to clarify the arena and efforts to create the product? Based on our discussion in this chapter we are now able to go a level deeper in our definition of a concept.

Definition: A **concept** is a proposal for a product's composition and issues that is detailed enough to justify it as a good answer to the task and intention. Further, the task and intention are justified with respect to the conceptual need satisfaction and the knowledge required, i.e. the probability of successful realization, need satisfaction, and success in the widest sense.

Note that this definition highlights the arena, the concept itself, and the knowledge accompanying the concept.

2.7 Conclusion

This chapter has led us through a breakdown of society's systems to this book's focus: conceptualization and design. Understanding 'the conceptual' means rethinking or imagining new situations, needs, and possibilities, and accepting the double task of clarifying the situation and conceptualizing the design. This leaves the question: how do the designers see and experience their situation? We will approach this by first explaining a brief history of design to give a foundation for understanding the rest of this book.

2.7.1 The Designers' Situation

The designers' role and situation have changed radically since the days of the lone craftsman, where one person cared for all aspects of the design and sale of a product. Today we face disintegration: design is separated from production, marketing, and sales. However, the designer still retains their key integrating role. Cantamessa (2011) describes this progression in three major steps, illustrated in Fig. 2.10:

- **1969–1979**: Fig. (2.10a) the 'Fordish' era toward the end of the 'American century' of innovation and technological enthusiasm. Products were feats of engineering and/or aimed at mass production in large, integrated firms. The Fordish paradigm of mass production and mass consumption is coming to a close.
- **1979–1999**: (Fig. 2.10b) the customer-centric post-Fordish era where products compete to win customers' attention and money. A new paradigm emerges: design is focused on the customer and their perception of the product.
- **1999–today**: (Fig. 2.10c) the 'yet to be named' era where firms move from supply chain to ecosystems, from products to systems, platforms, product-services, business models, and policies. This again moves into a new paradigm where the customer and innovator are blurred.

The message we draw from this progression is that we face an era of design where a designers' tasks are immensely expanded in scope and challenge. These point to key features of future design situations (McAloone et al. 2007). Design is becoming a **globally distributed** activity, with the attendant temporal, spatial, and cultural challenges. In this context, designers are increasingly expected to show **social responsibility**, taking into account sustainability and ethics, and requiring the designer to understand both **lifecycle** and **product**. Building on this holistic understanding there is also a shift towards **services** as key deliverables, not just products. Ultimately, these expanded responsibilities require designers to understand the context, complexity, and business potential of their and their colleagues' actions.

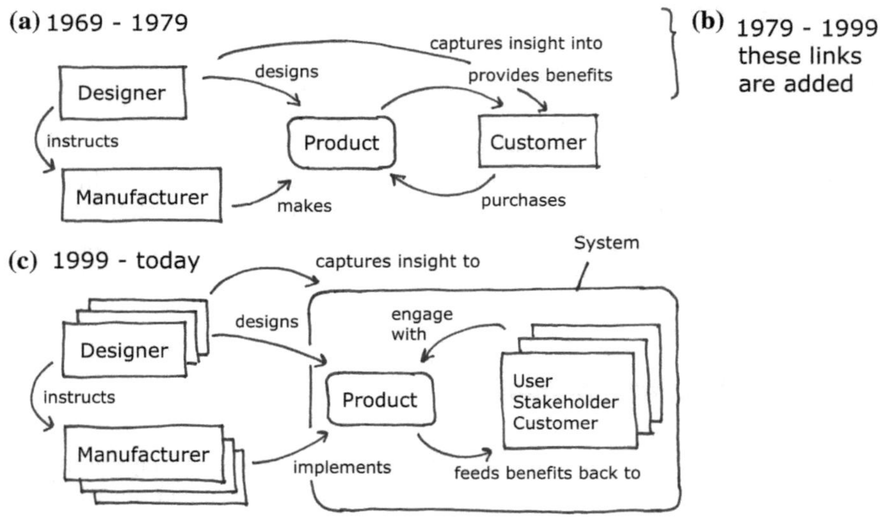

Fig. 2.10 Three phases of development in the twentieth century

2.7.2 *Needs and Challenges*

This progression in the design landscape and our initial discussion of conceptualization reveal a number of **needs and challenges** for the design practitioner, which we seek to address in this book. In particular there are **specific needs** related to the three groups of readers and their situation. In **design practice** methods and reasoning are becoming ever-more complex, with reliance on computer support always increasing. As such, there is a renewed need to better understand the logic behind design thinking and design practice. In **design education** diversity of design fields is a significant problem. As such, we go back to the core of all design, building on fundamental design reasoning and conceptualization (Andreasen 2011). Finally, in **design research** there is a similar need for consolidation and support in applying research to practice (Andreasen and Ahmed 2006; Andreasen and Wallace 2011).

In **conceptualization** key challenges stem from handling the many composed elements related to need, context, intention, possibilities, etc., which all feed into the 'good solution'. This can only be tackled by reasoning about functionality, properties, need satisfaction, and value across the whole product lifecycle. The challenge for the **individual designer** is to balance the demands of structured planning and methods against the individual and team's efforts to: **stage** the design activity to best realize their abilities, knowledge, and creativity; **'design the design activity'** to best realize strategy, procedure, and creative effort; maintain **creativity** and personal, entrepreneurial goals; and maintain **transparency** with respect to ecological, ethical, and legislative demands.

2.7.3 *From Here*

Bearing these needs and challenges in mind, and considering the fundamentals of conceptualization described in this chapter, our next step is to look at how we might answer some of these issues. **Part II** takes us into **the design machinery** where we will explore the designer's competences and skills, the staging of the design activity, and the creation of new products.

References

Andreasen MM (2011) 45 Years with design methodology. J Eng Des 22(5):293–332. http://www.tandfonline.com

Andreasen MM, Ahmed S (2006) Thoughts on design research consolidation. Unpublished presentation at DS's advisory board meeting, Heraclion

Andreasen MM, Hein L (1987) Integrated product development. Institute of Product Development, Technical University of Denmark Copenhagen. IFS (Publications)/Springer, Berlin (Facsimile edition 2000)

Andreasen MM, Wallace K (2011) Reflection on design methodology research. Key note speech at ICED 11 Copenhagen (Unpublished)

Arthur WB (2009) The nature of technology—what it is and how it evolves. Penguin Press, London

Cantamessa M (2011) Design ... but of what? (Chap. 20). In: Birkhofer H (ed) The future of design methodology. Springer, London

French MJ (1985) Conceptual design for engineers, 2nd edn. Design Council, Springer, London, Berlin

Hales C, Gooch S (2004) Managing engineering design, 2nd edn. Springer, London

Hansen CT, Andreasen MM (2010) On the content and nature of design objects in designing. In: Marjanovic D et al (eds) Proceedings of the 11th international design conference DESIGN 2010. Faculty of Mechanical Engineering and Naval Architecture, University of Zagreb and The Design Society, pp 761–770

Jørgensen U (ed) (2008) I teknologiens laboratorium – ingeniørfagets videnskabsteori (In the laboratory of technologies—the science of engineering). Polyteknisk forlag, København

McAloone Tim C, Andreasen MM, Boelskifte P (2007) A Scandinavian model of innovative product development. In: The future of product development (ISBN: 978-3-540-69819-7), Springer-Verlag, Berlin, pp. 269–278

Olesen J (1992) Concurrent development in manufacturing—based upon dispositional mechanisms. PhD thesis, Technical University of Denmark

Pahl G, Beitz W (2007) Engineering design. A systematic approach, 3rd edn. Springer, London (1st edn, 1977)

Roozenburg NFM, Eekels J (1996) Product design: fundamentals and methods. Wiley, Chichester

Part II
The Design Machinery

Chapter 3
Designers and Their Knowing

Designers' knowing is shaped by their education, training, experience, and research. In order to empower the designer we must be conscious of both what they *know* and what they *do* when they design. Professional and branch knowledge relate to technology, product, and market insights; our focus is on design knowledge, which concerns operation in the design activity. Designers generate and use a rich spectrum of models for creating solutions, insight, and predictions. These are often integrated into methods with the aim of rationalization, transfer of experience, and increased productivity. Thus their use is not just a question of knowing the method, but also possessing the correct mindset. This mindset is related to background, context, and method validity, and is key to transforming a method from a series of instructions into a truly productive means of design. In doing this the designer brings to bear their knowledge, skills, and abilities in design methods as a powerful actor in the design team.

3.1 Dimensions of Knowing

In many situations all that is needed for a designer to be creative is pencil and paper. However, situations in practice are typically much more complex, involving many aspects of design knowledge. As such this chapter aims to answer what is needed for designers to create previously unknown and unseen things, based on the following questions: what do we understand by **practice** and how is it composed; and what underpins **designers' knowledge** and the efficiency and effectiveness of their work? In answering these questions we will move closer to understanding the mechanisms behind design and design knowledge, i.e. one part of the 'design machinery' of Part II. Figure 3.1 gives a foundation for this exploration by showing a rough articulation of the various aspects of design knowledge.

M.M. Andreasen et al., *Conceptual Design*, DOI 10.1007/978-3-319-19839-2_3

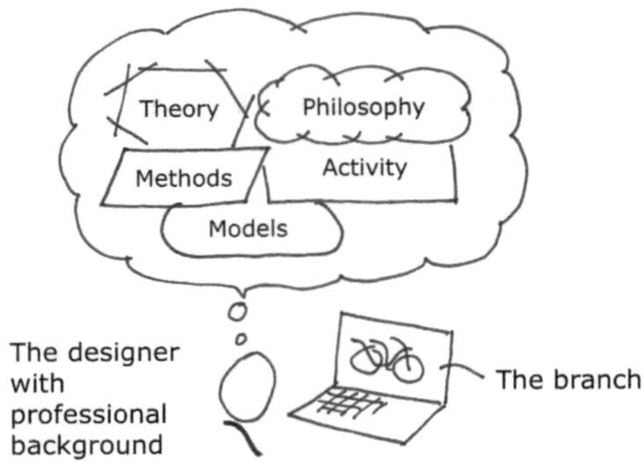

Fig. 3.1 Key knowledge elements in design

Figure 3.1's knowledge elements can be broadly grouped into three areas. **Design knowledge** is generally related to the rationalization and empowerment of designing. It is based on education and research inputs. **Professional knowledge** reflects education determined by the specific demands of industry and engineering domain. It is based upon education and experience. Finally, **Branch knowledge** concerns technology, market, specific products, and the customer. This knowledge is almost exclusively based on experience and is not normally shared between companies.

The distinction here is ambiguous. Some professions are supported by rich design knowledge and highly developed design theories, e.g. ship design, aircraft design, and the food industry. Other professions are characterized by highly developed professional knowledge and craftsmanship. However, all professions build on fundamental design knowledge. As such, we focus on these foundational elements in this chapter.

- Section 3.2 explores **designers and their practice**.
- Section 3.3 examines how designers create and use models to support this practice in **'models, the designers' language'**.
- Section 3.4 builds on these models to discuss **methods and tools**.
- Section 3.5 explores how **competence and skills** affect method use in design.
- Section 3.6 brings these elements together by looking at **design philosophy and principles**.
- Section 3.7 closes the chapter by discussing how the design process is influenced by the designers' **perceptions of design activity**.

3.2 Designers and Their Practice

'Practice', in the pursuit of craft or profession, is core to our ideas of design. In industry the concept of benchmarking is seen as 'best practice'. Benchmarking tries to determine if a procedure (method, technique) has consistently given results superior to those achieved with other means. This implies a type of comparative practice, which develops by experience, learning, and negotiation as new means

Fig. 3.2 Teamwork in an educational situation, where the students start developing their understanding of practice, courtesy Torben Lenau

are compared to old. However, in reality much of this progression is reliant on the designers and their ability to articulate their practices, Fig. 3.2. This is particularly important in industry where design is often erroneously seen as an 'art or craft' rather than knowledge that can be learned.

At the core of this book is the idea that design can be learned and constitutes a way of thinking similar to the sciences or humanities. This is also recognized by Cross (2006) who describes the values in each culture as (emphasis added):

- *"In the* ***sciences****: objectivity, rationality, neutrality, and a concern for 'truth'.*
- *In the* ***humanities****: subjectivity, imagination, commitment, and a concern for 'justice'.*
- *In* ***design****: practicality, ingenuity, empathy, and a concern for 'appropriateness'."*

Elaborating this, Cross (2006) describes three designerly ways of knowing, to which we add a fourth point.

- *"Designers tackle* ***'ill-defined' problems****.*
- *Their mode of* ***problem-solving*** *is 'solution-focused'.*
- *Their mode of* ***thinking*** *is 'constructive'."*
- They use **languages** to both 'read and spell' design entities, allowing abstract requirements to be transformed into concrete objects.

Designers show common traits in their skills and capabilities. In order to characterize these common "designerly approaches" (Kleinsmann et al. 2012) used a design game to interview 55 designers with experience in design and/or design management. From this she identified six 'images' of approaches:

- "***Design as product driven****: focuses on the actual product to be designed. A multidisciplinary team is formed to meet the complexity of demands and integrate all knowledge needed, while using designerly approach.*
- ***Design as solution driven****: focuses on the wicked problems within the product or system to be designed. An engineering design team is formed to solve the problems encountered in a designerly way.*

- ***Design as value driven****: focuses on creating value for the end-user. Contextual complex problems, with diverse stakeholders, are reframed to create better value for people within the particular context.*
- ***Design as business driven****: focuses on innovation as a strategy for a sustainable company. The designerly approach strengthens the product portfolio and business development.*
- ***Design as experience driven****: focuses on (social) innovation through the facilitation of a stakeholder dialogue. Through co-designing experiences value is created for all stakeholders.*
- ***Design as vision driven****: focuses on the designers' personal style. The designer is invited to make his statement through the design."*

Kleinsmann's images show the range of designers' beliefs, strategies, and approaches, which we also think reflect differences in their mental processes and reasoning. The images' common elements are articulated in the list of characteristics above. Throughout this book we explore this confrontation between structured, systematic work patterns and the designer's cognitive approaches.

3.2.1 Design Practice

Although all design builds on common principles it is still possible to identify typical design practices from the finished products. For example, Scandinavian, German Bauhaus, and American product styles are distinctly recognizable. Further, aircraft designers claim to be able to see different designers 'fingerprints' in the final planes. In spite of the different images and fingerprints we find it interesting to try and understand how a design practice is realized in teams and departments of companies.

> **Definition: Design practice** is a **work pattern** based on the type of product to be designed, the company—its past design activities and future aspirations, and an understanding of which approach leads to the highest probability of success.

Thus it is possible to describe common design knowledge elements linked to the realization of design in practice as illustrated in Fig. 3.3 (Andreasen 2009, 2011).

3.3 Models, the Designers' Language

Models are both tools and the documentation of the design results. Indeed, it is hardly possible to imagine design without some form of models being used! Modelling describes the activity through which designers utilize models, i.e. artefacts created in this activity. Models have multiple applications in design.

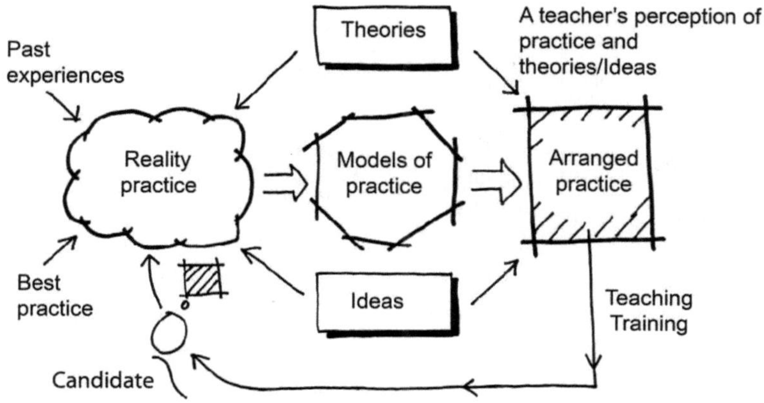

Fig. 3.3 Staging of industrial practice with respect to theories, ideas, and experiences (Andreasen 2011)

However, before we go on, let us clarify that *design researchers* are interested in *modelling **of** design,* i.e. creating theories or frameworks that explain designing; while *designers* are interested in *modelling **in** design,* i.e. creating and applying models to support the design activity. It is this second perspective that we focus on here, as illustrated in Fig. 3.4.

The figure above shows the following phenomena:

- Something is being modelled: the **phenomenon** or **object**. This is either physically present or situated in a designer's mind as a mental model.
- The **model** articulates the attributes of the phenomenon or object through its composition, representing what is modelled.
- **Creating the model** builds on **theory** or **vision** of the phenomenon or object combined with a suitable **way of modelling**.
- **Applying the model** results in a simulation of the phenomenon or object (Roozenburg and Eekels 1996). Results are then generated and consequences derived in a more or less active process, e.g. inputting data verses reading a drawing.

Models shape the designer's knowledge and are core to the articulation of the product and the design process. This is achieved via a wide range of models, both complete and intermediate.

3.3.1 Phenomenon and Model

The product does not exist during the design process, only finally materializing during production. Thus, designers deal with the 'product' as articulated through models. In this way models form a type of 'design language' (Tjalve and

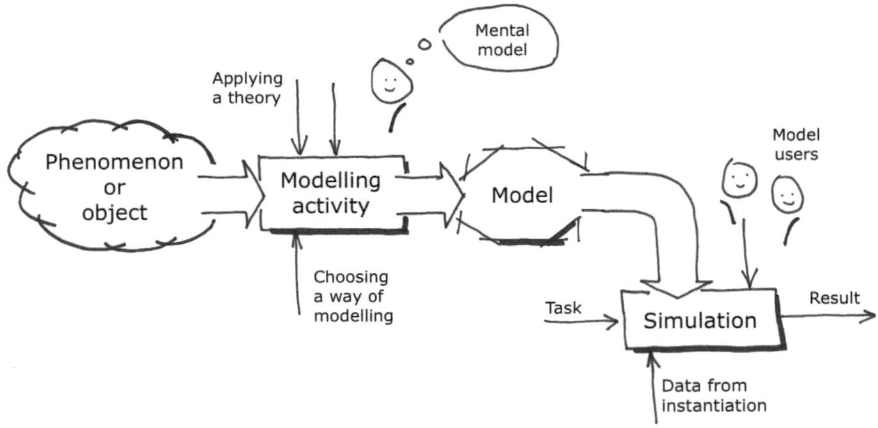

Fig. 3.4 Describing the modelling activity (Maier et al. 2014)

Andreasen 1975), with different models being applied for different purposes in the process, e.g. defining, synthesizing, communicating, clarifying, justifying, and documenting for manufacture. As models represent 'something' they share certain characteristics with this something, Fig. 3.5. The 'something' can be a phenomenon, i.e. an observable episode or articulation of an activity or operation, or an object, i.e. an 'operand' entity created during design, e.g. a part description. As such, we can define a model:

> **Definition:** A **model** is a human creation that carries attributes similar to the modelled phenomenon or object.

Models are normally described as 'representing, reproducing or visualizing', with definitions focusing on the models being 'in some other form' or 'using a different conceptual or representational system'. Thus, a model is a translation into another 'language', one of representation. This is normally coupled with simplification or reduction of complexity, i.e. models are partial. Models can be assessed with regard to, e.g. goodness, accuracy, relevance, profitability, clarity, or productivity. A clarifying discussion is given by Vermaas (2014).

The attributes related to the model can be both the characteristics shared with the object/phenomenon and the models own articulation of its properties, i.e. its explanative power (German: *Aussagefähigheit)*. A core characteristic of models is 'similarity', which relates to appearance, behaviour, and composition. This relationship between the model and the phenomenon is seen as a theory, see Fig. 3.4 (Tomiyama 2006), i.e. the modeller either makes assumptions or builds on known theories about the nature of the phenomenon. When a sketch is shown to designers they often see or reason about things that are not explicitly illustrated, for example, how the sketched object might be assembled, based on their own assumptions.

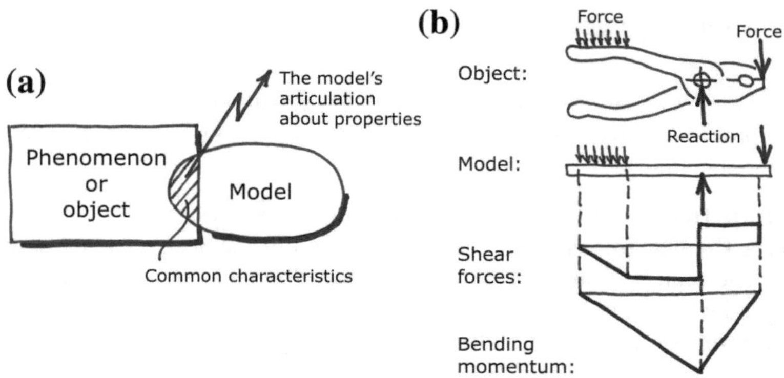

Fig. 3.5 **a** A model has characteristics in common with, or similar to, the object and is thus able to illustrate certain properties shared with the object **b** e.g. a pair of pliers is modelled as a beam, allowing shear forces and bending moment to be calculated, inspired by Samuel and Weir (1999)

3.3.2 The Model and Modelling Techniques

As defined above models are created to represent something. Thus a huge palette of model types is available, with no constraint on the medium: both model and object may be steel, clay, paper, etc. As such, the modeller may master many specific modelling languages, e.g. mathematics, logic, Simulink, Systems Theory or Entity Relation modelling. Further, depending on the language used this can dominate the modelling activity. Thus, phenomena are often decomposed as a chain of models with each transformation based on different theory as illustrated in Fig. 3.6. Here a phenomenon is transformed into an information model via, e.g. object orientated modelling. However, it is important to realize that these tools do not tell us about the nature of the design phenomena but rather the nature of information structures.

> The step of phenomenon modelling comes before any information structuring considerations.

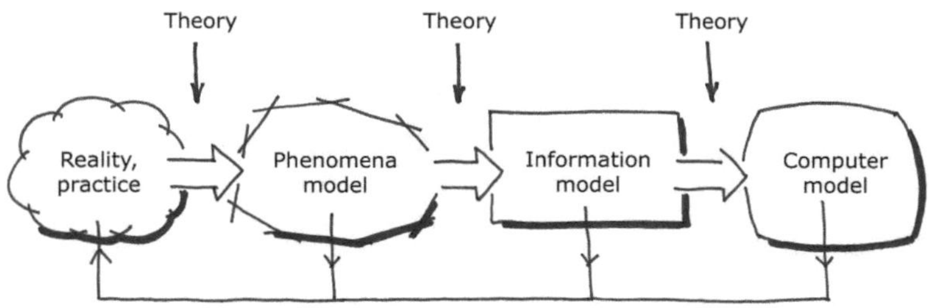

Fig. 3.6 From a design phenomenon to a computer model (Duffy and Andreasen 1995)

3.3.3 Model Applications

Models are typically applied in key satiations to support the progression of the design process as exemplified in Fig. 3.7. Based on this we can distil common model applications.

- **Capture the unknown**: supporting ideation and synthesis.
- **Define the design**: supporting the specification of the design such that its characteristics can be translated into other types of models and the specification of the product's manufacture.
- **Communication**: supporting the design activity between team members, stakeholders, specialists, users, and others. Here models can act as *boundary objects,* i.e. translators, or for *self-communication,* i.e. supporting a dialogue between 'mind and design'.
- **Obtain insight**: giving insight into the phenomenon through *simulation,* e.g. manipulating the model's parameters to observe the design's behaviour under different conditions.
- **Manage**: supporting the management of the design activity by, e.g. describing an ideal sequence of operations or helping facilitate coordination. As above the models parameters can be supplemented with, e.g. operation time and resources.

Although models may simultaneously serve a number of the roles outlined here they are all mediated by their application in the design team. Further, they all contribute to the gradual design **progression**, helping to synthesize, navigate, and make decisions leading to the final design (Maier et al. 2014). We will explore each of these roles in the following sections.

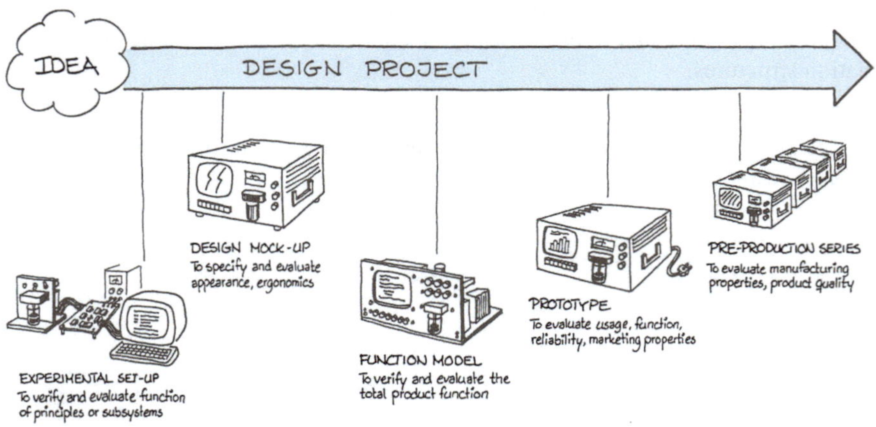

Fig. 3.7 Examples of mechatronic models and their purpose (Buur and Andreasen 1989)

3.3.4 Capturing the Unknown

Generating ideas and concepts based on perception of need, opportunity, intention, and task formulation means to create something unknown. This unknown may be a lack of known principles or a gap in our understanding of how the product will meet the use activity. Further, this unknown points to the endless possibilities confronting a designer. The designer explores the unknown based on **imagination**, and the ability to **foresee and predict**, coupled with theoretical and experimental insight, and supported by models. Initially models are simple and quick to produce, e.g. sequences of sketches articulating how the product interacts with elements in context. These sketched models allow the designer to read, interpret, and imagine how it might work, etc. Further, models allow prediction of the solution's properties, e.g. size, weight or speed, based on subjective experience, interpretation, and adding imagined details to the solution.

Example:
Physical models: Handpresso. We use the design of the Handpresso here and in two further examples on conceptualization Fig. 7.1, and design, Fig. 9.10. Here we will focus on gaining early insight into the product's functionality and the project's tractability. Handpresso is a portable device for making espresso coffee using standard coffee capsules. The inventor worked out different concepts: with a hand pump, with an electrical motor, and with an external energy source. Some of the physical models used for clarification of functionality are shown in Fig. 3.8. Note how 'off the shelf' components are used, giving the designer a good and early feeling of the embodiment and functionality.

Fig. 3.8 Early models of the Handpresso concept, supporting the designer's imagination of functionality and tractability (*Courtesy* Nielsen Innovation, France)

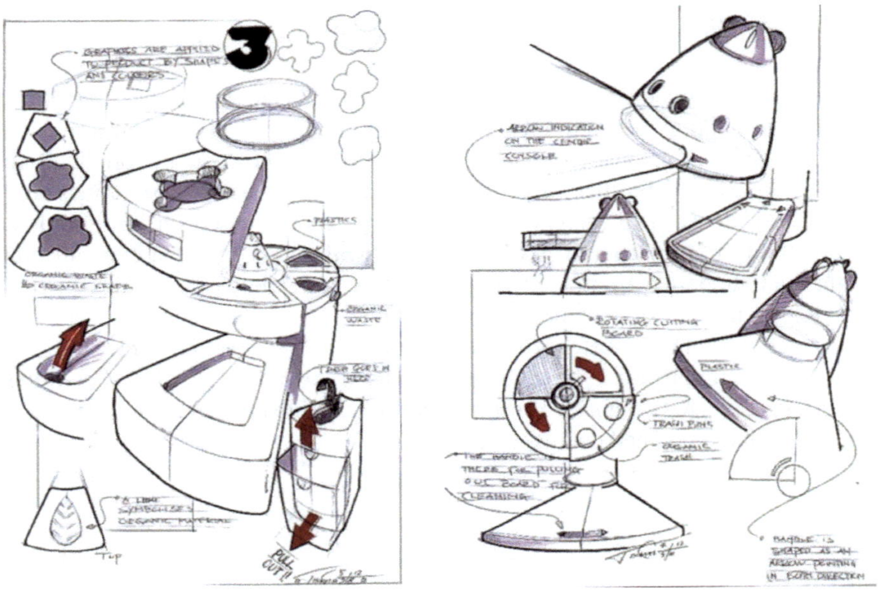

Fig. 3.9 A sequence of sketches made for a designer's self-communication and supporting the imagination. *Courtesy* Thomas Ulrik Christiansen

Ideas start as **mental models** supported by a mental image, i.e. our brain's ability to visualize imagined things and situations (Fig. 3.9). In this way sketching activity is often experienced as a direct conversation between the brain, the hand, and the pencil. Early models can also take other forms, such as 3D clay or foam sculptures, which can also be used to facilitate this reflective dialogue. Early models have no formalization as long as the designer finds them supportive. However, externalization and communication become important in teamwork, making choice of coding and understanding critical (see **communicating design** below).

3.3.5 Defining the Design

Design describes a gradual **concretization** and **detailing** of a product as illustrated in Fig. 3.10. Here, each part and assembly is documented such that the production can be planned and performed. There are many different types of models used for articulating criteria, for capturing use, form, functionality, for clarifying spatial layout and interfaces, etc. Each of these models constitutes one step in the gradual definition of the design but few are related to the structure of the parts. In particular, computer aided design supports later stage part definition but is not generally helpful in the early phases where geometries have not yet been defined.

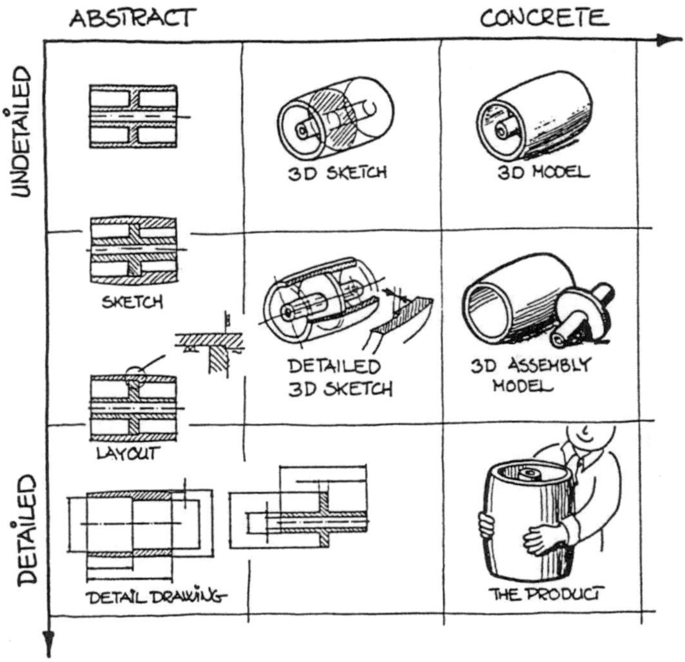

Fig. 3.10 Independent processes of concretization and detailing (Buur and Andreasen 1989)

Definition of the design is strongly supported by representing the design as a system (explored further in Chap. 8). When the design is defined as a system model it serves as a framework for bringing together other types of partial models, e.g. providing a structure for mapping product related knowledge to the product's life phases. Special types of models are used to define and communicate the final design for production, e.g. assembly drawings and bill of materials. In this way models are key to design communication.

3.3.6 Communicating Design

Cooperating in teams, dialogue with management, and interaction with users and other stakeholders all demand effective communication of the design. Models are central to such communication activities, which can be principally illustrated as in Fig. 3.11.

Models represent a **coding,** where something is articulated as a result of the chosen representation and the receivers ability to de-code this, e.g. their ability to read and interpret the model. The model contains a message and is created in a certain medium. Further, the communication can be disturbed by noise in, e.g. the aspects of the model not related to its purpose or loses in the transfer through, e.g. misunderstandings as illustrated in Fig. 3.12.

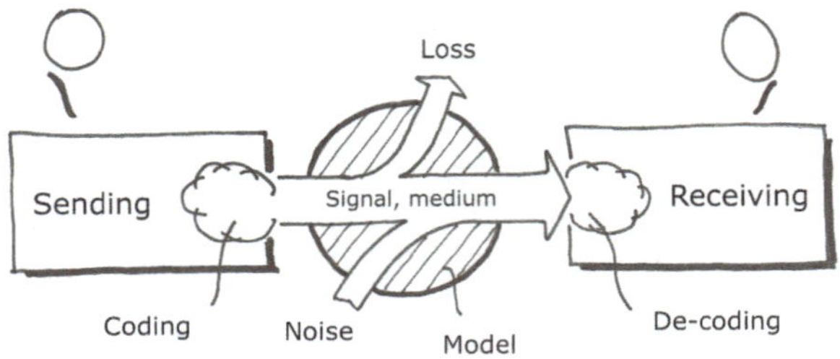

Fig. 3.11 Communication model showing the role of coding

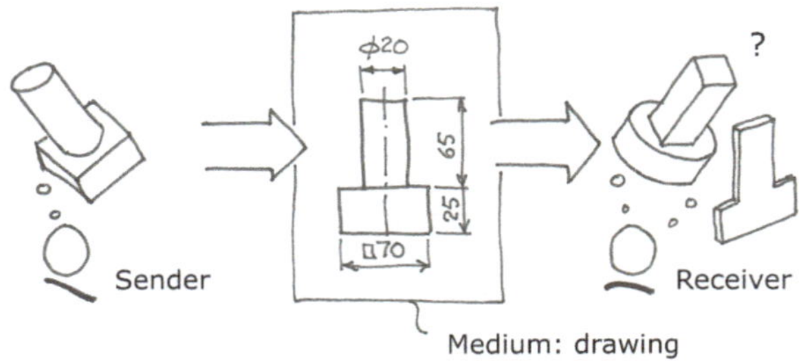

Fig. 3.12 Coding a standard mechanical drawing and de-coding with a misunderstanding

Example:
Models in communication. In Denmark pavements are traditionally a combination of concrete flagstones and smaller pave stones. Laying these is time-consuming and hard on the craftsmen's knees. In order to solve this a machine was proposed that would allow a standing work position. Figure 3.13 shows from left to right the original work situation, interim experiments on sliding mechanisms, and the final experiments with real craftsmen. Through these models and experiments the designer's ideas and the professional's tacit experiences were combined. Figure 3.14 shows a CAD drawing of the final machine design. Courtesy Andreas Lolle and Steffen Nielsen.

Fig. 3.13 Models and experiments leading to a concept for a paving machine

Fig. 3.14 CAD drawing of the final machine concept

Creating the best model in the actual situation is influenced by a number of aspects initially identified by Tjalve et al. (1979):

- The **object** to be modelled, its context, and/or its different perspectives, e.g. use, functional units, or part structure.
- The **property** to be illustrated by the model. This is related to the object's properties, e.g. aesthetics and complexity.
- The **purpose** of the communication, e.g. defining, showing ideas, embodiment, or verifying certain aspects.
- The **receiver** of the model, e.g. the designers themselves, colleagues, clients, etc., or communication with a computer.

Similarly, the communicating model to be produced is determined by the following characteristics:

- The **code** chosen for visualization, e.g. three-dimensional image, symbolic representation (mathematical terms, electrical or mechanical symbols, etc.) or just using language.
- The **medium** used, e.g. conversation, graphics (photographs, movie, video, computer display, etc.) or three-dimensional materialization (paper, cardboard, wood, etc.).
- The level of **abstraction** and the number of **details**.

Models are not just a translation or representation, but a way of seeing the world and making it visible and tangible. Models facilitate both individual thinking and interactive communication, Henderson (1999): *"The visual culture of engineers is not made up of school-learned drafting conventions but rather the everyday practices of sketching, drawing, and drafting that construct their visual culture—a visual culture that in turn constructs what and how designers see".*

3.3.7 Models for Insight

Design projects are high risk because of, for example, misinterpretation of need and market, or misfit between product and customer expectations. Designers incrementally reduce and clarify these risks as they progress and concretize the solution. Thus, models provide a means for giving insight into situations that cannot wait for this incremental clarification. Modelling is a substitute for true clarification when the product does not yet exist, when it is too expensive or time-consuming to create a prototype, or when the act of modelling directly supports the search for solutions. In this way models become a means for buying insight.

The question "how does a model actually model?" is difficult to answer generally because of the many types of model. Fundamentally, a model's characteristics can be manipulated and/or interpreted to reflect the phenomenon's/objects' behaviours and properties. This can be through physical laws or through apparent similarities between the phenomenon and the model.

Example:
Modelling use. Tjalve (1979) used models to choose between design alternatives, as shown in Fig. 3.15, for a piece of laboratory equipment. Here (a) shows one of four possible spatial arrangements for the functional units and operator's workspace. Later in the design process (b) was used to communicate the selected concept. This was modelled in 1:1 plastic foam together with an operator simulating the job in a real time. In this experiment the operator felt that the concept was 'oppressive' and subsequently helped identify key design changes for the final product (c).

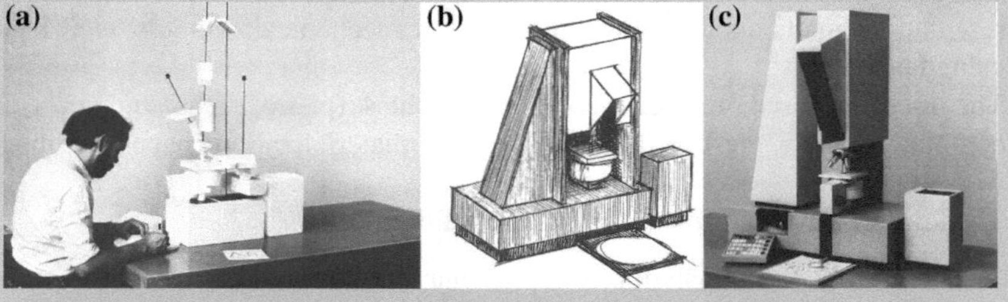

Fig. 3.15 Various models of a new laboratory device, from sketch to final product (Tjalve 1979)

3.3.8 Models for Managing

Project-oriented design activity is modelled in the 'design process' showing the planned activities, their relationships, and timing, as well as the allocation of teams, personnel, and tools. Companies add activities based on the nature of the company and task to create **procedures,** see Chap. 9. These procedures are then fitted to the specific project and can be used as the basis for time and resource plans. This model thus provides a means for comparing the actual activities and progress to the plan, supporting coordination and helping improvement via reflection and deviation analysis.

All the model types mentioned above can be used to support design management. System models or models showing the design's parts, functional units, etc. support **composition** activities. **Interaction** and **interfaces** between the design's elements can be captured and manipulated in design matrices (Eppinger and Browning 2012). Finally, design relationships can be used to manage the links between the tasks in a project. These clarify the transfer of design information and design **dependencies** in a design matrix that supports the scheduling of a project.

3.3.9 Creative Use of Models

Designers again and again face situations that can be seen as 'problem-solving' but which can also be seen as issues to be scrutinized, reframed, and envisioned. Here designers use models to support exploration and structuring of a problem or to help find solutions. The models used here are diverse and often invented by the designers themselves. As such, although these support teams in developing shared understanding and force designers to consider new perspectives, they can be of limited use in communication outside of the team. In particular, the exploration of new perspectives and fostering shared understanding highlight the power of model making, and that models, just like new products, are invented and applied in intermediate steps throughout the design process.

Time should be made in a project to allow designers to create and use models.

Sketching is an important tool for creating many types of models and ideas, fast, rough, and dirty. "*Unfortunately, in design education the sketching process is treated mostly as a skill, rather than a powerful tool to structure design problems and to develop solutions*", Restropo (2004). The use of models for conceptualization is treated in details in Chap. 7.

In the 1990s our team at DTU aimed to create a 'Designer's Workbench' (Andreasen 1992; Svendsen 1994; Jensen 1999; Mortensen 1999). In doing this we realized that a core challenge was creating a 'design language' or way of articulating the synthesis and modelling of a design. This language aimed to support the designer's cognitive operations as well as the computer's handling of the formalized information. We asked: What should govern such a language: the designers' cognition, the modelling, the methods or the computers' requirements? We found then, as now, that although models as discussed above are key, they work in conjunction with methods and tools. In this way they can adapt to and effectively support synthesis throughout the design process. We think it is important to get deeper insight into a 'designerly way of modelling' to maintain the power we see in the use of models: the drawings' surprising ability to let the designer '*see things not explicitly on the page*', the early models' role in linking mind, hands, and eyes, and the models' role in supporting synthesis.

3.4 Methods and Tools

A company's design practice is composed of a set of tools (for information handling, calculations, and graphics) and methods (for monitoring the design process and solving design sub-tasks). Despite this central role method/tool selection is often neglected in practice. This is in contrast to research, where methods and tools are dominant. As such, this section aims to extend our understanding of methods and how they can be used effectively in design.

3.4.1 What Is a Method?

Methods are at the core of many human activities: in the **sciences**: controlled experiment, classification, and analysis; in the **humanities**: analogy, metaphor, and evaluation; and in **design**: modelling, pattern formulation, and synthesis (Cross 2006). Methods represent a continuum from precise calculation with distinct numerical solutions to simple modes of reasoning where solutions are fuzzy. Further, Araujo (2001) found that the words method and tool are used synonymously. Roozenburg and Eekels (1996) see methods as composed of rules, with many design methods being of a heuristic nature because "*they aid in finding something, but there is no guarantee that it will be found and by everyone*". In contrast an algorithm is a sequence of operations leading to a result, e.g. the decision support method 'weighted criteria' is formulated as an algorithm but at its core it is heuristic. This leads to our definition of a method.

Definition: A **method** is a goal-oriented rationalization or simplification of engineering work in the form of a standardized work description.

Methods can originate from work practices or be created by designers or researchers. Some methods make seemingly impossible tasks more feasible by supporting creative work, while other methods simply make tasks more rational. In design, there is a broad spectrum of methods of very different nature, validity, and applicability. To reduce confusion we make a distinction between engineering methods and design methods, even if they overlap in reality:

Engineering methods originate in engineering insight into an artefact's nature. They are normally analytical and calculation based, giving a single metric output or set of parameter values as a result of different inputs/conditions. For this type of method it is possible to rigorously prove that the result matches reality with a certain precision. These methods typically build on an actual model and parameterization of a phenomenon. Thus, the validity of the result depends on the designers' ability to model and judge the method's applicability.

Design methods are normally synthesis and clarification oriented. They create a staging for creative activities or structured search, classification, discussion, evaluation, and decision-making. This type of method typically has few mechanisms for obtaining results and can sometimes have a very low probability of success. The application of design methods is critically dependent on the designer's known-how, mindset, and interpretation of how the method should be used. It is difficult to rigorously validate this type of method due to the many uncontrolled influencing factors.

Example:
Design method application
When designing a service the Customer Activity Cycle is an important model of the product's lifecycle elements and their potential service need. Figure 3.16 shows a situation where an 'ideal' model of a ship's lifecycle is on the table and various stakeholders identify possible services and networking set-ups. The method describes the application of the model and staging the situation.

Fig. 3.16 Design method application: customer activity cycle, *courtesy* Tim McAloone

In the Delft Design Guide we find the following classes of design methods (van Boeijen et al. 2013):

- **Creating a design goal**: for strategy and vision, design specification, and problem definition.
- **Creating product ideas and concepts**: for concept mapping, function analysis, scenario writing, morphology, etc. (see Chap. 7).
- **Decisions and selection**: for managing value and weighing objectives, etc.
- **Evaluation of product features**: for simulation, testing, concept selection, and analysis.

3.4.2 Methods' Origins

The earliest methods aimed to support engineering design details such as strength and cost, while later methods aimed to generally improve design work and education (Jones 1992). A recent, collection of design methods can be found in the mentioned 'Delft Design Guide' by Boeijen et al. (2013). A dominant area in design research is Design Methodology, i.e. the study of the work procedures designers follow (Roozenburg and Eekels 1996). Design Methodology gives us models of design, methods, and techniques and a system of concepts and corresponding terminology.

Design methods may have their origin in practice and experience (e.g. weighted criteria for evaluation), in understanding of human cognition (e.g. brainstorm-like methods), in understanding of an artefact's nature (e.g. morphology), in understanding cooperative teamwork, or in combinations of these (e.g. House of Quality). The method itself is normally articulated via verbal instructions related to a tool in the form of a scheme/model or to the staging of actions. As we shall see the formulation of this instruction is critical to the application of a method in practice.

3.4.3 Methods' Formulation and Application

A condition for using a method is that the user can come to the same insight as the method creator's interpretation of the use situation. Araujo (2001) illustrated this expected understanding of the designer's situation as shown in Fig. 3.17. This shows that beside the method's instructions there are a number of other critical elements (Andreasen 2011):

- Interpretation and understanding of the task.
- Interpretation of the context, judging appropriateness, and timeliness of using the method.

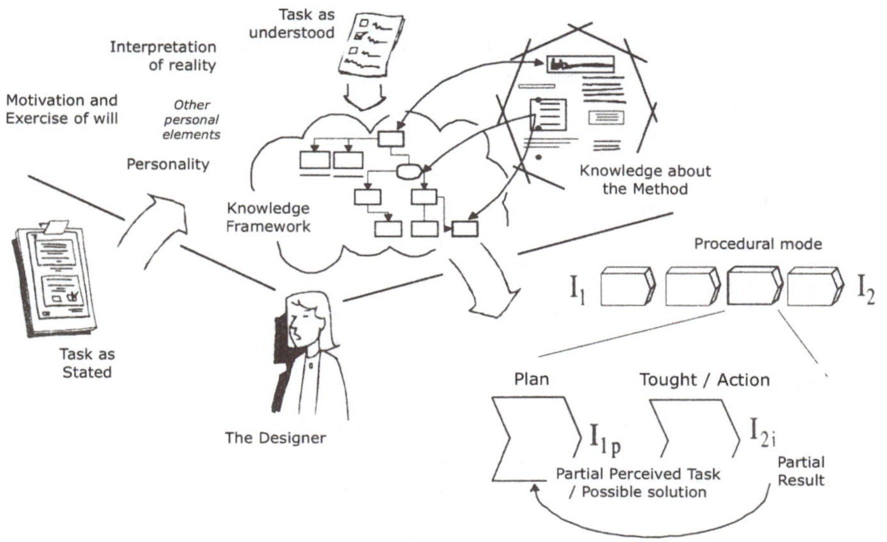

Fig. 3.17 Understanding the designer's situation when executing a method (Araujo 2001)

- Imagining and planning the use of the method based on knowledge, experience, and skill.
- Understanding the theory related to the phenomenon to which the method is applied, translating its concepts into the method's language and models.
- Understanding the use of the method and proper procedure to undertake when designing, judging the validity of the method and its results.

In practice it can be difficult to clearly isolate a method from a designer's wider practice. However, in any activity the goal, the result, the designer's knowledge, and the activity itself reflect the domain, e.g. designing a pump, as illustrated in Fig. 3.18. Although methods are key to design, design is not simply a sequence of methods, raising the question: What makes a good method?

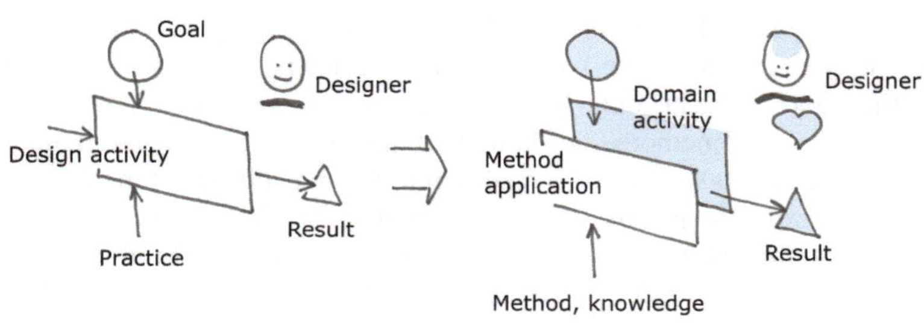

Fig. 3.18 Method's use in practice separating the method application and domain activity

A method should always be understood in combination with the design activity it will support. The result of the method is rarely the result of the wider activity.

3.4.4 What Makes a Good Method?

As noted above it can be difficult to isolate specific methods in practice. This has prompted significant study into the deficiencies of design methodology revealing three main issues (Cantamessa 1997; Araujo 2001; Birkhofer et al. 2005; Badke-Shaub et al. 2011):

- Difficulties in defining the **performance** of methods.
- Difficulties in how methods are **presented** and formulated.
- Difficulties in applying methods in a wider **process**.

Many of these issues stem from an erroneous belief that methods are rational 'machines' for producing results. Thus, when a method does not give the expected results the problem is attributed to the method itself (Jensen and Andreasen 2010). Andreasen (2011) sums this up by stating "*that even if designers often attribute their results to methods (also in situations where it is obvious that the method is not doing the work), we can only get proper explanations of their use by understanding how the designers talk and feel about methods, individually and collectively*". As such, methods cannot be seen as rational machines, instead they must be carefully understood in context. Thus, when a designer masters a method it simply becomes one part of 'how we always do it'. Daalhuisen (2014) states that a method works, when or because it '*extends the ability to design, enhances reflection on designing, and enhances learning to design*'. The robustness, flexibility, and execution of a method lie with the designer and cannot be transferred through a verbal instruction only. Similarly, Badke-Schaub et al. (2011) advocate a 'designer centred methodology' as well as '*methodology support in an individually and situation-oriented manner to meet the demands of the user, and thus increase design performance*'.

Validation relates to the actual effects and productivity of a method but also to a method's theoretical background. The results from applying a method depend on its timely and appropriate use. Here, appropriateness is not just a matter of the method and thus can only be truly judged retrospectively. The relationship between theory and methods, i.e. how methods build on theory, is a matter that only indirectly affects the methods quality through the abilities of the designer.

A typical example is this book's use of Domain Theory (Chap. 8). Based on this we have created powerful methods for modularization and platform approaches (Hvam et al. 2008; Harlou 2006). The theory itself is 'model based' and is composed of concepts and mental constructs that cannot be proven or falsified. Its 'truth' comes via its support of the designer's mindset and thus their successful application of the methods.

3.4.5 *Mindset*

When we look at design process models, which are generally seen as methods, it is evident that the model only contains a small percentage of the understanding necessary for completing the design. The same is true for most design methods descriptions: they only describe the 'machinery' of the method, not the basic understanding of the actual phenomenon or the 'why' dimensions. This, in part, contributes to the over-optimistic belief that methods can be used directly by simply following the instructions, and even more optimistically that they lead to a masterful, successful result on their first application. Understanding the proper context and phenomena related to method application we call **mindset** (Andreasen 2003). In particular this highlights socio-technical aspects and appropriate/timely application as key challenges.

Definition: A **mindset** is the proper understanding of a method's use in accordance with the designer's reality (interpretation of task, situation, execution, validation, etc.), and the method's background and proper use.

Daalhuisen (2014) explains mindset as follows: '*It is not only knowledge and experience that allows a designer to use a method beneficially; it is also belief in its added value, trust in its applicability and preference for its use*'. Mindset and method description unite in method application as illustrated in Fig. 3.19, together with the staging of the method (see Chap. 4).

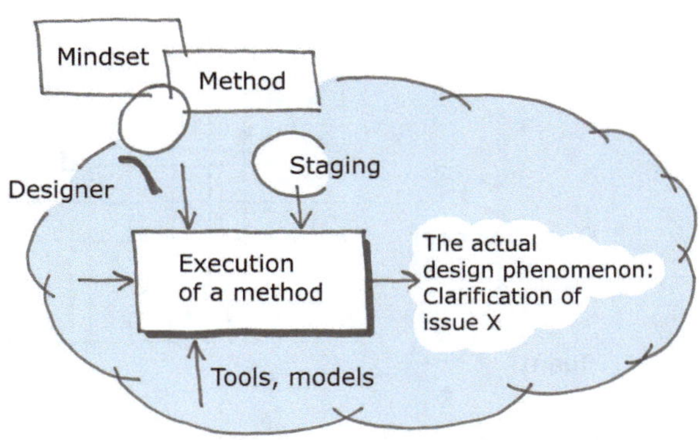

Fig. 3.19 The execution of a method based on mindset, method description, and staging

Example:
Application of the House of Quality (QFD) method. The basic idea of the QFD method is to compare customer reactions (called Voice of the Customer) to a launched product [(g) in Fig. 3.20] with the company's efforts to build good attributes into that product. Based on this mapping the aim is to identify improvements for the next generation of products. The core is the scheme illustrated in Fig. 3.20, which has the following elements: (a)—list of customer statements, (b)—list of product attributes, (c)—customer's articulation of the issues' satisfaction compared to competing products, (d)—mapping of relationships answering the question: What is the influence of the attributes on the customer's issues? The strength of this relationship is graded on three levels. (e) Calculates sum score for each attribute to show their relative importance for customer satisfaction, based on the strength of relationships and their importance. Finally, the scheme's 'roof' (f) maps interactions between technical parameters to support assessment of feasibility and possible trade-off areas (see Chap. 12).
QFD is basically a scheme, i.e. a tool plus a method mechanism, for confronting a company's supposed values with the true customer values. This mechanism is supported by the designers' mindset, i.e. understanding the transformation and confrontation of parameters, and the business effects of customer values. The staging of the method [combing the actors (q), (r), and (s) with the rest of the team (t)] aims to foster shared understanding of the success of past designs and the challenges faced by the new development and its goal formulation.

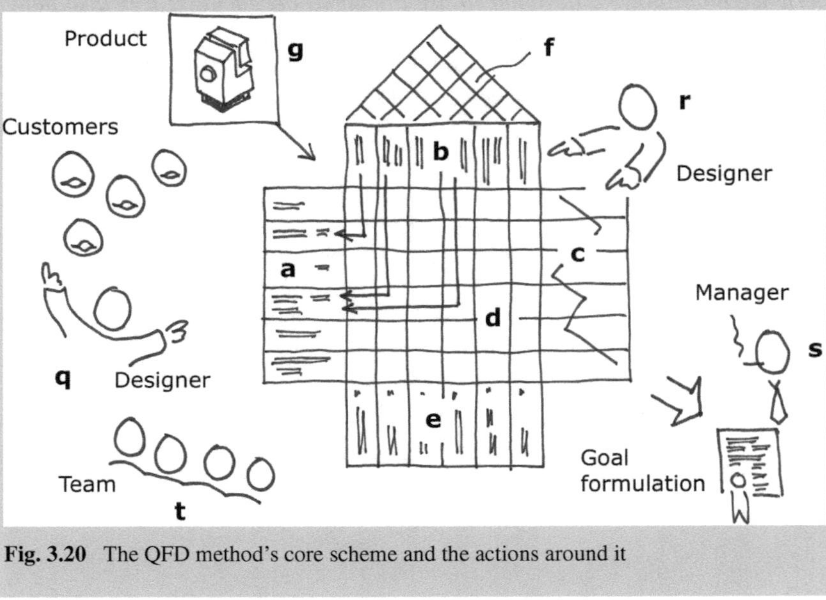

Fig. 3.20 The QFD method's core scheme and the actions around it

The heuristic and socio-technical aspects of QFD are also shown in Fig. 3.20 through the actors' roles. The designer's (q) role is to interpret and translate the voice of the customer into the list A. The engineer's (r) role is to map and grade the influences based on their interpretation of what technical parameters the product (should) possess. The manager's (s) role is to ensure that the goal formulation balances the company's strategy/policies and the need to create good business.

Bringing these points together leads to the following:

Both **design practice** in its use of methods and **design research** in its creation of design support need to build effective **mindsets** related to the design phenomena.

Our focus on mindset underlines the difficulties scholars and consultants face when articulating proposals for theories, methods, and models such that the designer can effectively use them. As such, we offer the general advice: *Dig one step deeper!*

- Methods are **fuzzy**: they can lead to a result even if used incorrectly, and often give poor/no results even if used properly.
- Methods are **narrow**: they treat a single issue and require the user to create the totality.
- Methods are **nonlinear** (even if many claim to be!): synthesis is fundamentally 'trial and error'. Thus even methods like algorithms must respect this.
- Methods must be **learned**: experience and rich understanding are required for effective use.
- Methods challenge the designer's **mindset**: is the proper insight present?

3.5 Competences and Skills

Design is characterized by specific knowledge and ways of thinking. Thus, we must understand what competencies are required in design and how we can understand design from this perspective. Throughout the book we deal with design knowledge, and find the following definition from Sim and Duffy (1998) powerful:

Definition: "***Knowledge*** *is a competence notion that promotes rational action depending on goal(s) to be achieved for a particular activity*".

From this we can see that knowledge is linked to action and understanding. Thus, it is only via a designer's actions that we are able to 'see' their knowledge in the form of competences and skills.

Definitions: **Competences** are the ability to actually realize knowledge as rational actions depending on the goal and the activity.
Definition: **Skills** are learned abilities and the capacity to carry out a task.

Key to realizing these actions is motivation and attitude. **Attitude** provides the driver for turning knowledge, professionalism, abilities, and competences into value based, rational activities.

A precondition for identifying the specific competences and skills characterizing designers is observing their behaviour. Developing these competences takes time and experience in design practice. We find it interesting that competences and skills seem to be a dimension of knowledge that is different from the 'tool-oriented' dimension of design.

In order to understand these terms in a design context it is helpful to consider the concrete example of **competences in conceptualization**:

- Show **awareness**: react to and sympathize with actions and people to build up insight into needs, values, and problems.
- Show **synthesis competence**: creatively seek solutions and systematically articulate concepts, embodiments, etc., in order to gradually concretise your ideas.
- Show **innovative thinking**: unfold perceptions of the solution space and seek out limitations and possibilities.
- Show **integrative reasoning**: balance the various design demands, establish 'sensible' goals for completeness, and identify possible consequences.

Whenever a person's competences and skills are evaluated the person's creativity interferes in a way that is difficult to identify and isolate. We treat creativity in Chap. 7 but will comment on the core competences relevant here. First, **competence in synthesis** is key to design. It brings together creativity, intuition, systematic thinking, technical insight, and imagination to drive productivity in problem-solving and synthesis situations. Second, **competence in visualization** is key to creativity, sketching, and modelling and thus core to design communication and understanding.

The combination of competences and skills is not only determined by the practice of the designer, but also the nature of the tasks. For example, innovative tasks demand more than the core competences mentioned above. **Competence in innovation** is often seen as entrepreneurial and related to individuals, technology, organizational interaction, and a company's culture. The nature of innovation means it relates to new tasks or expanding dimensions such as (Jørgensen 2004), cooperation (within a team or in wider networks), going beyond products and systems seen as simple physical entities, and structuring innovative work.

Our treatment of competences and skills is an attempt to articulate 'designerly' competences and contribute to awareness, respect for, and understanding of their importance in design. It also points to the designer orientated aspects of knowledge management. The richness of design knowledge is channelled into the design activity through the designers' endeavours. This knowledge contains 'wisdom' which can be descried in terms of design philosophy and principles.

3.6 Design Philosophy and Principles

This section aims to answer the question: what is the role of high-level design insight in the form of philosophy and principles? In the author's experience engineering design has not been heavily influenced by specific philosophers, rather design philosophy and principles have been developed from the ground up by designers and design researchers.

3.6.1 Design Philosophy

Philosophy generally deals with questions related to our perception of the world, our thoughts, and our understanding. Philosophers also deal with design and the basic concepts of designing, for example, in the Centre for Philosophy and Design (Galle 2010). Their goal has been to bridge philosophy and design research; one may say to supply design researchers and practitioners with a philosophical understanding of design's basic concepts. However, it is a long way from this initiative to the colloquial understanding of design described by designer's or company's attempts to articulate 'philosophies' in the form of metaphors, one-liners or buzzwords. Between these two extremes we find more or less explicit design philosophies. We refrain from trying to give a comprehensive overview but focus on the questions: What answers do we find in design philosophy; and what answers do we want? It seems that there are three dimensions of perception and understanding that individually or in combination provide a starting point for articulating a philosophy:

- What the product is and does in itself, i.e. the technical tasks leading to **function and structure**.
- What the product can be used for and its link to need/purpose, leading to **function and use**.
- What values are ascribed to the product, its usefulness and the **symbolic values** it carries. These are **socio-technical** dimensions and **semiotic** articulations.

"**Form follows function**" was a design credo formulated by the French architect Louis Sullivan (1924), who created high-rise buildings using iron, glass, and concrete. Sullivan was inspired by the simplicity that was created when engineers'

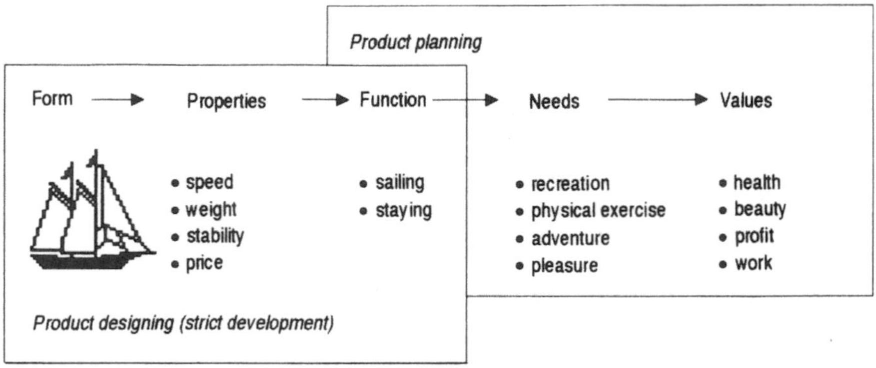

Fig. 3.21 Function seen as the link between form and value (Roozenburg and Eekels 1996)

dimensioning was used to determined functional form (Faber 1962). Heretically one might ask if Sullivan's credo leads to good solutions. Similarly, "**Space, construction, form, function**" is the articulation of the Danish architect Harboe's (2001) design philosophy. Here, form also creates internal and external spaces, based on a task and composition plan for the totality and the details, social and economic issues, and the creation of frames for human life.

Current design philosophers use the concepts function and form, e.g. Papanek (1971) or Roozenburg and Eekels (1996), to build links between a product's aims and value as shown in Fig. 3.21.

Here, form determines the product's properties and function, while functionality leads to need satisfaction and value perception. This philosophy is built upon in this book's Domain Theory and Link Model (Chaps. 11–14).

"**Structure and function**" are the basic concepts of systems thinking (Chap. 8). The basic aim is to understand composed product's behaviour by decomposing their ingredients and their relationships. In Gero's (1990) FBS framework (Function, Behaviour, Structure) structure carries a product's functions and behaviour, and is thus synonymous with 'form'. In our book we see this structural definition of products as the core of design.

Example:
Synthesis from function to structure in a bilge pump Progressing from what we want the product to do, i.e. its function, to its way of doing this, i.e. its structure, is often described as a creative leap, exemplified here. Sailing boats that are moored in a harbour collect water from rain and leakage, and can be damaged or even sink because of this. Thus, there is a desire for a product with the function "remove water" or "move water out of the boat". One idea might be to use a type of wick to passively suck up the water based on wind and warmth as energy sources driving evaporation. Another structure could be a conventional valve-based pump driven by the boat's

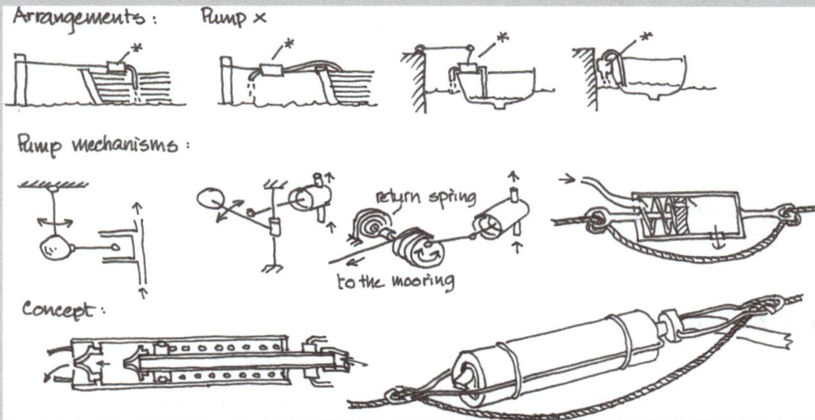

Fig. 3.22 Sketches of a pump driven by forces on the mooring cable

movements or forces in the boat's mooring wires, as illustrated in Fig. 3.22. Concretization, detailing, and experiments allow us to determine if the expected functionality is achieved by either of these ideas.

Next consider **semiotics and symbols**. A product is both a symbol in itself and can carry symbols. To read and understand symbols is the domain of semiotics. This is driven by insight into culture and social context. The values that users assign to a product's symbols can be diverse, including pride of ownership, boasting value, indication of social belonging or political, moral or environmental position. However, we also find more neutral symbols serving as identification, operation instructions, and visualization of functions.

Example:
Identity of bicycles. Bicycles, even after many years of development and many digressions (big/small wheels, more wheels, and different transmissions) have not found a single standardized design; customers ask for individualization (Fig. 3.23). For example, in towns we find that bicycles are designed to symbolize and signal the owner's identity, rather than provide the optimal transport experience.

Fig. 3.23 Various bicycles: racer, Christiania bike, and lady's bike

In the bicycle example (Fig. 3.23) the product's symbolism is carried by their structure, primarily in the design of the frame. While graphics on the product might enhance the bicycle's symbolic values (speed lines, logos, colour, and text) they are not the main carrier in this case. Because of the malleability of software the form of products, such as mobile phones, iPods, PCs, etc., are less influenced by their internal functionality. Therefore, designers must invent characteristic, iconic forms while balancing mainstream conformity and originality. This leads us to design principles.

3.6.2 Design Principles

Just as with philosophies design principles are broad, multidimensional concepts that are fundamentally valid. Principles can be used in practice by confronting them with design proposals and asking 'is this principle satisfied'? However, principles can also address ethical, professional, and practice aspects of design.

The Royal Academy of Engineering in the UK formulates what they call design principles (Armstrong 1980): *"Engineering design encompasses three key stages of realization:*

- ***Need**: all design begins with a clearly defined need.*
- ***Vision**: all designs arise from a creative response to a need.*
- ***Delivery**: all designs result in a system, product or project that meets the need".*

These three declarations aim to communicate the essential elements necessary for design decisions to produce desirable results and have been derived from experience, practice, and pragmatism. Several scholars have also formulated principles of design, e.g. Pahl and Beitz (2007) 'basic rules':

- **Clarity**: Ensure user's can reliably predict a design's functionality and performance.
- **Simplicity**: Avoid unnecessary components and production processes to ensure a simple structure.
- **Safety**: Ensure peoples safety by focusing on strength, accident prevention, and protection of the environment.

We can elaborate on the simple principles by defining them in terms of a **strategy**. Strategy has a more unified meaning than design philosophy or principles because it describes the way we attack a certain task or problem. It is a situation dependant approach, which can sometimes look similar to philosophy or principles. For example, many design process models are based on the advice that 'the problem be verbally articulated' before any search for solutions is undertaken. We see this as a strategy but it is very close to a principle in nature.

Here we focus on the design strategies related to project and team, which are subordinate to product development and business strategies. Mastering design strategies is an important dimension of design knowledge. The many proposals for design process models can be seen as strategies, claiming that there is a good, general way to approach a design task. However, these are often of limited validity despite claims go the contrary. As such, we will revisit principles and strategy throughout the book with the aim to ultimately understand what makes a good product.

We think that the perception of philosophies, principles, and strategies relates to reflective practice as defined by Schön (1983): *"Reflective practice is the capacity to reflect on action so as to engage in a process of continuous learning"*. Schön sees this as a core characteristic of professional practice. Bolton (2010) writes that it involves *"paying attention to the practical values and theories which inform everyday actions, by examining practice reflectively and reflexively. This leads to development insight"*. We see our treatment of philosophy, principles, and strategies as concrete patterns of insight found in practice and representing valuable understanding. However, as with all insights, the designer's own perception of the situation affects their interpretation.

3.7 Perceptions of Design Activity

The key knowledge element in design is the designers' explicit and implicit understanding of the design activity. This includes their mental models of the activity and the pattern of activities that best support their knowledge. Activity models provide us with methods when we use them in a normative way—they describe structured ways of proceeding. However, this still does not answer what mindset we should use when designing and what perspectives we should consider.

Let us start with the researchers' explanations. Hubka and Eder (1996) propose three basic, elements for explaining design:

- **Theory of Technical Systems**: containing descriptive statements on the nature of the artefact to be designed.
- **Design knowledge about objects or systems**: the branch and domain dependant know-how related to the ways and means for creating technical systems.
- **Theory of Design Processes**: an explanation of the actions of design in a socio-technical and industrial context. This is coupled with **Design Process Knowledge** regarding the operations and means for performing the design process, e.g. methods, procedures, strategies, and tactics.

Hubka and Eder's (1996) book was epoch-making in defining the existence of design science, seen as the collection of knowledge about designing in an encyclopaedic style. They viewed the design process as a system with activities as elements. Cross' book "Designerly ways of Knowing" (2006) was just as epoch-making in its identification of design as a general dimension of human cognition. Cross combines insights from design processes and design artefacts to describe ways of thinking unique to design: *"Design has its own distinct 'things to know, ways of knowing, and ways of finding out about them'"*. Here we can take a more specific look at these insight areas.

- **Design processes**: these integrate strategy (how to tackle a task), finding satisfactory solutions, coping with the ill-definedness of design situations, and the designer's *"confidence to define, redefine and change the problem-as-given in the light of the solution that emerges from their mind and hands"* (Cross 2006). Here, Cross describes designer's ability to translate ideas concerning individual, organizational, and social needs into artefacts, as a coding or language.
- **Design artefacts**: Cross states that *"there is a great wealth of knowledge carried by the objects of our material culture"*. Learning from past designs is important and designers are characterized by their *"ability both to 'read' and 'write'"* in our material culture.

Hubka and Eder use system thinking as a framework for their treatment of the design process, while Cross is more concerned with the designer and how they tackle specific situations and tasks.

3.7.1 A Cybernetic Understanding of Design

Cybernetics is a science that aims to understand natural and artificial systems as functional, dynamic, and goal oriented. Central to this are control and communication, the system's ability to observe and react to changes in the surroundings, and the system's ability to influence the surroundings in a controlled way. We use this perspective to further explore the 'machinery', its dynamics and functionality.

Design teams operate in a cybernetic way as highlighted by Wynn et al. (2010) and Maier et al. (2012). These authors focus on the design activity's methods and models, and how they can be seen from a utility perspective. A Cybernetic system's behaviour and functions are described by a set of abilities and functions, which are comparable to a team's necessary behaviours and functions in solving a design tasks in dynamic surroundings.

While socio-technical models focus on the actors (human or non-human) as carriers of knowledge they do not describe what they are doing. Cybernetic models fill this gap.

From a cybernetic perspective design is as a goal-oriented, perceptive, and self-governing system, which highlights the **active functions** to be performed by the team.

The cybernetic parallels with design underline the self-organizing, self-propelling, and goal-searching nature of design in a team seen as a system. Maier et al. (2012) highlight the role of models, which the team uses to reflect its perception of the surroundings, situation, and design.

3.7.2 A Coordination Understanding of Design

The socio-technical and cybernetic models of design do not really show what is synthesized and manipulated when designing. This is instead found in design coordination, a high level concept for supporting design: *"The design of complex products involves the coordinated organization of multi disciplinary groups, activities, and information, which continually evolve and change during the design activity"* (Andreasen et al. 1994, 1997). Core to this is the Design Coordination Framework, which has nine frames each of which can be seen as a coordination dimension, Fig. 3.24. Each frame is a dynamically evolving aspect of design, which links all the frames. For example, a progression in the design activity may lead to an added entity in the artefact model, followed by a goal break down, a new task definition, and a call for new resources.

The linking lines in Fig. 3.24 represent the coordination actions required to manage progression (synthesis) or changes in the frames. The means for coordination are manifold: teamwork, modelling of the different frames, use of experiences from past projects, and team members' knowledge and imagination. The complexity of coordination relates to the question of design entities, i.e. what should be designed in a project?

The design coordination perspective describes a high level framework balancing control and coordinated efforts, underlining of the complex **dimensions of results** in design.

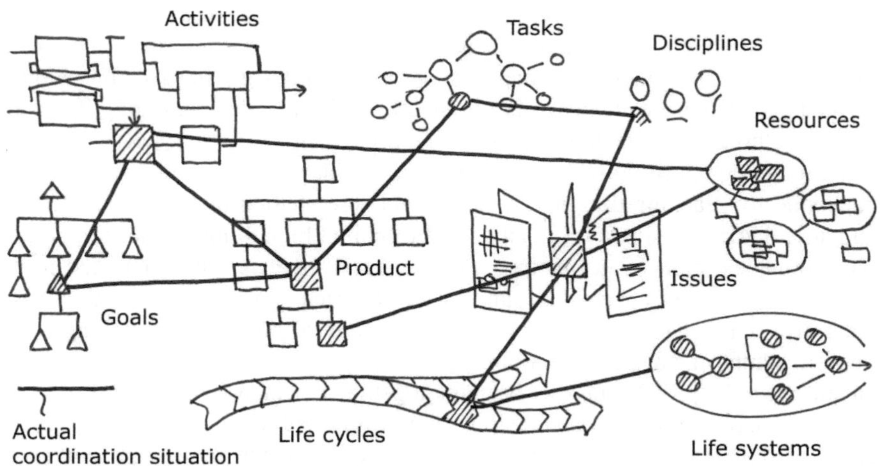

Fig. 3.24 Design coordination framework

3.7.3 Textbooks' Mapping of Knowledge

The cybernetic and coordination views outlined above both build on a systems perspective in their mapping. A less systematic insight into design is related to human cognition and the nature of the object to be designed. Design textbooks normally aim for an 'all inclusive' understanding of design. These combine knowledge from design, specific professions', and examples from various branches. Thus, we synthesize these various insights here to get a general understanding of designers' knowledge. This can be understood by looking at the history of design education.

In the mid-1900s design teaching as we know it today did not exist. Instead students were taught the underlying sciences, e.g. thermodynamics, solid and fluid mechanics, and control theory. This gave an incomplete view of design. As engineering design evolved design was split into its own field that shunned many of these elements. Design textbooks used few, fragmented examples of concrete products, and generally did not treat the underlying sciences. Thus they also gave an incomplete view of design. A true and comprehensive understanding of design brings these views together, combining aspects such as complexity, dynamics, coordination, cooperation, trade-offs, and the role of practice.

The textbook view of design knowledge is split between **design** with few product and science examples, and **engineering sciences** with a focus on analysis. Both must be considered.

3.8 Conclusion

This chapter deals with our fundamental understanding of design and how it is practiced. Although this is rarely discussed by designers 'because they know what they know' the insights in this chapter underpin all of design. In particular without understanding these elements it is impossible to fully understand and effectively utilize methods or tools, strategies or design principles.

Scholars compose the elements presented in this chapter into curricula as the backbone for teaching design. These elements are studied by design researchers and articulated as insights for practice and teaching. In addition to theory, philosophy, strategies, models and methods, design knowledge is also articulated in the form of case stories, episodes, examples, and the rich explanations found in many textbooks. This gives an overwhelmingly rich body of information, making it difficult to synthesize a specific school, practice or personal philosophy. Our book is an attempt to help in this task by bringing together our understanding of design.

These elements make designers powerful actors with the ability to create influential, sustainable, and useful products. Thus, this chapter provides a foundation for understanding the machinery of design—the designers, their knowledge, and their staging of the teamwork and design process. The role of staging is to facilitate design and support the designer in expressing their competences and skills, and thus forms the focus for the next chapter.

References

Andreasen MM (1992) Designing on a "designer's workbench", 9. WDK Workshop, Rigi, Switzerland. Unpublished

Andreasen MM (2009) Complexity of industrial practice and design research contributions—we need consolidation. In: Meerkamm H (ed) Proceedings of 20. Symposium design for X, Neukirchen. Erlangen, Lehrstuhl für Konstruktionstechnik

Andreasen MM (2003) Improving design method's usability by a mindset approach. In: Lindemann U (ed) Human behavior in design—individuals, teams, tools. Springer, Berlin

Andreasen MM (2011) 45 Years with design methodology. J Eng Des 22(5):293–332. http://www.tandfonline.com

Andreasen MM, Bowen J, MacCallum KJ, Duffy AHB, Storm T (1994) Design coordination framework. In: CIMMOD/CIMDEV Workshop at Torino, Italy unpublished

Andreasen MM, Duffy AHB, McCallum KJ, Brown J, Storm T (1997) The design coordination framework: key elements for effective product development. In: Duffy AHB (ed) The design productivity debate. Springer, London, pp 151–172

Araujo CS (2001) Acquisition of product development tools in industry:—a theoretical contribution. Ph.D. Dissertation, Technical University of Denmark

Armstrong JA (1980) Design principles—engineer's contribution to society. Royal Academy of Engineers, London. Estimated publishing year

Badke-Schaub P, Daalhuizen J, Roozenburg N (2011) Towards a designer-centered methodology: descriptive considerations and prescriptive reflections. In: Birkhofer H (ed) The future of design methodology. Springer, London

Birkhofer H, Jänsch J, Kloberdanz H (2005) An extensive and detailed view of the application of design methods and methodology in industry. In: Samuel A, Lewis W (eds) Proceedings of ICED 2005, engineering design and global economy, Melbourne

van Boeijen A, Daalhuisen J, Zijlstra J, van der Schoor R (2013) Delft design guide. BIS publishers, Amsterdam

Bolton G (2010) Reflective practice, writing and professional development. SAGE publications, California

Buur J, Andreasen MM (1989) Design models in mechatronic product development. Des Stud 10(3):155–162

Cantamessa M (1997) Design best practice at work: an empirical research upon the effectiveness of design support tools. In: Riitahuhta A (ed) Proceedings of ICED 97, international conference on engineering design, Tampere, Finland

Cross N (2006) Designerly ways of knowing. Springer, London

Daalhuisen J (2014) Method usage in design—how methods function as mental tools for designers. Ph.D. thesis, Delft University of Technology

Duffy AHB, Andreasen MM (1995) Enhancing the evolution of design science. In: Hubka V (ed) Proceedings of ICED 95, Praha, Heurista, Zürich

Eppinger SD, Browning TR (2012) Design structure matrix methods and applications. MIT Press Mass, USA

Faber T (1962) Rum, form og funktion (Space, form og funktion). Berlingske leksikon bibliotek, Copenhagen

Galle P (ed) (2010) CEPHAD 2010: The border land between philosophy and design research. Plenary sessions and supplementary items vol 1 and 2. The Danish Design School Copenhagen

Gero JS (1990) Design prototypes: a knowledge representation schema for design. AI Magazin 11(4):26–36

Harboe KE (2001) Funktion—Form—Konstruktion (function—form—construction). Arkitektens Forlag, Copenhagen

Harlou U (2006) Development of product family based upon architectures. Contribution to a theory of product families. Ph.D. Dissertation, Technical University of Denmark

Henderson K (1999) On line and paper. Visual representation, visual culture, and computer graphics in design engineering. The MIT Press, Cambridge

Hubka W, Eder EW (1996) Design science. Springer, Berlin

Hvam L, Mortensen NH, Riis J (2008) Product customization. Springer, Berlin

Jensen T (1999) Functional modelling in a design support system—contribution to a designer's workbench. Ph.D. Dissertation, Technical University of Denmark

Jensen TE, Andreasen MM (2010) Design methods in practice—beyond the "systematic approach" of Pahl & Beitz. In: Marjanovic D et al (eds) Proceedings of DESIGN 2010, 11th international design conference Dubrovnik, Croatia

Jone JCh (1992) Design methods—seeds of human future. Reinhold Van Nostrand, New York

Jørgensen U (2004) The design engineers of the future have innovative competences. Produktudviklingsdagen 2004, IPU, Technical University of Denmark

Kleinsmann MS, Valkenburg R, Sluis J (2012) Demystifying design thinking: images of a design thinker within an organizational context. In: Marjanovic D et al (eds) Proceedings of DESIGN 2012, 12th international design conference Dubrovnik, Croatia

Maier A, Wynn AD, Howard T, Andreasen MM (2014) Perceiving design as modeling. A cybernetic systems perspective. In: Chakrabarti A, Blessing LTM (eds) An anthodology of theories and models of design. Springer, London

Maier AM, Wynn DC, Andreasen MM, Clarkson, PJ (2012) A cybernetic perspective on methods and process models in collaborative designing. In: Marjanovic D et al (eds) Proceedings of DESIGN 2012, Dubrovnik

Mortensen NH (1999) Design modeling in a designer's workbench. Contribution to a design language. Ph.D. Dissertation, Technical University of Denmark

Pahl G, Beitz W (2007) Engineering design. A systematic approach. 3rd edn. Springer, London. First edition 1977

Papanek V (1971) Designing for the real world. Pantheon books, New York

Restropo J (2004) Information processing in design. Ph.D. Dissertation, Delft University of Technology

Roozenburg NFM, Eekels J (1996) Product design: fundamentals and methods. Wiley, Chichester

Samuel A, Weir J (1999) Introduction to engineering design. Modelling, synthesis and problem solving strategies. Butterworths Heineman, Oxford

Schön D (1983) The reflective practitioner. How professionals think in action. Basic Books, New York

Sim SK, Duffy AHB (1998) A foundation for machine learning in design. AI EDAM 12(2):193–209

Sullivan L (1924) Autobiography of an Idea. Press of the American Institute of Architects New York, New York City

Svendsen K-H (1994) Diskretoptimering af sammensatte maskinsystemer (Discrete optimization of composed machine systems). Ph.D. Dissertation, Technical University of Denmark

Tjalve E, Andreasen MM (1975) Zeichnen als Konstruktionswerkzeug (Drawing as design tool). Konstruktion 27 Nr 2

Tjalve E (1979) A short course in industrial design. Newnes-Butterworths London. Faximile edition (2003) Systematic design of industrial products. Institute of Product Development DTU Copenhagen

Tjalve E, Andreasen MM, Schmidt FF (1979) Engineering graphic modelling; workbook for design engineers. Butterworths, London

Tomiyama T (2006) A classification of design theories and methodologies. In: Proceedings of the ASME IDETC, Paper No DETC2006-99444, ASME

Vermaas PE (2014) Design theories, models and their testing. On the scientific status of design research. In Chakrabarti A, Blessing LTM (eds) An anthology of theories and models of design. Springer, London

Wynn C, Maier AM, Clarkson PJ (2010) How can PD process modeling be made more useful? An exploration of factors which influence modeling utility. In: Marjanovic D et al (eds) Proceedings of DESIGN 2010, Dubrovnik

Chapter 4
Staging Conceptualization

Staging describes the process of understanding a team's situation and fitting it to the project at hand. Both mental models and physical entities in the team's surroundings contribute to this shared and gradually shaped articulation. Staging builds on, and helps shape, a team's shared understanding, values, and insights. As such, it is a key element in any community of practice. Core to this is developing the link between a team and its surroundings to support both awareness and insight. Finally, we explore how staging also helps build up interaction with users, often using boundary objects to facilitate understanding and communication.

4.1 The Role of Staging

Chapter 3 focused on the designer, however, a lone designer is seldom able to fully realize a new product without support. Thus, we now expand our view to consider the team and its composition. Arranging a team's cooperation and project work is called staging; a relatively overlooked area but an important part of design nevertheless. Staging has two key dimensions: empowering the team activity and supporting innovation (Hansen and Andreasen 2006). In this chapter we lay the groundwork for staging by seeking to answer the following questions: what is the scope of staging, how is it adapted to specific design situations, and how can it help a team work with its surroundings?

M.M. Andreasen et al., *Conceptual Design*, DOI 10.1007/978-3-319-19839-2_4

A condition for innovation is that things are done differently from routine, thus we should always seek to reflect on and improve the 'design for design', i.e. the staging.

In this chapter we deal with the following topics.

- Section 4.2 sets out the core concepts underpinning **staging and its challenges**.
- Section 4.3 looks at how designers can use **the design space and staging elements**.
- Section 4.4 applies these insights to design **teamwork** and how this can be supported.
- Section 4.5 finally examines how teams can use staging to **interact with their surroundings**.

4.2 Staging and Its Challenges

The terms *staging* and *space* both come from socio-technical design (Bijker 1995). Space describes how designers' work, know, and operate, in a team (Clausen and Yoshinaka 2007). This deals with both the physical and mental frame in which a team operates. Staging is the subsequent arrangement of this space to best support the team.

Definition: Staging is the act of establishing and fitting a team's space to a project, to best support the design activity.

Staging is closely related to design practice, the process, its conditions, and its surroundings. Here 'the surroundings' describe both the social and technical environment where the product is produced and deployed. This is also where the designer might draw information from when solving the task. Interaction with these surroundings is a critical element in design. Conversely, 'the conditions' describe what the design team controls, including the project, its staff, and the design practice used.

Staging brings together individual designers into an effective team. This is achieved by formulating an agenda: setting perspectives, defining contributions, clarifying roles, and creating motivation. As such, staging is as important for a team's collaborative work as it is for supporting innovation. Further, staging is on a level where it is relatively independent of broader organizational efforts, such as management or productivity approaches, e.g. agile or lean.

The word staging is of course taken from the world of theatre and film. It is about interpretation of the play, formulation of the script, direction of the actors, staging the scene, etc. This runs parallel to its meaning in design and is an apt

Fig. 4.1 A 'product life gallery' workshop in Steelcase, USA, *courtesy* Tim McAloone

metaphor. This similarity is illustrated in Fig. 4.1 where a design workshop is being used to explore a product's lifecycle. The staging is here the physical arrangement of the room, the illustration of the product lifecycle in a poster gallery, and the interaction of the design team and lifecycle actors.

Although staging is relatively independent of management efforts it is still influenced by the organization. In particular, if a team is not permitted to develop its own approach to the design project and tasks allocated to it the opportunities for staging are very limited. However, when a team is allowed to stage effectively, in harmony with the organizational management, it is possible to realize significant benefits. The project leader is a key mediator between these two domains, interpreting management and project objectives to allow the design team to stage their work as effectively as possible (Andreasen and Hein 1987).

Staging is most easily realized in small independent teams and is not a natural consequence of industrial organization. Care should thus be taken to harmonize the staging and management activities to maximize the benefits in larger organizations.

4.2.1 Key Challenges

Hopefully we have clarified 'what' staging is but we also need to ask: what is it used for?

- To support **practice** by helping to best arrange task, knowledge, and work processes.
- To support team **interaction** in terms of both mindset and physical project execution.
- To facilitate the **design process** by fitting it to the actual situation.
- To help **develop individuals** intellectually, in terms of both design knowledge and ability.

Staging influences creativity, inspiration, and well-being through the team's surroundings, as discussed by McKim (1972). In a physical sense 'creative' workspaces can be extremely valuable in creating a mindset, however, this is only one (rather static) aspect of staging. A key challenge in the more dynamic aspects of staging is adapting to each specific project. Here, staging must articulate a specific agenda, formulate perspectives for the team, identify roles, contributions, and motivations for each designer, and establish trust. Further, these should be adapted as the project progresses to ensure each team member is working to their full potential and doing the 'right things'. An example of this type of evolving staging comes from a train carriage company. The company was experiencing problems with quality and timely product delivery. In response to this the design and supply teams were relocated to the assembly site. Further, their uniforms were changed to match those of the assembly personnel. This put them in direct contact with the workers, the product, and problems, resulting in greatly improved quality and delivery performance.

From this example we can distil some key challenges facing staging. First, the **social process** aspect of design work means staging must strike a careful balance between structure and openness in its **community of practice**. Structure is required to help create a common agenda and goals, while openness and rule breaking can be key to inspiring innovation. However, too much structure kills creativity and too much rule breaking can cause chaos. Also in this social mix is the need to support designers in developing their **awareness** of the situation and task, through motivation, active and systematic search, and interpretation. Successful design builds on effective **need and value interpretations**. Thus staging should also support interaction with users, and integrate knowledge about them into the design activity.

Similar to the value interpretation above, staging also plays a key role in the parallel development of **synthesis and clarification** (Andreasen and Hein 1987) by facilitating knowledge flow and productivity. Part of this knowledge comes from the business dimension. Here, a key challenge is balancing the team's and company's **vision** to align time and resource allocation, particularly in ideation and experimentation. Finally, as design practice is fundamentally a **learning** process, staging is challenged to support the continuous development of the designers. Thus, although staging can be made the more or less explicit responsibility of a team or project leader, it is also the responsibility of each designer in their professional development.

4.3 The Design Space and Staging

The design space is a mental model of the actors and design activities, where staging is introduced via both humans and artefacts. Figure 4.2 shows a model of space and staging inspired by the works of Clausen and Yoshinaka (2007), and Andreasen and Clausen (2011).

Here, '**actors**' describes the designers and their **mental models** of the design and situation. These are influenced by the team's **agenda**, **motivation**, and **roles**. Further, the designers carry the **competences** and **knowledge** used to execute the actual **practice**, supported by **methods**, **tools**, and **staging objects,** e.g. models, plans, and prototypes. In this model we can see staging as the team creating, arranging, choosing, and deploying these various elements. This gives an idealized image of the space and the team's interaction with its surroundings, partly through **boundary objects**.

> Although all these entities (actors, etc.) are brought into a space when a team is formed they require staging to bring them together to best solve the actual task.

Space gives us one way of describing a designer's world and thus 'what it takes to design'. However, although this model seems comprehensive there are many open questions regarding the importance of each actor's worldview, the company's

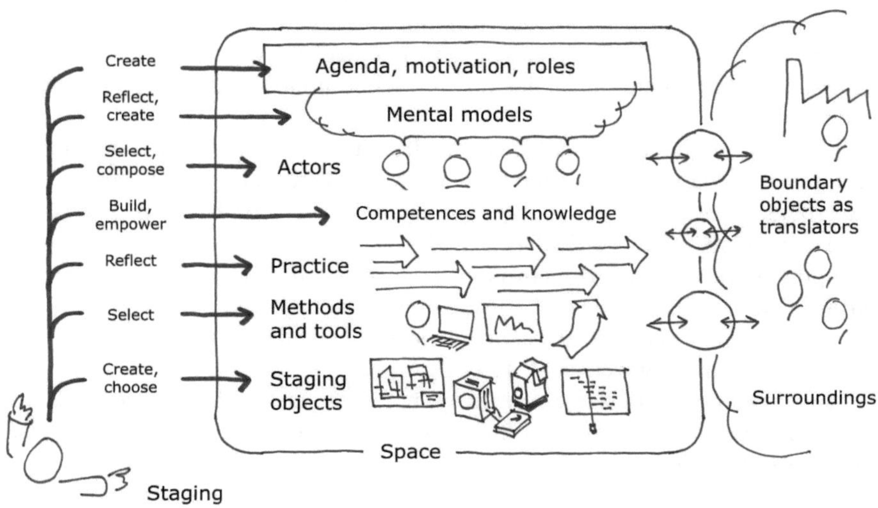

Fig. 4.2 Staging in a socio-technical design space

values and norms, and the influence of design philosophies. In order to better understand the role of these staging elements we will explore each in turn.

4.3.1 Staging: Agenda, Motivation, and Roles

Using the parallel to staging a theatre play, the play should be explained along with the staging idea and director's interpretation, and the roles distributed. Design projects need similar clarification of their specific angle of attack, need interpretation, importance, and ambition. Everyone should agree 'what we are up against' and be confident that the team can tackle this. These staging dimensions have no specific carrier or visualization but are realized through goal formulation, activity planning, and daily work practices. A formal project leader might facilitate the creation of these entities but they can also grow from the team members and their interaction.

4.3.2 Staging: Mental Models

Mental models describe our insight into and understanding of something. These are fundamentally individual but can be partially shared by creating common models or ways of thinking that the team all understand and agree on. Mental models can describe any subject imaginable and are similarly widely used in the design context. Common subjects include: planning, issue prioritization, idea organization, basic product framing or structure, cost reduction strategy, product properties or features, and competitive strategy. Mental models describe and articulate our understanding while mindsets describe our attitude and beliefs.

Typically, project failure can be traced back to a lack of shared mental models or common understanding of the design space (Kleinsmann 2006). Mental models provide "*sense making devices that establish the parameters of a problem*" (Valkenburg 2000). In particular, shared mental models are important because they provide "*knowledge structures held by members of a team that enable them to form accurate explanations and expectations of the task, and in turn to coordinate their actions and adapt their behaviour to demands of the task and other team members*" (Cannon-Bowers et al. 1993).

A key element of design is creativity, however, this is hindered by both over and under constraint (Onarheim 2012). In this context a team's mental models can be essential in setting their own game rules for creativity, putting them in the sweet spot of constraint. For example, when a team decides to ignore a major requirement in order to allow them to brainstorm more freely.

Just as with creativity different models can be produced and combined to form a team's overall understanding, empowering all of their design activities. These collective help to define a team's identity, role, and understanding of its position in a company. This is usually complemented by reference frames, such as a company's scheme of responsibilities, dependencies between functional areas, and factors influencing dynamics, reaction time, and results. A larger scale reference frame is the **arena** in which a company is operating. This can be defined graphically, by a company's network, or by the type of product produced; and is populated by competing companies and products, market mechanisms, advertising, and sales, as well as customers and their interpretation of brands, product quality, value, and preferences.

Example:
Understanding design
Understanding and creating a model of design that in useful for coordination, planning, and discussion is a challenging proposition. Many authors (including the authors of this book) describe design process models. Some of these models are 'rhetoric', i.e. they show a mindset or single aspect of design. Such models can be useful as the basis for 'mental models'. For example, Fig. 4.3 shows one such rhetoric model focusing the gradual concretisation of the final design.

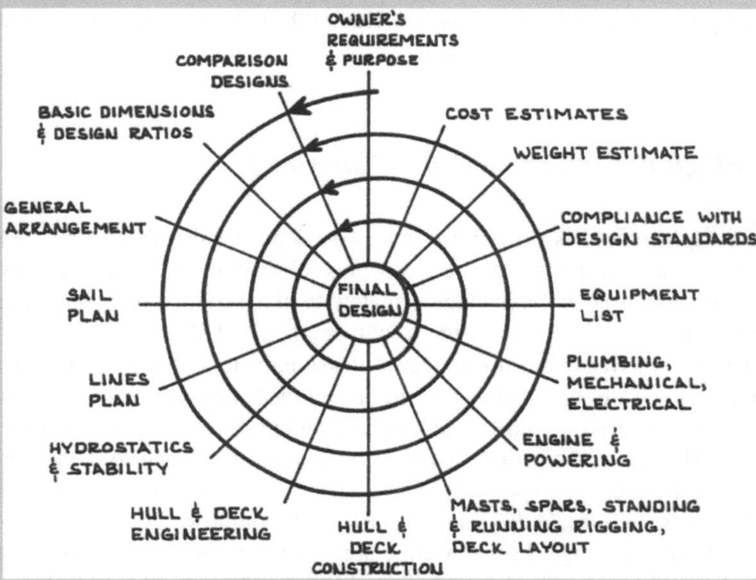

Fig. 4.3 A rhetoric model of design that could be used as the basis for a shared mental model, *courtesy* Eric W. Sponberg

4.3.3 Staging: Actors

Design is still a fundamentally human centric activity. Even the most detailed procedural design models only explain a tiny fraction of the insight needed to complete the design. Thus, a key element in staging is effectively organizing the actors and their insight. Actors come from a range of specialist job roles and it is the staging that allows them to incorporate their diverse knowledge into a cohesive whole—the space does not organize itself! Key tasks are (Jørgensen 2004):

- Identify and **coordinate actors** with different knowledge based on their perspectives, ideas, and innovative contributions.
- Motivate actors to be '**translators**' so that relevant contributions and knowledge are brought into the space.
- Balance **task composition** to take advantage of both detail-oriented and broadly skilled actors.
- Monitor, analyse, reflect on, and improve **coordination** in the space to increase performance.

In a team both the formal and informal leaders play an important role in creating and maintaining the community of practice. This typically links to the roles informally defined by the team, e.g. trainer, idea generator, and entrepreneur (Sant 1988) or the creative product developer, entrepreneur, sponsor, information handler, and project leader (Andreasen and Hein 1987). Team management is critical to people's interaction and motivation.

4.3.4 Staging: Competences and Knowledge

A first thought would be to build a team based on the member's knowledge. In Chap. 3 we made a distinction between general design knowledge, professional knowledge, and branch knowledge (related to the product type and industry). However, we also linked knowledge to competence and skills, and it is these that allow actors to realize their knowledge, professionalism, and abilities in practice.

Competences are not static but are developed through reflective practice. Actors' competences and skills need to be challenged and applied if they are to be honed. Therefore, the project leader or stager must balance to what degree an actor should be seen as a specialist with a specific area or should be brought into new learning situations. A key staging task is to compose the necessary competences and skills, arrange the knowledge areas, and promote a constructive a conflict of ideas. These should also be weighed against the individual development of the team members. A planned introduction to company practices and continuous learning are key means for developing and balancing competences.

4.3.5 Staging: Practice

Practice describes how the team works, i.e. when we observe a team we see its practice but the practice's composition and nature may be hidden to us. Practice is composed of all the elements in Fig. 4.2, as well as being influenced by past design activities and products, successes, and failures. Practice also points to the need for mutual understanding and interpretation of goals and results as the cornerstone of strategy.

Practice is developed through what Kolb (1984) calls experiential learning and Schön (1983) calls *reflective practice*. Both are learning processes that occur in parallel with practice in reality (explored further in Chap. 7). Thus, an important dimension of practice is the designers' ability to work with physical artefacts in their design activity, i.e. 'working concrete', see Fig. 4.4. This concretization practice is clarified in IDEO's five-step procedure (Brown 2009): Observation, Brainstorming, Rapid Prototyping, Refining, and Implementing. This builds on intensive use of visualization, mock-ups, prototypes, and scenarios, particularly in communication with the client.

The Dutch designer Jan Glas described a characteristic aspect of his practice as arranging the design process to give the best possible 'contact with the matters', i.e. allowing him to work in the workshop, seeing ideas materialized and prototypes grow. This gives him the feeling of mastering the synthesis and allows an early dialogue with clients in a co-creative, co-design approach.

Fig. 4.4 The workspace of mechanical designers, *courtesy* Jakob Parslov Radiometer Medical ApS

A key element in reflective learning and development is to finish a design activity with a specific closing action where the team reflects on their performance, results, experiences, cooperation, etc. Some scholars, e.g. Ulrich and Eppinger (2004), show this explicitly in their design process models.

A planned and consciously reflective closing activity in a project is a powerful tool for improving a team's practice.

4.3.6 Staging: Methods and Tools

The role of methods in a project varies widely. Sometimes, mastery of a specific method is required to take part, while at other times the team barely uses methods at all. In situations where practice is dominated by specific project support tools, standard procedures, and standard documentation the designer might have no choice in how to work. Here these tools may 'arrange' the daily work to such a degree that reflecting on appropriate method use becomes redundant. However, this is rare and it is typically preferred that design teams are able to stage their own work practices.

Even when designers plan their own practice the question of method application is fuzzy. We discussed this in Chap. 3, where we highlighted the rather soft nature of methods. Araujo (2001) made several studies of product development methods in industry and found that decisions regarding their use were not easily traceable. It is also typical that nobody is responsible for methods' strategic deployment. It is up to the designer and their team to make them work in a given situation and project. This is in contrast to the false belief that methods are machines that will automatically create useful results if only the instructions are followed. Similarly, the design team must also answer how methods are combined and used in sequence for their specific project. As such, designers should be wary of texts that claim a rigid application sequence. We compare using design methods with learning to play a musical instrument. Nobody would expect a reasonable result on the first try; although most people could probably make a noise on a piano there is a long way to go before we could tackle Rachmaninoff.

- Part of good practice is to **master methods**, being able to understand them, and adapt them to the specific project. This almost always builds on some unsuccessful attempts.
- **Systematic reflection** on method use (successful or not) is a powerful means for reaching this mastery.

Methods and tools should support the team based on a designer-oriented approach, i.e. staging the application so that the method is discussed, planned, and reflected on. You might note that application is only one small part of the method's purpose—methods play a large role in facilitating team dialogue and creating shared understanding. For example, a checklist was introduced for goal formulation and requirement documents in a Danish company. Based on this checklist these documents were semi-automatically created and circulated via email. This replaced the previous method where the team facilitated a workshop gathering all company stakeholders. Ultimately, the automated approach created fast and complete documents, but there was no shared understanding in the team or shared vision of the project in the company. The key contribution of the method had been lost!

4.3.7 Staging: Staging Objects

Staging objects need particular attention. In Chap. 3 we quoted Cross (2006) in pointing out "*the great wealth of knowledge carried by objects of our material culture*". A company's and its competitors' products are extremely important objects to confront the design team with. They can be in any form, e.g. drawings, prototypes, etc., and can be used in any context, e.g. brought into the office. Objects carry knowledge and help realize the material reality. An aircraft designer said: "*In the old days we had so many components in the office that we might assemble them and fly home. Today I only see computer screens and children's drawings as decoration*". The statement is a reminder that the design will be realized in the material world!

Important staging objects are anything that brings past experience into the design space: drawings, mock-ups, reports, prototypes, explorative results, and sales numbers from past launches, etc. Sim and Duffy (1998) highlight the potential for systematic reuse of past solutions in terms of saved man-hours and reduced risk. Unfortunately, this is more easily said than done as past projects are rarely completed with reuse in mind and designers are normally reluctant to reuse, seeing 'creation from scratch' as superior. When creating and choosing staging objects it is important to look for this learning effect.

4.3.8 Staging as Innovation of the Organization

Change and innovation in a company's organization aims to improve capability and performance. A company's portfolio of competences and capabilities may be articulated as a radar diagram, where competences are the 'slices of cake' and capability the radius, see Fig. 4.5. Change and innovation can be formulated as desired 'lifts' and 'turns', i.e. moving to a new focus and leaving old focus areas behind (Kirkegård 1988).

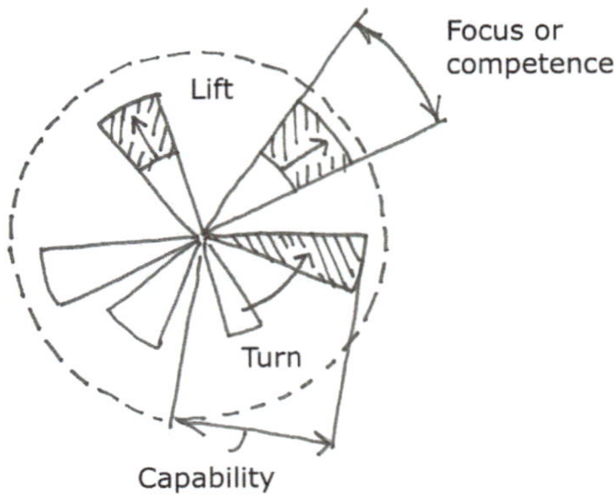

Fig. 4.5 Lifting capability and turning to a new focus as dimensions of team related innovation

This type of change requires strategic insight into the areas of professional or branch knowledge where competitive advantage is to be found. When new areas are established it is necessary to restructure the staging of the team: new methods, new competences, new worldview, etc.

Examples:

Change dimensions

When Bang and Olufsen wanted to launch integrated systems (e.g. a music system located in one room but with speakers in several other rooms) they found that they lacked knowledge in software and communications, and that their existing electronics knowledge was no longer able to deliver the competitive edge they desired. Similarly, when 'movement' became the focus of a new product lines (e.g. Fig. 4.6) mechanical knowledge gained renewed importance. With each strategic shift they also changed the companies staging to maximize their competitiveness.

Fig. 4.6 Bang and Olufsen changed their organization in order to realize these mechanized designs, Beosound 3000 and 9000, *courtesy* Bang and Olufsen A/S

4.4 Teamwork

This section explores the nature of teamwork and how staging influences it. Teamwork is positioned between the organizational and the individual. The organization provides the team with knowledge, data, human resources, and managerial guidance sometimes overlapping with teamwork; and at the individual level designers work with and for the team. It is widely recognized that teams are powerful organizational entities, especially in design and development.

Definition: "*A* ***team*** *is a small group of people with complementary skills who are committed to a common purpose, performance, goals, and approach for which they hold themselves mutually accountable*" (Katzenbach and Smith 1993).

The organization of teamwork can be characterized as "*cycles of distributed design combined with collaborative design phases*" (D'Astous et al. 2004). D'Astous et al. also underline the reciprocal relationship between individual work and teamwork.

4.4.1 A Community of Practice

A team's practice belongs to a wider community of practice that glues the team together.

Definition: A **community of practice** is a socio-technical pattern that evolves in a team and its space as a result of experience, cooperation, learning, knowledge creation, and sharing.

Historically, practice has been the carrier of abilities from building houses to brewing beer. This is still true today, where even highly refined, formalized descriptions of design must be tempered through practice. We authors have experienced the power of ingenious individuals and of expert teams and respect both. Thus, we must consider both formalized practice and the development of the designer. Figures 4.4 and 6.13 show how a group of designers have staged their community of practice.

4.4.2 Design Dynamics

The cybernetic view of design highlights the *active functions* actors should perform to make the design activity goal-oriented, perceptive, reactive, and self-governing. Awareness, perception, and reaction are related to several levels: societal change and development, technological and industrial product development, and opportunities, possibilities, and conditions relevant to the actual project. This is reflected in the Design Coordination view, which deals with the complexity of the design operation and describes the dynamic control and focusing of the many dimensions required to stop the design work growing out of control. From this perspective the key object is the synthesis result. This demands staging the team's perception of the task (goal, plan, and coordination of resources), of the synthesis activity (the pattern of sub-activities and contributions to the design), and of the team's composition and support (work resources, disciplines and knowledge). Duffy and Andreasen (1995) uses design coordination to 'design the design', i.e. systematically stage all the elements needed to support the best possible process, see Fig. 5.5.

4.4.3 The Team's Collaboration Dynamic

Kleinsmann (2006) identifies a set of collaboration elements, Fig. 4.7, that are a useful way of decomposing a team's interaction. These turn basic collaboration into successful teamwork, if properly supported by the staging.

Collaboration is an act of mutual motivation and joint process. This is the foundation for cooperation, which drives the development of shared understanding and the realization of collective goals (Kahn 1996). Based on Kleinsmann (2006) we define collaboration as:

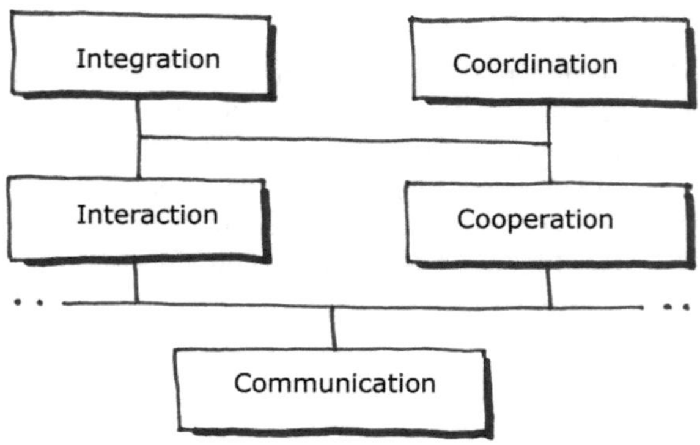

Fig. 4.7 Collaboration dimensions in organized teams and projects

Definition: **Collaborative design** is the process through which actors from different disciplines share their knowledge about the design process and the design itself. This creates shared understanding related to both process and artefact, helps integrate their knowledge, and helps them focus on bigger common objectives—the final product to be designed.

From this we can distil three building blocks (Kleinsmann 2006):

- **Knowledge creation and integration** between actors from different disciplines and functions.
- **Communication** between the actors about both the design content and the design process.
- The creation of a **shared understanding** of both the design content and the design process.

Kleinsmann's studies have shown that collaboration is most difficult during the conceptual stage, especially with respect to shared understanding. The barriers and misunderstandings at this stage are not immediately obvious but grow through the project to cause serious issues in later stages. Thus the goal of collaboration is knowledge creation and integration. This can be hampered by actors from different disciplines being unable to understand each other. For example, Andreasen and McAloone (2001) found that many Danish companies found it difficult to articulate multidisciplinary concepts, e.g. mechatronics (a mechanical, electronic, and software-based product). This lack of shared understanding is a problem because it influences the quality of the design (Valkenburg 2000). Bucciarelli (2003) highlights that "*since no actor has at any stage of the process, a comprehensive, all encompassing understanding of the design, actors have to share knowledge*".

4.4.4 Communication

Communication is the more *or less* direct and synchronous activity of conversation in the design work. Factors here are what to communicate, how it is done, and what media to choose. Olson et al. (2001) found that 20 % of the time in design meetings is spent planning and staging the process, 30 % discussing progression (and thus design coordination), and only 40 % on the actual design content (goals, synthesis, evaluations, etc.).

The communication of design content follows from disintegration, i.e. a team distributes tasks or activities. The challenge here is to create and consolidate the task understanding, conditions for synthesis, evaluation and choices, and coordination. Design communication is primarily verbal. Perry and Sanderson (1998) highlight that the most effective form of communication is face-to-face. The communication process becomes more explicit and challenging when a team is geographically distributed. A key element here is visual communication, e.g. text, drawings, slides, posters, reports, models in hardware, photo, video, CAD-drawings, diagrams, and demonstrations, often followed by verbal explanations. The importance of effectively communicating design work is highlighted by Henderson (1999): "*Sketching and drawing constitutes the communication in the design world*". In particular, the use of drawing-based communication is key in teamwork, thus we briefly discus their nature and challenges here.

Drawings are often based on conventions, codes, and standards that facilitate communication and foster precision but only for those people who know the standard. This can be a significant barrier to experts from other domains (Henderson 1999). Another limitation of drawings is that to do not intrinsically communicate their importance or validity. A computer rendering can be very convincing while at the same time being meaningless.

Even if drawings are basically static, they can still be used in dynamic modelling because the viewer is actually "*able to see things which are not there*", (McKim 1972). For example, by imagining sequences of assembly, misuse of the product, failures influencing reliability, number of machining operations, or tolerance problems. This is compounded by the fact that designer's intentions, rationale for choices, and interpretation of what to design, are not shown in drawings. Thus, leading to problems when there are frequent changes encountered during the design activity. Finally, when graphical communication is used the visualization defines both what we see and implicitly what there is to see, i.e. in choosing what to show we also chose what not to show and hint at the scope of the possibilities. As such, although drawing and visual communication are essential tools, a designer should be familiar with their limitations and possible pitfalls.

4.4.5 Interaction and Integration

Interaction can be seen as direct (or not so direct) contact between actors. The most important mode of interaction is dialogue where opinions, body language, jokes, irony, and seriousness mix. Integration, on the other hand, has a number of meanings. At the organizational level integration is "*…the quality of the state of collaboration that exists among departments that are required to achieve unity by the demands of the environment*" (Lawrence and Lorsch 1967). In product development Andreasen and Hein (1987) see integration as the kernel of efficiency. Dorst (1997) defines integration at the team level as: "*Someone is designing in an integrated manner when he/she displays a reasoning process building up a network of decisions concerning a topic (part of the problem or solution), while taking into account different contents (distinct ways of looking at the problem or solution)*". The three types of reasoning discussed in Chaps. 11–13 all represent different aspects of design integrating.

Key to good communication and integration is continuous reflection leading to recognition of goal satisfaction, progression, quality, and goodness in the team's efforts. Decision-making activities are also closely linked to integration as they help foster alignment between team members.

4.5 Interacting with the Surroundings

The team's interaction with their surroundings is illustrated in Fig. 4.2. Key to this is the flow of information giving insight into opportunities, competition, user needs, and new technologies. Another type of interaction stems from the team being embedded in a larger organization. This gives insight into and opportunity to contribute to, e.g. strategy, policies, plans, and tasks. Underpinning these are human-to-human interactions between the team, external actors, and users.

Above, we saw how staging draws together actors from different knowledge domains in order to bring contributions and knowledge into the space. Actors external to the team can be seen as 'specialists', contributing specialist and branch knowledge. Similarly, users contribute composed and detailed articulations of needs, values, and insights from life conditions. For both, boundary objects are an important means of facilitating knowledge transfer.

4.5.1 Use of Specialists

Specialists are any person who adds to the team's insight, e.g. stakeholders or consultants. Design and product development has many issues that are most efficiently solved using specialists. An extreme example is car design, which needs specialists in collision testing, embodiment design, aerodynamic simulation, internal noise and sound, ergonomics of users and in assembly, environmental impact analysis, and many more.

Specialists' role in design differs from team members. Specialists are not responsible for or involved in the design, instead they make proposals. It is then the project leader that decides if the proposals are acceptable, sufficiently clear, timely, and appropriate for the design progression. Specialists are not design oriented and are therefore outside the team's professional practice.

4.5.2 Users in the Design Activity

There are many varieties of user-orientated design (e.g. empathic design, market pull, customer oriented, social innovation, co-design, co-creation) but these all come down to some form of user involvement. The main idea is that users are 'specialists' in the product's use. It is interesting to look at the history of user involvement in design. It was early recognized that understanding user needs was crucial. This led to the idea of listening to the 'voice of the customer'. Here, selected groups of users were involved in the design process through, e.g. observing their behaviour or engaging them in co-creation. Thus, it is now normal to include users in the design activity or even in the proposal of new products. This reflects the need to balance technological push and user pull. This balance is dictated by opportunities in the market and the value perception of the company or team.

The design team does not always appreciate users: "*If we had involved users in our product development, no Bang and Olufsen product would have seen the daylight*" is a statement from one B&O development manager. However, it is well known that 'lead users' (users who formulate requirements, demand new features, and push ideas) are a key source for product innovation.

A novel development in this area is open-source design where companies offer design documentation to users, allowing customers to create and develop their own products. Here the idea is that customers form an active group sharing designs and providing the company with innovative ideas.

Example:

3D-printing as open-source design

Consumer 3D printers have become very popular, with more than 180 types on the market today. One open source printer, Bower (2005), was the result of researchers tinkering with self-replicating machines—and how they can evolve—with the first prototype emerging in 2006. The designs were shared and well documented from the start of the project, allowing the idea to attract significant public interest and engagement. In particular, the open source nature of the machine has allowed users to create a huge variety of new product options. Figure 4.8 shows one version of RepRap in operation together with one of its printed components.

Fig. 4.8 The RepRap 3D printer in action and an example of a printed component, *courtesy* Ali Gürcan Özkil

4.5.3 Focus Dimensions in User Involvement

Staging design and innovation is much more than just supporting the team; we must also integrate the user. In order to simplify the many descriptions and types of user involvement we propose the following focus areas.

First, **system dimensions** are commonly part of public projects, e.g. a metro, where admission, visual design, information system, safety, etc., all need to be integrated to satisfy both the authorities and individuals. This focus area has substantial elements of use, societal value, and utility.

Service dimensions are related to supporting a user over time via, e.g. service offers, professional control or certification programs. This supports service-based innovation, i.e. the joint development of product and service, and builds on an intimate integrated view of the needs and problems confronting the user.

Socio-technical dimensions deal with the interaction between systems, products, and service, either planned or believed. Clarification of this area can drive innovation in, for example, the product's domestication (acceptance, use, and deployment), branding, and reputation.

Network dimensions allow collaborative groups of companies to innovate based on knowledge, technology, cultural/user insight, and closeness to market, which are not present in the individual companies. Similarly, **market dimensions** draw insight from access to established markets, customer relationships, cultures, and traditions or by establishing new markets with new sales and service opportunities.

Finally, **user dimensions** describe a wide area: user behaviour, man/machine interaction, usability, use activities, ergonomics, experience, value perception, cognitive processes, sustainability, social interaction, etc. All of these deal with features related to 'user driven' design (Janhager 2005).

Based on these areas we can see that user involvement plays a number of roles: as *design strategy*, *focus area* in exploration, and a distinct *type of staging*. For example, these three elements are brought together in the design of computer systems. Here, bad experiences, unsatisfied expectations, and problem understanding led to research by Beyer and Holtzblatt (1998) and van Hippel (1978), and resulted in a general theory of work space design.

4.5.4 Boundary Objects

Boundary objects (also called *intermediary objects*) describe a concept from sociology related to shared understanding. A boundary object is something that is understandable to multiple parties, without any one party being required to share the full understanding. The concept has won popularity in the design domain;

being used for communication and translation between, e.g. designers and users, software and mechanical engineers or engineers and government. Common boundary objects are models, sketches, drawings, or components that can be interpreted by different disciplines whilst maintaining a common identity. Such staging elements have to be created throughout a project to support coherence, integrity, coordination, and decision-making.

Example:

User involvement with a full-scale boundary object

The Institute of Product Development at the Technical University of Denmark was asked to design and deliver a test-tube filling machine with semi-automatic sterilization, filling, and stopper insertion. The machine was one-of-a-kind and user interaction was a critical criteria. Thus a full-scale model was produced to gather user feedback before finalizing the design (Fig. 4.9).

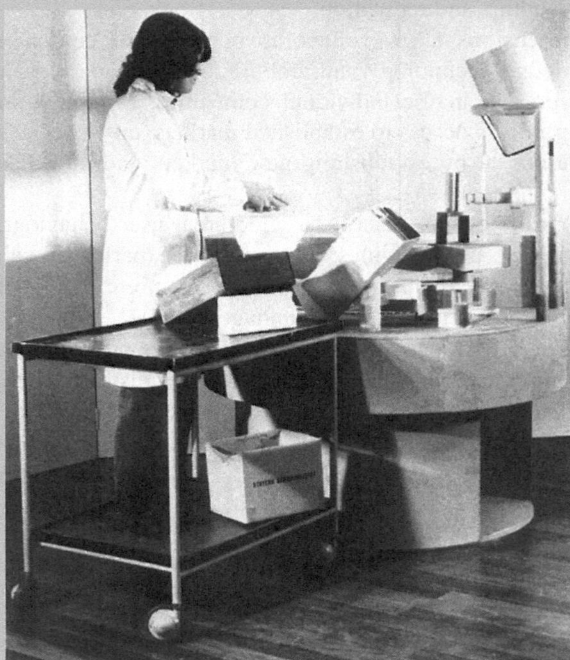

Fig. 4.9 A full-scale model used as a boundary object between designers and laboratory personnel (Tjalve 1979)

4.6 Conclusion

At the heart of this chapter is the idea that staging is critical to design and an important skill for the designer to master. Further, by building from team staging up to product development and company organization we are able to create the best possible conditions for design success. This is a fundamentally different perspective from the top down view of management. With this in mind we are able to turn our focus to the final aspect of the design machinery: the design process.

References

Andreasen MM, Clausen C (2011) Staging competences in design. Presentation at "Design Engineering Education Entrepreneurship", DTU (not published)
Andreasen MM, Hein L (1987) Integrated product development. IFS (Publications)/Springer, Berlin
Andreasen MM, McAloone TC (2001) "Joining three heads"—experiences from mechatronic projects. In: Meerkamm H (ed) 12th design for X symposium, Neukirchen, pp 151–156
Araujo CS (2001) Acquisition of product development tools in industry: a theoretical contribution. Ph.D. dissertation, Technical University of Denmark
Beyer H, Holtzblatt K (1998) Contextual design—defining customer-centered systems. Morgan Kaufmann Publishers, San Fransisco
Bijker WE (1995) On bicycles, bakelites and bulbs—toward a theory of socio-technical change. MIT Press, Cambridge
Bower A (2005) The RepRap project. www.reprap.org. Accessed Feb 2014
Brown T (2009) Change by design: how design thinking transforms organizing and inspires innovation. HarperCollins, New York
Bucciarelli LL (2003) Designing and learning: a disjunction in contexts. Des Stud 24:295–311
Cannon-Bowers JA, Salas E, Converse SA (1993) Shared mental models in expert team decision making. In: Castellan NJ Jr (ed), Individual and group decision making: current issues. Lawrence Erlbaum Associates, Inc., Hillsdale, NJ, England, pp. 221–246
Clausen C, Yoshinaka Y (2007) Staging socio-technical spaces: translating across boundaries in design. J Des Res 6(1–2):61–78
Cross N (2006) Designerly ways of knowing. Springer, London
D'Astous P, Déticeme F, Visser W (2004) Changing our view on design evaluation meeting methodology: a study of software technical review meetings. Des Stud 25:625–655
Dorst K (1997) Describing design—a comparison of paradigms. Ph.D. dissertation, Delft University of Technology
Duffy AHB, Andreasen MM (1995) Enhancing the evolution of design science. In: Hubka V (ed) Proceedings of ICED 95, Praha, Heurista, Zürich
Hansen CT, Andreasen MM (2006) Conceiving product ideas in an initial and uncertain design situation. In: Jónsson MPh, Uumphersson R (eds) Proceedings of NordDesign 2006, University of Island, pp 32–41
Henderson K (1999) On line and paper: visual representation, visual culture, and computer graphics in design engineering. The MIT Press, Cambridge
Janhager J (2005) User consideration in early design stages of product development—theories and methods. Ph.D. dissertation, Royal Institute of Technology, Stockholm
Jørgensen U (2004) The design engineers of the future have innovative competences. Produktudviklingsdagen 2004, IPU, DTU Copenhagen

Kahn KB (1996) Interdepartmental integration: a definition with implications for product development. J Prod Innov Manage 13:137–151
Katzenbach J, Smith D (1993) The wisdom of teams. Harvard Business School Press, Boston
Kirkegård L (1988) Fastlæggelse af udviklingsopgaven (Determination of the product development task). Institute of Product Development, DTU Copenhagen
Kleinsmann M (2006) Understanding collaborative design. Ph.D. dissertation, Delft University of Technology, Industrial Design Engineering
Kolb DA (1984) Experiential learning; experience as the source of learning and development. Prentice-hall, Englewood Cliffs
Lawrence PF, Lorsch JW (1967) Organizations and environment. Harvard Business Press, Boston
McKim RH (1972) Experiences in visual thinking. PWS Publishers, Boston
Olson EM, Orville CW, Ruchert RW, Bonner JB (2001) Patterns of cooperation during new product development among marketing, operations and R&D: implications for project performance. J Prod Innov Manage 18(4):258–271
Onarheim B (2012) Creativity under constraints: creativity as balancing 'constrainedness'. Ph.D. dissertation, Business School, Copenhagen
Perry M, Sanderson D (1998) Coordinating joint design work: the role of communication and artifacts. Des Stud 19(3):273–288
Sant K (1988) Fastlæggelse af Udviklingssystemet (Determination of the product development system). Institute of Product Development, DTU, Copenhagen
Schön D (1983) The reflective practitioner—how professionals think in action. Temple Smith, London
Sim SK, Duffy AHB (1998) A foundation for machine learning in design. Int J Artif Intell Eng Des Anal Manuf (AIEDAM) 12(2):193–209
Tjalve E (1979) A short course in industrial design. Newnes-Butterworths, London. Faximile edition (2003) Systematic Design of Industrial Products. Institute of Product Development, DTU, Copenhagen
Ulrich KT, Eppinger SD (2004) Product design and development, 3rd edn. McGraw-Hill, New York
Valkenburg RC (2000) The reflective practice in product design teams. Ph.D. dissertation, Delft University of Technology, Industrial Design Engineering
van Hippel EA (1978) Users as innovators. Technol Rev 80:3–11

Chapter 5
The Design Process

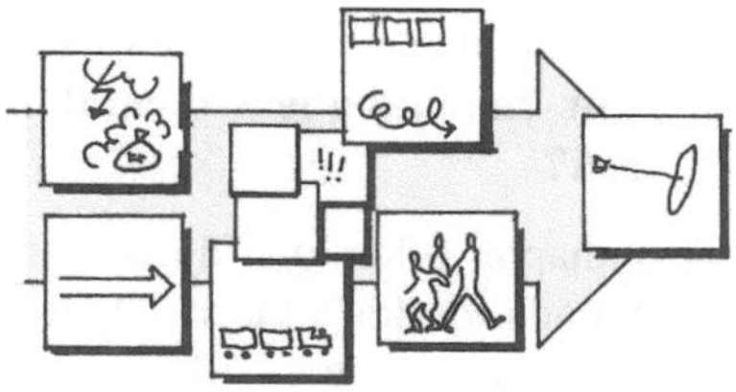

The third element in the design machinery can be seen as part of the staging, but has such a major role that it needs its own treatment here in this chapter and further detailing in Chaps. 6–10.

Understanding and mastering an effective sequence of activities normally leads to a preferred, explicit design process as part of practice. In this chapter we highlight what really influences the composition of a process fitted for the actual task, context, and organization. We emphasize its dependency on human thinking and the artefacts' nature, its role in supporting and managing the design activity, and its shaping into procedure. Our proposal for a design process, the Encapsulation Design Model, is a guide to understanding design's nature, a framework for creating a fitted procedure, and a backbone for the application of models, approaches, and methods.

5.1 The Design Process' Composition

The third element of the design machinery outlined in Part II is the design process. Here the design processes refers to both the actual activities undertaken by the designer, and the models used to describe and guide these activities. This dominates the staging activity and is by far the most complex element. However, the design process itself provides little prescriptive support for the designer.

Taking a step back, the idea of a design process as a support for the designer originated in the 1950s, but really found its modern form in the 1960s. In 1962 Asimow created the Design Morphology model, providing a rich description of new product development in the industrial context. Independently, Hansen (1966) created his model combining problem solving and gradual product synthesis. In this model Hansen set the foundation for the modern conception of the engineering design process. However, in order to transform these abstract design processes into useful support for the designer we need to answer three key questions—explored in this chapter.

M.M. Andreasen et al., *Conceptual Design*, DOI 10.1007/978-3-319-19839-2_5

1. **How can the design process support the designer?** Here, we seek to understand how the process can support and instruct given a specific task, the nature of the design problem, and the wider context of the design activity.
2. **What do the various design process models tell us?** What is their role, are they valid and how can they be applied?
3. **How can designers formulate their own design processes to best fit the task?**

This chapter offers preliminary answers to each of these questions in order to lay the foundation for a more detailed examination of their implications in Chaps. 6–10. As such, we deal with three main topics. First, Sect. 5.2 explores the various ***factors influencing the design process***. Here we seek to understand how these factors impact the actual situation, and how the designer can incorporate them in their design process. Expanding on this multifaceted view of the design process, there are hundreds of theoretical process models, and even more procedures in industry. As such, Sect. 5.3 contrasts these two perspectives—***literature's models and industry's procedures***—in order to help unite insights from theory and practice. Finally, Sect. 5.4 describes the ***Encapsulation Design Model,*** which both brings together a detailed explanation of a design process, and also provides a framework for the reader to understand design processes and their composition.

5.2 Factors Influencing the Design Process

Fundamentally, there are two main sources of influence on the design process: the nature of the artefact to be designed, and the nature of human problem solving and cognition. This is reflected in the previous chapters where we described product development as about both understanding the design needs, and synthesizing the product in line with the designer's thought process. These dual influences pose a serious challenge to model creation—how can we decompose the influences affecting the designer and the design process?

One approach is the use of *design process models*. These are widely used and have convincing benefits in practice; however, they often meet strong academic critique. This originates from a conflict between the sequential view of artefact development as a series of stages, and the less linear cognitive processes underpinning designers' creativity. Our philosophy is that we cannot expect models to prescribe human behaviour, but they are needed for navigation and cooperation. As such, we see models from two perspectives: one articulating the procedural and co-coordinative aspects, and one in which the designers drive the process and operate creatively. Here, we draw an analogy to one of our favoured Wednesday afternoon pastimes, Jazz and beer in the cafes of Copenhagen. If jazz were to be modelled, it would also need these two perspectives, respecting the composer (supported through time signatures, structure, and chord progression) and the improviser (supported through harmonies, scales, and modes). The beer has no

Fig. 5.1 Stepping stone methodology and creative freedom together

relevance but has other benefits. Describing this dual-perspective mindset, captured in Fig. 5.1, is our ultimate goal in this chapter.

Reflecting on this dual-perspective we have discussed some of the different views on design Chaps. 3 and 4. Here we saw four main themes emerge:

1. The '***how to tackle a task***' view. This considers how to handle the ill-defined design situation, the perception and translation of needs and intent into an artefact, and balancing the interaction between the perceived problem and the emergent solution.
2. The ***socio-technical*** view. This considers the impact of the design on individuals and society, as well as understanding the design team as a community of practice.
3. The ***cybernetic*** view. This considers a team as a goal-oriented living organization that observes, reflects, and reacts to create the best possible conditions for solving the task.
4. The ***design coordination*** view. This considers the complex nature of designing through the composition of goals, activities, actors, competences, results, etc.

Bearing these views in mind the following sections aim to answer the question: what other factors influence the design process? Here each section explores one of the six major factors highlighted in Fig. 5.2.

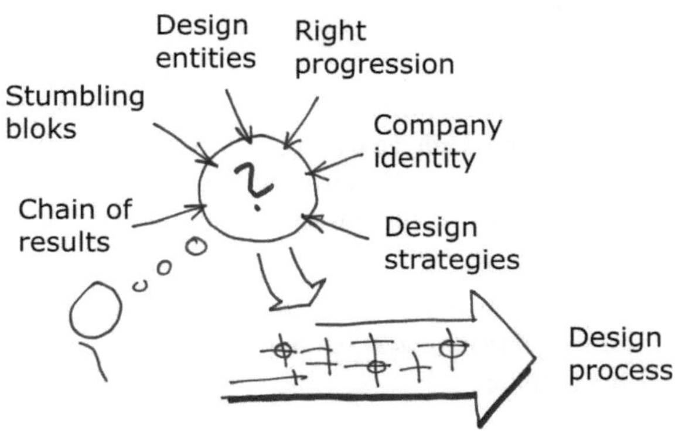

Fig. 5.2 Factors influencing the design process' arrangement

5.2.1 Chain of Results

Design is only one part of the wider product development activity; however, it forms a key integrating mechanism for the many disparate elements found in industry. In particular it helps bring together task identification, manufacture, sales, and business considerations—outlined below. This results in a process where design is realized through multiple parallel and sequential tasks, which ultimately produce a timely, integrated solution.

- ***Task identification***. Identifying, planning, and coordinating tasks with respect to a company's aims and overall strategic goals usually falls to specialists or management. Further, task plans are critically related to a company's existing portfolio of ideas, projects, products, and procedures. Despite this, a typical task formulation only vaguely sketches intention, need, opportunity, and relations to past or future products. As such, the design team plays a key role in enriching and clarifying these dimensions in order to achieve proper conceptualization aligned with the wider needs of the company.
- ***Manufacture***. Manufacture is normally both geographically, and temporally distant from the designer in the early stages of the design process. This can result in substantial misalignment between the product design delivered and the capabilities of the manufacturing process. As such, the designer is responsible for having a good understanding of the manufacturing technologies available and their fit with current products, processes, and capabilities.
- ***Sales***. As with manufacturing, sales are often far removed from the core design activity. However, the successful designer continually seeks to incorporate the influences of market information, customer analysis, and user needs into his design process. Further, this integration of sales and design should be mirrored by feedback from the designer to the specialists determining the sales strategy, price policy, and follow-up sales.
- ***Business***. Here the designer is often detached from the value perception of the product, seldom being the intended user. As such, the design process provides a means for integrating usability experience, as well as identifying the true needs of the user. This is particularly true of approaches where users are actively integrated into the design process as a driver for improved satisfaction and product customization.

One question you might ask yourself here is, how concerned or responsible should a designer be with achieving this harmonious integration, and where does the design process end? Unfortunately there is no easy answer—although we offer the following heuristic for allocating your limited resources:

> The sequence Task/Concept/Design/Business/Use/Life forms the backbone of the product development process and thus also the priority of concerns for the designer.

Based on this, the designer's understanding of the design process is critically linked to these five elements. This is captured by the fact that just as they influence the design activity and its results, the designer also influences the execution of the tasks throughout this chain. As such, a designer can have a range of different roles in completing this chain. As project leader a designer might be involved throughout the process from idea identification, through product development, to sales and manufacture (Andreasen and Hein 1987). More usually, however, a designer will be involved with respect to specific tasks or phases. Despite this, many designers dream of an ideal where their ideas give them a central position in a company or even allow them to establish their own. Here, ideas may be the ticket for admission but the professionalisms needed for manufacture, distribution, and commercialization cannot be omitted or escaped from.

In practice this means that a designer's primary concern is his contribution as a pathway to influencing the whole chain of elements. Designers are valuable and powerful when they know what matters in the arena where the company operates, what matters for the company, and what it means to have a good solution:

> Conceptualization has two dimensions: *creating an idea* and *understanding the demand* and the relative power of the values, which tell the designer if this is a good idea!

The goodness of the idea relates to the totality of the concept—integrating all the elements discussed in this section to give a holistic understanding of a concept's merit in a given context.

5.2.2 Stumbling Blocks

The nature of design means that there are many stumbling blocks waiting to trip the unwary designer; here we begin to explore some of these. The sequence *Task/Concept/Design/Business/Use and Life* seems a logical one at first glance, with each topic's clarification and realization being the precondition for the next. However, in order to totally fulfil each element it is also necessary to draw information from those elements following it, making the whole process iterative. For example, a successful concept incorporates insights from the use activity, but before this is possible at least some concept is needed for imagined use. Thus, the designer is continually faced with incomplete knowledge at each step, forcing him to iterate to successfully fulfil the overall process. In order to understand what this means for our conception of the design process let us consider the main stumbling blocks emerging from the above sequence.

The first and most significant stumbling block is that the above sequence, as well as the majority of design models, portrays a goal-orientated process that appears to offer a stable and relatively linear path to the top down determination of an artefact. In other words, despite some effort to recognize the complexity of actual design activity, process models often fail to adequately portray the iteration and messiness inherent in real world design work. Decomposing this we encounter four horsemen: ***co-evolution***, ***emergent design***, ***ill-definedness***, and ***intractability***.

First, ***co-evolution*** describes a world where task and solution mutually and gradually clarify each other. This can lead to substantial revisions as well as significant uncertainty throughout the process, demanding numerous corrective loops. Co-evolution is the main reason for the feedback loops often depicted in design models, although these belie its fundamental relationship with the design activity. Similar to co-evolution, **emergent design** describes the process by which a design emerges from a creative process while displaying properties and values which could not be predicted from its detailed parts. This can lead to changing of goals or the redefinition of a problem creating a feedback loop into the design itself. Although this emergence can be both positive and negative, it ultimately means that through much of the design process the designer faces an ***ill-defined*** product. Ill-definedness describes the fact that design essentially concerns products that do not exist yet. In this context, designing must proceed from sparse insight and information, gradually establishing the parameters of the design as the process progresses. This lack of knowledge is often disguised behind seemingly relevant information, extrapolated from past designs, which presents a danger for a designer seeking new opportunities. Finally, these stumbling blocks can combine to make a design problem seem ***intractable***. Intractability is the perception that there are no feasible solutions or sub-solutions, and that a design will incur unexpected or unacceptable problems, such as spiralling costs. This demands careful consideration to identify when a problem is truly insoluble or merely apparently so. Ultimately this can lead to substantial changes of goals and plans, or termination of the project.

Compounding these stumbling blocks is the constraints defining and limiting the solution space. Constraints are both necessary for and limiting to creativity (Onarheim 2012). The design activity needs constraints in order to be defined, e.g. in terms of task and goal formulation, company conditions, and market and societal limitations. However, where a situation is either over or under constrained creativity can be stifled. This links to the dynamic nature of design, where changes in target, markets, and technologies are commonplace. In this context, assumptions, goals, methods, and decisions all need to be considered dynamically as constraints propagate; any synthesis step adds to the constraints for those following.

Complexity can be experienced through these constraints, with the designer perceiving the compound impact from, e.g. problems, effects, criteria, stakeholders, and their various interrelations. As such, the development process demands flexibility and opportunism, requiring dynamic coordination (Chap. 3) and navigation. This dynamism makes each design project **unique** in terms of focus, procedure, strategy, management, staff, decisions, etc. (Andreasen and Hein 1987).

Thus we warn against blindly adopting a 'typical' design procedure and expecting the best outcome.

5.2.3 Design Entities

The design entity describes the composition of the final design outcome. We simplify this by using the label 'product' when discussing need satisfaction or the overall aim of the design activity. However, during the course of design activity several intermediary design entities will inevitably be created in the form of both artefacts and activities. In order to identify these entities, three perspectives should be considered, adding to our discussion from Chap. 2:

- ***Design and realization***: concepts and issues related to the product. These lead to new sub-systems or ways of realizing the product, e.g. a new supply chain or different manufacturing and assembly processes.
- ***Business***: concepts and issues related to the distribution, sales, and service. These lead to entities, such as new service offers, distribution systems, or accessories.
- ***Deployment***: concepts and issues related to the use activity, system integration, maintenance, and renewal of the product. These lead to entities dealing with a user's perception of 'what they have bought' and what the product actually does, and the need for service.

Figure 5.3 shows a simplified model of the design entities that can be envisaged at the conceptualization stage. Olesen (1992) links conceptualization and design entities in his model Fig. 13.12 via several parallel task activities.

Design entities are the dominant identifiers of design activity. When a design is based on previous similar projects, design entities are easily identified. However, in other situations the design process is often made up of entities closely linked to

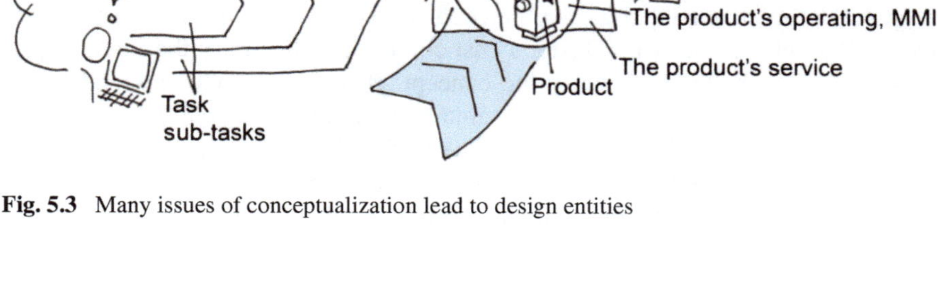

Fig. 5.3 Many issues of conceptualization lead to design entities

experts from other disciplines, such as business. As an example, we can imagine a company producing canned food. Here, the final product is made up of several design entities: the prepared food ready to be packaged, a new can type to be manufactured, a new label to be designed and marketed, and the development of all of these into a refined business plan for the product launch. Further, new sales channels and advertisement campaigns give rise to additional design entities. In this example, no one person or expert is able to deal with all of the entities associated with the project. Instead a wide array of experts need to be involved and managed in parallel in order to produce a successful outcome.

5.2.4 The Right Progression

Many models in the literature articulate the design process as a logical progression from functions through to detailed design, arguing that this progression is the 'right way'. This is especially prevalent in the mechanical engineering domain. However, we argue for a more pragmatic approach, where the designer is not afraid to challenge this ideal pattern depending on the situation arising in a project. In particular we highlight four situations that demand a different approach. The first is where the design strategy is incremental design. Here the design is based upon substantial *reuse* of previous work, demanding a more 'copy and paste' type process. Second, when the design is based on a platform strategy, there is a need for a *configuration* approach where the process and the product are aligned with the wider platform. Third, when the design faces significant changes in either the problem or solution definitions there is often a need for substantial *backwards steps*. Here the designer might revisit earlier process phases in order to re-evaluate previous decisions or tasks in the light of the changed design parameters. Finally, some products demand a *detail first* strategy where the designer must invert parts of the process in order to set the groundwork for conceptualization.

Overall we advise that design process should be geared towards traceability and low financial risk. In this context it is important to identify critical issues, e.g. unreliable technology, and clarify their possible impact up front (Andreasen and Hein 1987). Here, Baxter's (1995) textbook "Product design" describes a relevant, if traditional, design process. Baxter explicitly explores the dependency between decided costs and the progression of the design process, and suggests a shallow curve in the early design where changes are easy. This leads to an interesting proposal where financial risk is minimized. Figure 5.4 shows the Encapsulation Design Model together with Baxter's cost curve. This illustrates the iterative process used at each stage to clarify the concept, check the technology, and finally approve the product prototype whilst maintaining designer flexibility.

Rather than emphasize a timeline as the main driver for the design process we instead use a causality line, linking design activities. This focuses the designer on a preferred sequence of activities without constraining them to a linear time ordering.

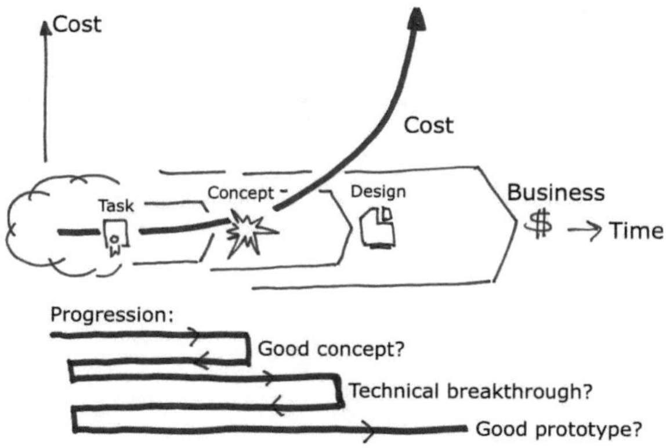

Fig. 5.4 The development activity related to cost and compared to Baxter's (1995) progression for low financial risk

In a sense the 'timeline' conceptualization of the process only becomes real when the model is transformed into real tasks undertaken as part of a design procedure.

5.2.5 The Company Identity

Let us not forget that the purpose of design, as seen from the company perspective, is the creation of products that lead to sales, and ultimately profit. Thus, the identity the company projects has a significant impact on the design process. Duffy and Andreasen (1995) described the model shown in Fig. 5.5 called "design of design". This challenges the designer to change one or more of the four boxes (normally considered as fixed conditions) in order to better fit the 'design machinery' to the situation. Figure 5.5 mirrors the company identity as articulated by its choices in the four boxes:

- ***The chosen design tasks***: the arena in which the company operates and how it interprets needs and opportunities in order to define its plan of action.
- ***The chosen design strategy***: the way the company chooses to tackle the actual project.
- ***What we are and know***: the company's knowledge, experience and access to technologies, as well as its perceived and real position in the arena.
- ***The way we design***: the tacit knowledge and unwritten rules governing how design is undertaken during new product development at the company.

The message of Fig. 5.5 is that the designer has the power to change *the space* in which they work; namely the company, organization or team's traditional way of perceiving and solving design tasks. This mental space is a consequence of past design activities and emerged practice, meaning that dynamic development or innovation in the company will be closely related to changes in product development:

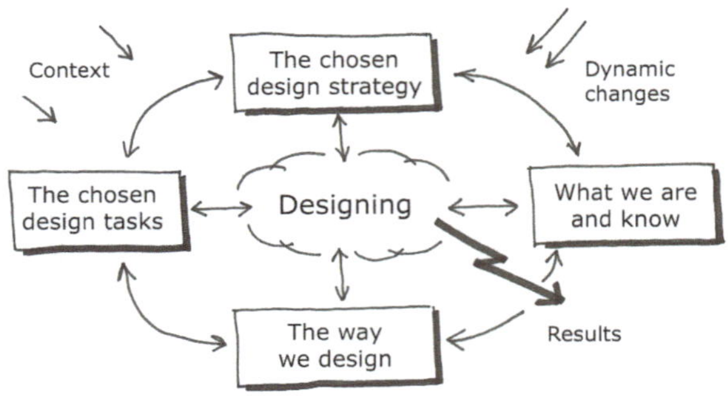

Fig. 5.5 Using 'design of design' to fit the design process to the situation (Duffy and Andreasen 1995)

Changing or innovating company identity means to change the fundamentals of development: strategy, tasks, knowledge, and way of designing.

From this we can see that innovative changes to a company's identity can have a strong influence on the design activity. However, the opposite is also true, these types of identity changes often originate from new product or design developments.

Example:
Model for 'identity design'. A consulting company used the model in Fig. 5.6 in its sales material. This was used to communicate information about the procedure for creating a new name, branding, strategic, and graphic identity for the customer. In the end, although the model was key to the project and important for the customer, it was not actually used for planning, with one manager stating that 'it is in our minds but stays in the drawer'. As such, it is important for understanding the efforts necessary for changing a company's identity.

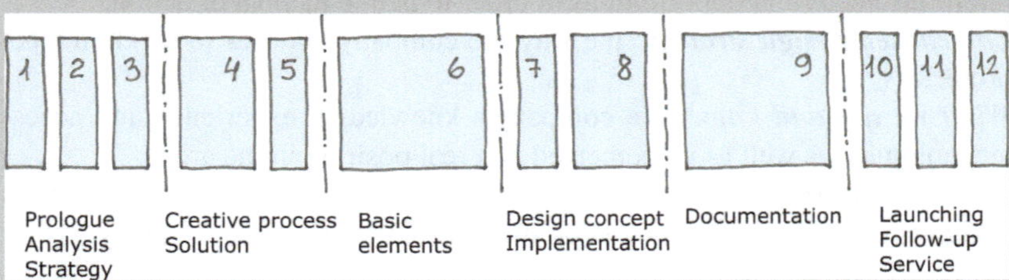

Fig. 5.6 Design model showing the structured progression of a consultancy's service, where customer dialogue is arranged. The figure shows relative time distribution between the steps

5.2.6 Design Strategies

From the four boxes highlighted in Fig. 5.5 company management often focuses more strongly on design strategy. This defines the approach for managing what is going on in the development organization. This organizational unit constitutes the product board, development department, and other functional areas, e.g. marketing and distribution (Andreasen et al. 1989). Further, this partially defines how the company's development portfolio is managed. The development portfolio is the sum of new development, upgrades, and customer orders, etc. This is a complex task that is not only about launching the right projects, but also composing these activities to support short- and long-term innovation and consolidation, resource optimization, and controlled reuse.

Parallel to the product design strategy are a number of related strategies, such as marketing and production. These all need to be incorporated in a development project such that the designer can answer: how will a new product influence sales and production, and how do sales and production influence the new product? Roozenburg and Eekels (1996) use this to describe product development as a loop, depicted in Fig. 5.7. Here, the company observes the effects of a product launch on its market in order to refine its overall goals, strategies, and policies. In particular, Roozenburg and Eekels distinguish between two levels of development: business search and strict development.

In general the design strategies outlined above aim to optimize **impact on the market**, while also **executing the project** successfully. In this context, both the literature and industry agree that generic design models are preferred. For example, Ulrich and Eppinger (2004) define six types of generic product development process, which Oja (2010) have supplemented with two more, shown in Table 5.1. There are two different types of project strategies illustrated in the table: the core way of creating products called 'description', and the characteristics of the design process called 'distinct features'. Ultimately the message of this section is:

> If a standard design process model or procedure is applied, we have to carefully adjust this to align with the design strategy.

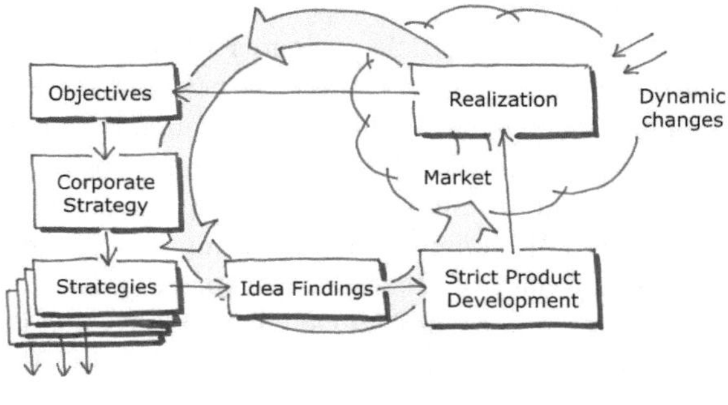

Fig. 5.7 Product development as a strategy-driven loop

Table 5.1 Generic product development process variants (Oja 2010)

Process	Process type	Description	Distinct features
Generic product development process	Market pull	Market opportunity recognition and selection of technology to meet customer requirements	Planning, concept development, system-level design, testing and refinement, production ramp-up
	Technology push	New technology introduction and evaluation of market	Matching the technology and market in planning, concept from technology
	Platform products	Application of established technology sub-system	Concept with proven technology platform
	Process-intensive products	Production process constrained product	Utilizing an existing process or development of a new (process)
	Customised products	Slight variation of existing configurations	Streamlined and structured development process
	High-risk products	High risk of failure due to technology or market	Early identification, analysis, and testing
Spiral	Quick-build products	Utilization of rapid prototyping and testing cycles	Repeated design and test phases
Complex systems	Complex products	System decomposition into sub-system components	Separate parallel teams, system integration and validation

Section 5.2 discussed six factors influencing the design activity and how we can model this. Next we bring these together with concrete design practices by contrasting them with industrial design procedures.

5.3 Literature's Models and Industry's Procedures

There are many proposals for design process models in the literature; at the time of writing a recent article compared 124 (Gericke and Blessing 2012). A key finding from this comparison was that these models fail to integrate the different disciplines encountered in industry. Although the models share common stages, this commonality only applies at a very high abstract level. Overall, this leads to a body of mono-disciplinary models that cannot be fitted to different types of design processes (Birkhofer 2011). However, despite these shortcomings, models are an essential part of process management, useful for researchers and practitioners alike. They support problem solving, aid decision-making, and provide a common platform for communication (Maier and Störrle 2011).

One of the major problems here is that these literature models are often mistaken for being directly transferable to industry. Clarkson and Eckert (2005) discuss several literature models and conclude that although they span a range of design disciplines, they are too general to properly help project planning and daily decision-making. As such, we argue that in order for literature models to be effective in practice they must always be adapted to the specific conditions and strategy necessary for solving the task. This is supported by studies where models have been successfully adapted in practice. Rückert (1997) sums this up by stating that "*A self-determined application of design methodology leads to better results compared to strict application*". We also believe this to be true and thus emphasize the importance of adapting the five modules based on the influences discussed in this chapter and, more generally, throughout the book.

5.3.1 Industrial Procedures

Industry's ideology of 'cost, quality, and time' is believed to encompass the main challenges in developing and delivering products. However, cost and quality are often only implicitly dealt with in design process models. In the context of this book we explicitly discuss their inclusion through design reasoning in Chaps. 11–13. In industry we seldom find literature's models used directly. In practice these models are changed into practical procedures, which can be seen as an instantiation of the design process model.

> **Definition:** A design procedure is a design process model fitted to a specific context. It is used as the basis for a procedural plan when a design project is executed.

Creating the procedure is part of a 'design of design' activity and is part of staging of design. As such, let us explore how this incorporates the influences described in Sect. 5.2.

The **chain of results** is addressed by defining the scope of the design model, e.g. creating a new concept versus creating a new business. Here, models can range from simple problem solving, to product development and engineering design. We discuss how these differences in scope can be dealt with through our Encapsulation Design Model described in Sect. 5.4.

The **stumbling blocks** are closely related to the human factor. Here, experience is often the best approach; however, risks can be reduced through reuse and the systematic consideration of design rational and information from past projects. For example, this can be achieved using the sign posting approach (Clarkson and Hamilton 2000).

The number and type of **design entities** to be created determines the number of synthesis activities needed as well as their nature. Further, the design strategy determines how they are sequenced, and how many parallel processes need be considered.

The **right progression** is a core project management problem concerning the gradual concretization and detailing of, for example, the design entities, need, opportunities, and legislation. Current practice ranges from specifying all sub-activities in a project to specifying the expected results of each design phase.

Finally, the **company identity** deals with the optimal utilization of company resources to fit the design result and design process to the company identity. Here, the development of company innovation should also be considered as part of the design activity.

5.3.2 Creating a Procedure

Fitting a process model to the actual circumstances of a company and designers demands is the core of the procedure creation activity. Here, information technology plays a dominant role; however, IT should not be seen as a readymade, ideal solution. Instead care should be taken to tailor IT support to best facilitate the required design activity. In order to do this, and create a procedure, we need to answer two key questions:

1. What must be accounted for in order to detail and fit a model to best support a specific **procedure** and design project?
2. What is 'left over' and must be understood and dealt with through the designer's **mindset**?

These questions reflect the dual-perspective philosophy illustrated in Fig. 5.1. In reality, industrial design procedures display many different applications and virtues. On one end we find rhetoric models. These inform the design philosophy used by the designers, and are followed by the company for the benefit of its staff, as well as communication with clients. On the other end we find procedures primarily intended for management purposes. These define the management of both project activity and corporate control of the process, resources, and results. As previously noted, procedures are often articulated as either detailed instructions and controls, or as expected results. These different articulations are primarily influenced by the maturity of the organization (Andreasen and Hein 1987).

Here you might ask, who makes these decisions and who should be in charge of installing a particular procedure? Unfortunately this is a rather fuzzy subject as in most cases it is very difficult to trace a procedure's origin. This is compounded by the fact that procedures are rapidly evolving documents, usually without a single responsible person overseeing their development (Araujo 2001; Jensen and

Andreasen 2010). However, this does not diminish the demand for these procedures, particularly as a company matures as an organization (Andreasen et al. 1989).

Example:
Installing Integrated Product Development. The book, Integrated Product Development (IPD), written by Andreasen and Hein in 1987 proposed one possible remedy for Danish industry's problems: lack of integration and effectiveness. IPD offered a model showing parallel activities across marketing/sales, design, and production. This explicitly developed a parallel **strategy**, integrating development, and three groups of **design entities**: business and sales, design and production, supply and distribution. The problem of **right progression** was dealt with by defining milestones based on the questions: what to clarify, and what to deliver from the three synthesis activities. Although this could be seen as constraining, the response from marketing and production was positive: '*Now we understand our role and contribution to new product development*'.
The procedure was fitted to the company through a consultancy driven series of courses. Typically, a course was arranged with a mix of lectures on the book's topics, and talks on the company's strategies, and plans. In particular, the different functions' expectations for new product development process were elicited and incorporated. This focused on what each function felt it could contribute or perform better (Fig. 5.8). During these courses we observed how the members of staff involved gradually improved their ability to articulate an 'improved' product development process. This resulted in the courses becoming a kind of negotiation with the management about change initiatives. In this way a new community of practice developed from the bottom up (Andreasen and Hein 1998).

Fig. 5.8 During his consultancy experience it was not uncommon for the first author to use cartoons to facilitate discussions (Andreasen and Hein 1987; Andreasen et al. 1989)

Sections 5.2 and 5.3 have laid the foundation for our proposal of a generic design process model, composed of five distinct modules.

5.4 The Encapsulation Design Model

Our aim with the Encapsulation Design Model is not just to create 'one more model', but instead to distill out a model that captures the core essence of design. We do this through five entities, merging two disparate phenomena:

- An ideal, partly causal, **progression** of the design activities: task/concept/design/business/use and life.
- An ideal articulation of the design process' **encapsulating and embracing nature**.

With regard to the second phenomena we emphasize two foundations for our thinking. First, early design activities are encapsulated through a gradual broadening view. Second, the activity includes a number of elements concerning goal and scope.

In order to clarify this model we must fully understand each of its five constituent elements. These five interrelated patterns of activity are introduced here, but are explored in depth in the subsequent chapters. This progression is illustrated in Fig. 5.9, and the five patterns are summarized here. The first element is **Exploration**. This aims to search out and find support for the product justification, key assumptions, and any prerequisite conditions necessary for project completion. This then leads into solution exploration where solution elements, product ideas, and possible tasks are considered in order to formulate the final design task process to be used. Based on this exploration the next process to be introduced is **Concept Synthesis**. This starts to transform the information found by the designer

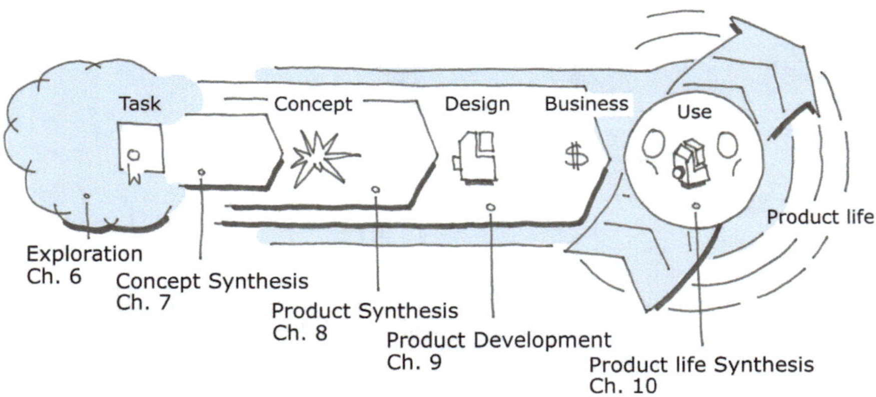

Fig. 5.9 The *Encapsulation Design Model* illustrated as a sequence of design activities. The following chapters explain each sequentially

into concept proposals. These concepts serve a dual purpose. First, they aim to answer the need, intention, and dreams articulated during exploration. Second, they help clarify the argumentation for the development activity's tractability, risks, and consequences. As such, concept synthesis leads to a final concept and an accompanying clarification of the space.

Based on the idea selected in concept synthesis, **Product Synthesis** focuses on establishing the design as a fully detailed and specified output. This activity alternates between synthesis and justification of the functionality and properties. Here, the activity can include conceptualization in order to ensure unbroken development as well as alignment with production and the other following phases. This ultimately leads to decisions about realization of production and sales. Based on this foundation the next activity to consider is **Product Development**. This establishes the company's ability to produce and market the product. This encapsulates product synthesis as the source for the product dimension in the multi-dimensional synthesis of, at least, sales, production, distribution, finance, and quality assurance. Similarly, the verification and justification activities in product development concern marketing, sales, production, distribution, competition, branding, maintenance, ethical responsibilities, and many more. These activities finally lead to product launch and thus close the development process.

With the development process concluded we transition to the **Product Life Synthesis** where the results from the development are deployed and adapted by the user. Here, every product experiences its life phase activities. Earlier phases mirror this by focusing on creating a fit for life. The product life synthesis embraces everything by being the ultimate result, leading to user satisfaction.

Concept Synthesis, **Product Synthesis**, and **Product Development** differ in scope by respectively creating concept, design, and business. Further, they also differ in relation to the product, transitioning from the abstract concept, to the more concrete product definition, to the final business case that encompasses market, sales, production distribution, etc. The difficulty in managing these processes is in setting an appropriate scope. Here, it is necessary to balance narrowing the focus—supporting concretization of the product, against broadening—supporting the wider appreciation and integration of secondary issues.

The key to understanding these complexities is to realize the nature of design activity, as a number of embedded processes. This is captured in the Encapsulation Design Model where each process is embedded within the subsequent phases as illustrated via a Russian Doll in Fig. 5.10. For example, exploration is both the basis for all subsequent activities, but also persists throughout the entire project, supporting the whole process.

The Encapsulation Design Model focuses on each entity's specific role, aim, professional content, and result; however, the sequence is tied to the overall progression of the embedded elements. Thus, depending on the type of design work undertaken, the product, and the organization, the model should be adapted, as we have discussed in this chapter. The model's aim is to serve as an eye-opening guide for understanding conceptualization and the surrounding design activities, without prescribing one rigid approach.

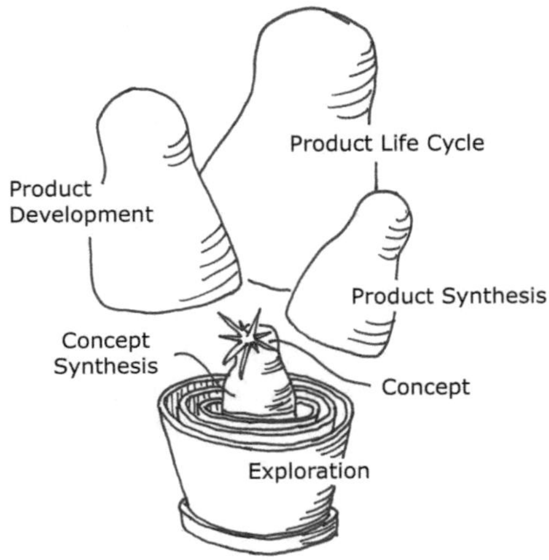

Fig. 5.10 The encapsulation design model's nature in a metaphor

5.4.1 It Is not Only About the Product!

All design activities are launched with unknown or uncertain aspects. These fall into two main groups, on the one hand the arena situation: need, competition, sales, branding, and market success; and on the other hand the necessary insight, technology availability, and networking as a condition for the product design. As such, it is risky to exclusively consider only one part of the five entities. For example, it is common for product synthesis to be undertaken while forgetting the clarification, i.e. the key underlying information and justifications. In this way information needs to flow between all five entities in order to make the overall process work. We characterize three types of information flow, which occur throughout the design activity:

- **Feed forward flow**: This describes the flow from clarification of the earlier phases into later design elements. This includes task information, decision rational, and information regarding the business opportunity, and risk reduction.
- **Situational**: This describes the immediate collection of information necessary to support the progressing synthesis or decision processes.
- **Feed back flow**: This describes the continuous imagination, forecasting, and consideration of past experience to ensure that the process is progressing in the correct way and that no anomalous activities have been identified.

Figure 5.11 shows ideal information flows. Here, Fig. 5.11 (a—situation) shows the situation during the search and task formulation phase, where the project is based on the results of the exploration work coupled with foresight, imagination, and projections based on experience. Figure 5.11 (b—situation) shows the general clarification situation of continual exploration feeding the design activities. This also includes research, forecasting, and experiences concerning the remaining activities.

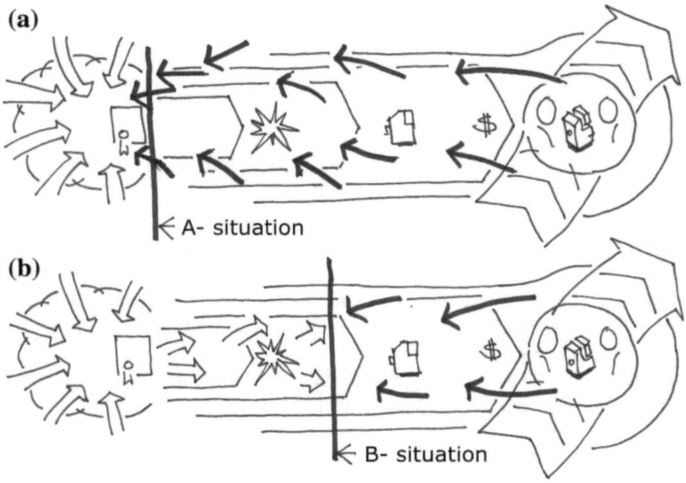

Fig. 5.11 Information flows in a development project: **a** feedback of ideas, forecasting, and experiences, **b** feedforward of actual clarification and experience

Considering these flows in conjunction with the Encapsulation Model (Fig. 5.10) we emphasize the fact that this only describes part of the situation. We do not show the designers' mental work, reasoning, or decisions, which play a substantial role in all design activity. As such, we warn against a blind belief in 'following the model' as this is not our intent. As a designer you never want to have to ask "*Why do we think the customer will buy our product*?" and find that no one on the team knows the answer. Thus, we offer the following heuristic:

The Encapsulation Design Model is a stepping-stone methodology as illustrated in Fig. 5.1. The stepping stones supply a map, but not the exact route or decisions to take.

5.5 Conclusion

The line of reasoning linking Part II together has been that the team and the designers' efforts can be seen as the machinery of design. We have brought this to the fore in this final chapter of Part II where we have used models to explain how this machinery is actually realized in performing the design activity and creating the desired outcome. In doing this we introduce the importance of positioning any model in the actual context, establishing its raison d'être, and its goals before proceeding with the process. In order to bring this together we propose the Encapsulation Design Model and its five entities. This is now explored in detail over the course of the next five chapters—Part III. Here, five partial design models are described along with the distinguishing traits linking them together and key to overall understanding.

References

Andreasen MM, Hein L (1998) Innovating the product development organisation. In: Birkhofer H (ed) Designers,—the key to successful product development. Springer, London

Andreasen MM, Hein L, Kirkegård L, Sant K (1989) Udviklingsfunktionen—basis for fornyelse (The product development function—foundation for innovation). Jernets Arbejdsgiverforening København

Andreasen MM, Hein L (1987) Integrated product development, IFS (Publications)/Springer, Berlin (Facsimile edition (2000) Institute of product development, Technical University of Denmark, Copenhagen)

Araujo CS (2001) Acquisition of product development tools in industry: a theoretical contribution. Ph.D.-dissertation, Technical University of Denmark, Copenhagen

Asimov M (1962) Introduction to design. Prentice-Hall, Englewood Cliffs

Baxter M (1995) Product design—practical methods for the systematic development of new products. Chapmann & Hall, London

Birkhofer H (ed) (2011) The future of design methodology. Springer, London

Clarkson J, Eckert C (2005) Design process improvement. A review of current practice. Springer, London

Clarkson PJ, Hamilton JR (2000) Signposting: a paradigm-driven task-based model of the design process. Res Eng Des 12(1):18–38

Duffy AHB, Andreasen MM (1995) Enhancing the evolution of design science. In Hubka V (ed) Proceedings of ICED 95 praha. Heurista Zürich

Gericke K, Blessing L (2012) An analysis of design process models across disciplines. Proceedings of the 12th International Design conference, Dubrovnik, Croatia

Hansen F (1966) Konstruktionssystematik (design systematics), 2nd edn. VEB Verlag Technik, Berlin

Jensen TE, Andreasen MM (2010) Design methods in practice—beyond the "systematic approach" of Pahl and Beitz. In Marjanovic D et al (eds) Proceedings of DESIGN 2010, 11th international design conference Dubrovnik, Croatia

Maier A, Störle H (2011) What are the characteristics of engineering design processes? Proceedings of the 18th international conference on engineering design: impacting society through engineering design Vol. 1: design processes. Design Society, pp 188–198

Oja H (2010) Incremental innovation method for technical concept development with multi-disciplinary products. Ph.D.-thesis Tampere University of Technology

Olesen J (1992) Concurrent development in manufacturing—based upon dispositional mechanisms. Ph.D.-thesis, Technical University of Denmark

Onarheim B (2012) Creativity under constraints. Creativity as balancing 'Constrainedness'. Ph.D.-dissertation Copenhagen Business School

Roozenburg NFM, Eekels J (1996) Product design: fundamentals and methods. Wiley, Chichester

Rückert C (1997) Untersuchungen zur Konstruktionsmethodik—Ausbildung und Anwendungen. (Investigations in design methodology—application and education) VDI-Verlag Düsseldorf

Ulrich KT, Eppinger SD (2004) Product design and development, 3rd edn. McGraw-Hill/Irwin, New York

Part III
Design Process

Chapter 6
Exploration

The design activity is normally based on perception of human needs. However, other factors from the designer's exploration or imagination might provide the initiator and driver. In this chapter, we discuss the front end of innovation based on awareness and imagination in our Exploration Design Model. This links five feed chains, each a potential starting point and all leading into the concept synthesis activity treated in Chap. 7. Through these feed chains our model brings together need interpretation, the designer's preferences, task, problem formulation, technology, and ideas.

6.1 Exploration: What and Why

Creating a new product is a synergy of at least three factors: business, need, and ideas. Someone must have the intent and ability to create a new product or business. There must be a need or at least willingness to buy. And there must be some ideas, knowledge, or technology which, coupled with design competences, can be

M.M. Andreasen et al., *Conceptual Design*, DOI 10.1007/978-3-319-19839-2_6

used to create a product. Our Exploration Model, introduced in Sect. 6.2, provides a framework for supporting this synergy. Exploration activity is critical to ensure that the correct product is developed! This has a number of dimensions:

- Right **product**: a need exists, the market is ready, and company effort is effectively distributed.
- Right **knowledge and technology**: the product is based on thorough accurate input from the knowledge and technology domains.
- Right **development**: the design process, production, sales, and distribution are aligned with the product, ensuring alignment between the company's functions and its resources.
- Right **lifecycle**: the final product is 'fit for life' across its whole lifecycle and gives the best possible conditions for the actors in these phases.

In dealing with these dimensions, exploration supplies information for the project, as well as supporting the search for new ideas and opportunities within the company's domain. Changes in competition, technologies, market conditions, legislation, and patents must all be included. Exploration ensures awareness of how to react to changes and what can be utilized in the company's surroundings.

Definition: Exploration is the upfront design activity that leads to the initiation and argumentation for a project. It is also the continuous process that supplies data, information, and knowledge.

This is clarified in the example below.

Example:
The Post-it. It is often said that Spencer Silver invented the post-it by accident. The truth is rather 'by incident', during his process of developing adhesives for the aerospace industry. Silver focused on the technology for many years, supported by 3 M's 'permitted bootlegging' policy. Further, he visited every 3 M division in his quest to find a business opportunity. Thus his colleague, Art Frey, originated the idea of pads for his hymnbook and introduced the notepad and sticky bulletin board concepts. Finally, a new production technology was developed for attaching 'non-sticking' adhesive to paper. This example of technology-based innovation shows the necessary interaction between resources (money, knowledge), technology development, and recognition of an opportunity; Frey was a user with a need (Koen 2004). Figure 6.1 shows post-its in action.

Fig. 6.1 Post-its used for classification in a professional design project, *courtesy* Jakob Parslov radiometer medical ApS

6.1.1 The Importance of Exploration

Unique, innovative products are created by individuals and teams, who explore effectively, finding new ways to, e.g. sell, distribute, or combine service and product. This ability is key to a product's success. It is not just the composition of the product that carries innovation.

The role of a goal formulation in pointing to the 'good product' is ambiguous. In particular, when the goal formulation, rather than the reality, becomes the target, creativity can be constrained. As such, we distinguish between good ideas and good solutions, based on the criteria proposed by Tjalve (1979) (detailed in Chap. 12). This links to the work of Pugh (1991), who describes how design work should be cyclical "*in all phases are considered all 'aspects' relevant for the development.*"

Many design process models focus on what might be called '*front end loading*'. Here, the task is fully clarified and defined, and a supply of information acquired upfront in a combined technical and marketing operation. However, this does not fit the reality of what is called the *fuzzy front end,* i.e. how the process is initiated. Koen (2004) uses the concept *Front End of Innovation* defined as "*activities that come before the formal and well structured New Product Development (NPD) portion*". Our interpretation of Koen's proposal for the creation of concepts is shown in Fig. 6.2. This shows a cyclical search and conceptualization activity

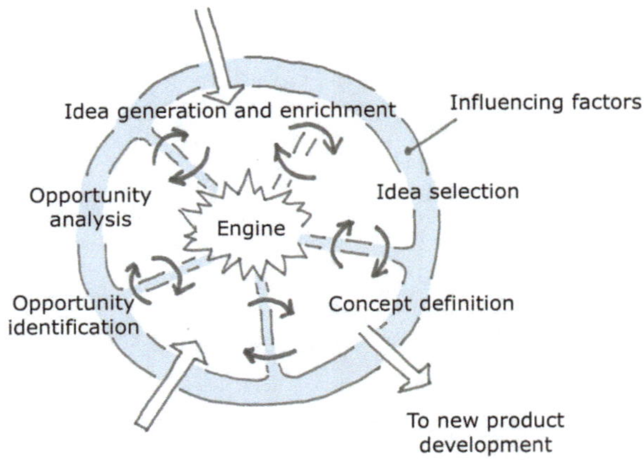

Fig. 6.2 Search and conceptualization 'machinery' in the front end of innovation

incorporating opportunity identification, opportunity analysis, idea genesis and enrichment, idea selection, and concept definition. There is no sequence merely a cycle.

Figure 6.2 highlights the synergistic nature of conceptualization but not the sources or search areas. We have separated these elements to form the Exploration Loop and Concept Synthesis. The purpose of Exploration is not only to identify opportunities but also supply rich information regarding technical conditions, users' belief systems and norms, and constraints (Lehtonen et al. 2011).

Our model, introduced in the next section, has visual similarities with Koen's but is based on a fundamentally different philosophy. Our intention is both to inspire a broad-spectrum search and to supply the project with necessary knowledge and information.

6.2 Our Exploration Model

Our Exploration Model illustrates an idealized view of the exploration work and its relationship with conceptualization. Our model is shown in Fig. 6.3 and illustrates how Concept Synthesis (Chap. 7) takes its starting point in this research loop.

This model articulates our perception of idealized exploration work. As such, it provides a mindset, a model of understanding, focusing on five main **feed chains** supporting conceptualization. Most other design models are articulated as activity models, i.e. they can be realized as a plan of tasks. That is not the aim with our Exploration Model.

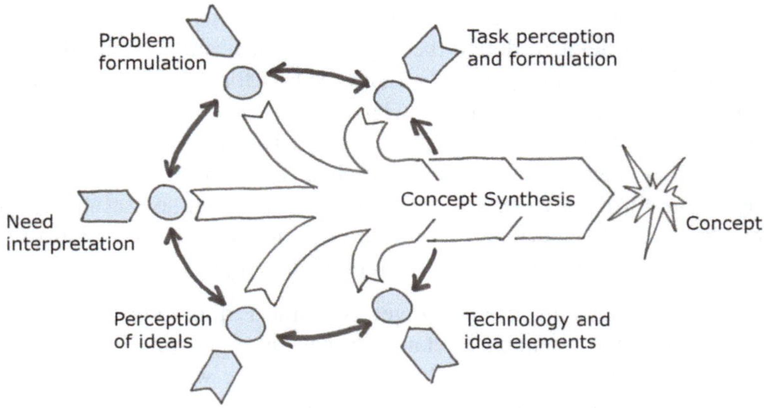

Fig. 6.3 Our exploration model: an idealized view of the explorative and clarification elements in the research work

6.2.1 The Five Feed Chains

Our model builds on the idea that conceptualization is not only based on product ideas but can take any of the feed chains as a starting point. These feed chains are summarized below and then discussed in more detail.

- **Need interpretation**: the designers' interpretation of a need can be triggered by any situation, e.g. tasks and problems, market knowledge, critique of existing solutions, customer reactions, or statements from critical users. The need articulation details the users' values, use activities, and situational identity. This insight can be captured through discussion, interviews, or observations of users, actors, and stakeholders. The end goal is a reliable, original, and insightful need description.
- **Perceptions of preferences**: what constitutes a 'good solution' should be captured and agreed amongst all stakeholders. Further, this should be described in terms of qualities and values related to the solution and articulated in the goal formulation. It can be productive to foster a constructive conflict between different perspectives, solution ideas, and goodness criteria.
- **Problem statement**: the problem describes the kernel of a task or key obstruction to creating a solution. The perceived problem is often the starting point in searching for solutions. Although this perception forms a starting point it is difficult to know, upfront, what the real problem will be or if there are a chain of problems. Basically, often the 'problem is a problem'.

- **Task perception**: when the task is formulated, i.e. its identity and range are decided, it is difficult to define the degree of clarification required. An apparently promising idea might be locked into the task formulation or the task might be open to explorative search for new business.
- **Technology and ideas**: a concept is always based on some concrete ideas regarding how a technology could be applied, fitted, or developed. These technologies can form the main driver for the project or can be useful solution elements from previous designs. For example, incremental and platform-based design intensively re-use elements from previous projects.

These five feed chains are brought together in the Exploration Model, although there is no causality or sequence related to their arrangement. Thus, how should the model be interpreted? In the following sections, we will treat each node individually. However, we will first give an overview of Concept Synthesis for context.

Our **Exploration Model** is a palette for idea search and the clarification of conditions for development and success.

Concept Synthesis is the main user of the output from exploration and ideally consists of: goal formulation, synthesis, and evaluation and choice. The concept search is based on 'someone's' intention, i.e. decision to do something. The starting point and procedure is explorative, building on experiment, or a belief that you have identified a great concept or epoch-making idea.

Each of the nodes in the Exploration Model can be seen as a **possible initiation** point for conceptualization and clarification.

Further to this the utilization of the research loop does not end once a concept is created. There is an on-going need for information from the feed chains in Product Synthesis and Product Development. This is needed to clarify and check the current state of a project's basic assumptions and consequences.

6.3 Feed Chain: Need Interpretation

Need interpretation is the process of identifying and understanding the actors related to the product's life cycle in order to satisfy their quality or value experiences. This interpretation leads to insights and decisions that define the project. In the following we introduce concepts and phenomena leading to insights that support productive thoughts on need.

6.3.1 The Need Satisfaction Process

The way needs are satisfied by new products takes so many forms that any attempt to prescribe this activity is almost hopeless. Instead we focus on what actually happens, Fig. 6.4.

Figure 6.4 shows an idealized model of the transformation leading to need satisfaction. We use this model to explain key terms and their relationships. Most importantly **need satisfaction** describes the new situation where the new product is offered, bought, and used. As such, **the need** is relate to the **users** in the **market** and is satisfied by **products and/or services** that lead to new utilization activities or better quality or value. Need recognition is thus a mental construct of the designer that, together with a client or company's intentions, initiates a **product development** project, as articulated in the Link Model. Users can choose substitutions using other means (bicycle instead of car), buy competitor's products or let themselves be surprised or convinced by new products. The need recognition leads to **the tasks** formulated by the company or client in cooperation with the team. This describes the activities as a project including the results to be created, the **product** and/or **service**, and the new **business**. The task is transformed into an activity when it is executed. The task also articulates the conditions for, and success criteria of, the activity.

Finally, we have the problem, concept, and solution. **The problem** is a construct formulated by the **team** as statements articulating what is perceived to be the key obstruction to be dissolved, the missing mechanisms to be created or the challenge to be met. As such, the problem belongs to the designer or team, not the users! The intermediary results for satisfying this problem and task formulation are described as **concepts**. These are justified with respect to tractability and probability of successfully leading to a new product and new business. Finally, **the solution** describes the final artefacts created: product, service, module, system, component etc.

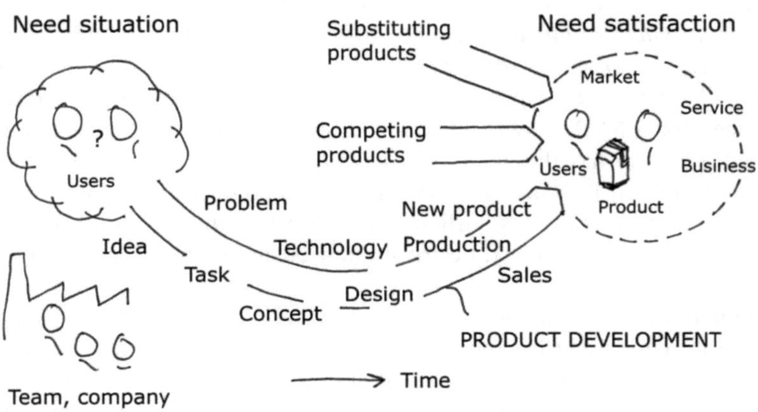

Fig. 6.4 From unsatisfied need to satisfaction via new product development

6.3.2 What Is a Need?

Based on our experience we build on two key works on ideation and need: Asimow (1962) and Alger and Hayes (1964). These older publications provide the foundation for much of modern design research; the role of need recognition is fundamental. In the first line of his book Asimow states that "*Engineering design is a purposeful activity directed towards the goal of fulfilling human needs, particularly those which can be met by the technological factors of our culture*". He concludes his Chap. 1 by formulating 14 principles. We highlight selected principals relating to need here.

- **Need**: design is a response to an individual or social need that can be satisfied via technology.
- **Physical reality**: the design or service is materially and physical possible.
- **Economic worth**: the design or service must be of value to the consumer equal to or greater than the sum of the costs required to create it.
- **Financial feasibility**: design, production, and distribution must be financial sustainable.
- **Design criteria**: the ideal outcome must be established relative to design criteria that represent the designer's compromise between conflicting value judgments, including those of the customer, the producer, the distributor, and his own.

Based on this we can see that needs are subjective and do not have an objective, independent existence. Thus, what makes a product attractive can be different for the various stakeholders: the producer (production, financial feasibility) and the user (need satisfaction, value, excitement). These two stakeholders have a mutual interest in what is economically agreeable: the trade. Alger and Hayes (1964) elaborate this by describing a need as follows: "*The engineer is concerned with creating material objects to serve human needs... Mankind recognizes a problem of human existence* (*often prosaic—such as how to dispose of garbage!*). *An engineer* (*or a person doing engineering work*) *then attempts to solve the problem*". However, we are left with the question: what qualifies a need for a NPD activity? The first response to this is to properly capture the need, i.e. who has the need, how new, how complex, and how strong is it (Thomas 1993). At the core is the designer, entrepreneur or producer's relationship with the need, i.e. it is sufficient to motive them to action. This motivation can be urgency in society, desire for new business or personal ambition and curiosity.

Although this articulation of need is widely adopted another version is described by Ulrich and Eppinger (2004) who talk about customer needs as single statements on product features and properties. We see such statements as the 'voice of the customer' to be translated into product requirements and goal formulation. As such, when we talk about need we talk about a total perception of an unsatisfied situation, e.g. the need for quick textual communication, which may be satisfied by internet messaging. Needs are satisfied by new or better means in terms of quality and value.

We can also differentiate between 'true needs' stemming from a basic or fundamental need (Maslow 1943) and 'artificial needs' created by marketing or sales slogans (see Chap. 12 for more). These two types create a contradiction between an idealistic focus on user's actual needs and creating a need for a new product through sales and advertising. Here, the customer is sold platitudes and exaggerations that are so common that they become internalized ('spoil yourself', 'you deserve it', 'more car for your money'). This type of advertising tries to instruct you, through life style illustrations, what class or tribe you will belong to if you buy the product.

Example:
Drug reminder device. Remembering to take medicine regularly is essential to many people's health. However, may people either forget to take their drugs or forget that they have just taken their drugs. One way to combat this is to supply a device that reminds the user to take their drugs and only dispenses the correct prescription. The aim is to both ensure correct treatment and to save costly home nurse visits. Here, an actor network model is interesting because it reveals problems of responsibility and financing, see Fig. 6.5.

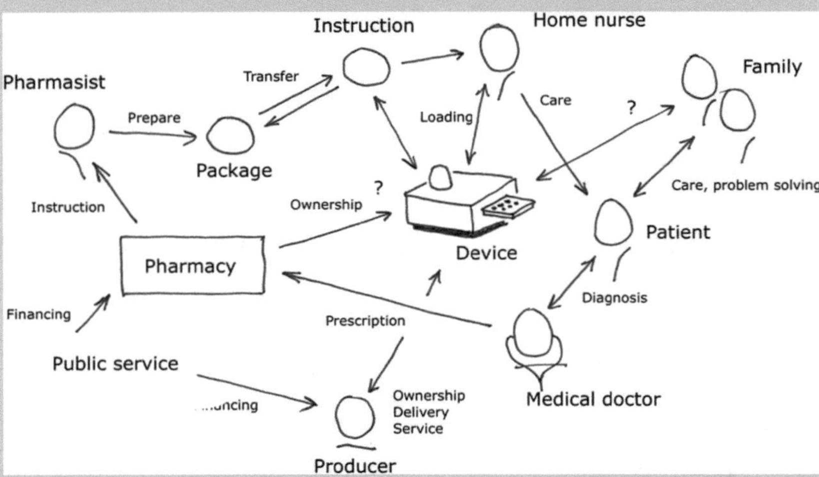

Fig. 6.5 An actor network model for a drug reminder device

These discussions lead to the following two heuristics regarding needs.

- A **need** is a designer's perception of an unsatisfactory situation or solutions related to specific users and use situations. Satisfying these is the core argument for new product development.
- A **need** can be articulated with respect to the unsatisfactory aspects plus characteristics of and conditions for their satisfaction.

6.3.3 Need Identification

The interaction between Exploration and Concept Synthesis can be seen as a gradual parallel concretization of the need and its satisfaction. The designer imagines, interprets, and defines ideal users and a speculative market, gradually justified via contact with actors and market research.

Vagn Aage Jeppesen (the creator of the authors' research group and one of the first design researchers) formulated a design process model in the 1960s. The first phase of this model was "Confrontation with the need situation" as illustrated in Fig. 6.6. It was his philosophy that the designer should work in the field, together with the users, in order to better understand existing products in use and thus be able to better 'read a need'—*here is something that can be changed.*

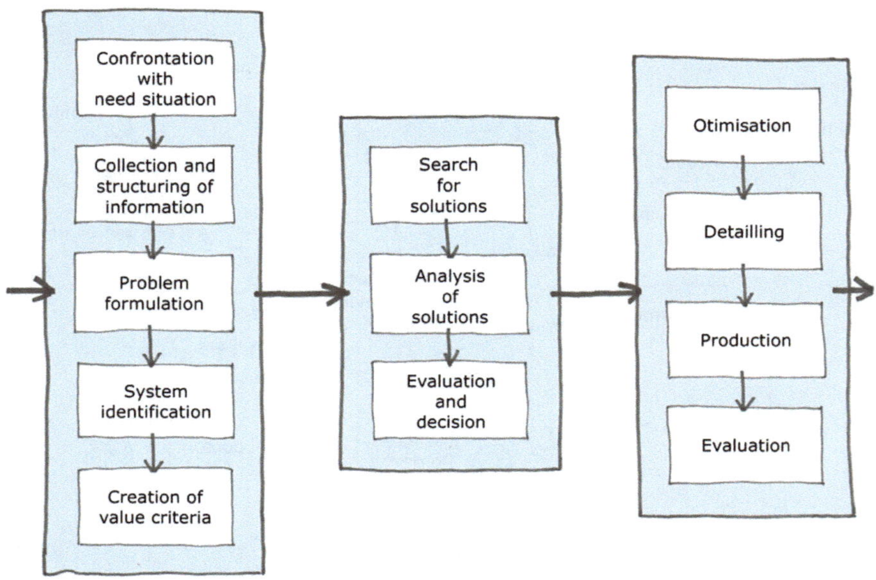

Fig. 6.6 Vagn Aage Jeppesen's process model (Andreasen et al. 1973)

Example:
Inventing a casting machine. Vagn Aage Jeppesen was educated as a mechanic in a machine factory and foundry company. As such, he was confronted with traditional foundry processes on a daily basis. In particular, he became familiar with stamped sand moulds in steel flasks, which required extensive handwork and many manipulations of the flasks. His invention, patented in 1957, was the principle of using a vertical parting plane instead of the horizontal, which was translated into a new machine. This machine produced a string of sand moulds formed with cavities and cores for the 'to be founded' product, Fig. 6.7. This mechanization allows for high precision and productivity in an automatic process supported by robots. This automation also led to substantially better work environment. Today 50 % of the world's cast iron products are produced on these machines.

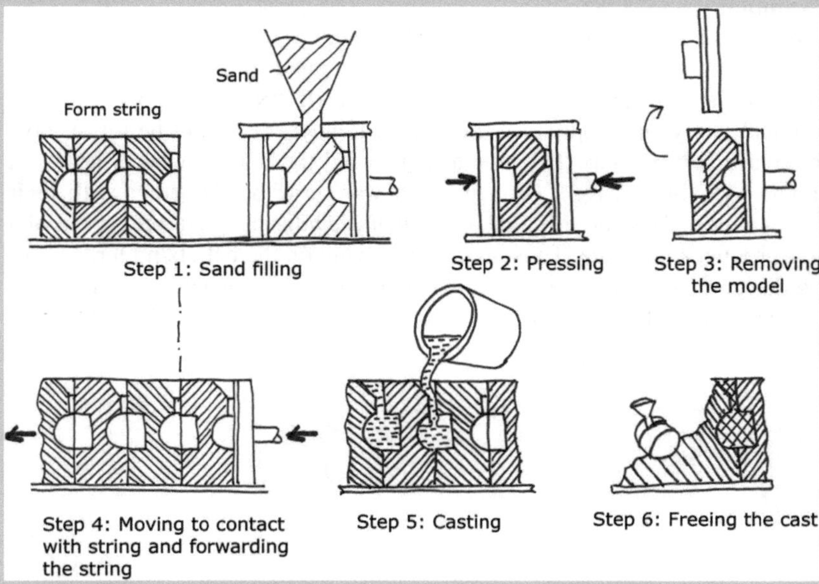

Fig. 6.7 Foundry process with vertical parting plane, explained in steps

The philosophy of Vagn Aage Jeppesen is the basic core for our view, retaining the key idea that 'someone' must have awareness, familiarity, and empathy in order to see possibilities. The starting point can in the products use and the users' value experience as found in, e.g. Workspace Design (Beyer and Holtzblatt 1998) where interviews and design games are used to identifying areas of improvement.

Dym and Little (2000) identify three roles in the design of a product. Other than the designer there is the "*client, the person or the group or company that wants a design conceived*" and the "*user, the person or set of people who will actually use the device or artefact being designed*". There is a tendency to think about the user as an individual and thus overlook actors belonging to, e.g. a networking client, different phases of the product lifecycle or at the societal level in the form of legislation, ethical rules, communal regulations etc. A key challenge in need recognition is identifying relevant, specific actors. This can be supported by *actor network analysis*, which combines field studies, interviews, questionnaires etc. On the other hand user behaviour and habits are at core of marketing and *market analysis*. This can lead to a market report, i.e. a summary of need interpretation, users, competition, market size, sales, and risk. Finally, we can try and understand specific users by articulating *user characteristics* for a group of people who broadly represent the wider user base.

Example:
Personas and user characteristics. The users related to the installation of a heating system in a house can be represented by the four personas in Fig. 6.8: the installation manager Hans (left): "New technology means trouble", the young apprentice electrician Benny: "It should be a fast job", the house owner Inga: "It is fine as it is", and the house owner Martin: "It should look impressive". Each persona was described together with their personalities. These descriptions were then used during the development of a new product's interface. Thus instead of relying on the beliefs/preferences of the design team discourse was based on the personas: what would Inga say or what would Martin mean?

Fig. 6.8 Illustration of five personas related to a heating system project (Hede Markussen 1995)

As we shall see in Chap. 7 the use activity is key to design; it is the use of a product that satisfies the users' need. Scenario techniques of all kinds are useful for capturing insight into the desired product's use, its interaction with the user, and its use properties. Use process models thus include interaction, functions, instructions, failure possibilities, and so on. Need interpretation feeds into decisions as minor as small updates to existing products and as major as need satisfaction in the various design proposals, i.e. a substantial step of synthesis.

6.4 Feed Chain: Perceptions of Preferences

We believe that designers, independent from focus (engineer, architect, industrial, software etc.), feel that they have a mission. Specifically, that they can solve societal problems and satisfy and create excitement for users. Thus designers' ideals and interpretation of society's needs has a significant impact on the design process. The feed chain 'Perception of preferences' deals with these ideals and how they relate to the identity of the design team.

The designer and the team stage themselves (see Chap. 4) in a **role** by formulating a vision and mission. Example roles include arranging citizens' involvement in urban innovation, establishing a network group or creating a service development. The task is then attacked based on the team's **ambitions**. They are aware of leading figures in their field and have pride in their work but also curiosity and playfulness. In a team, these ambitions should be balanced to form a shared perception of preference, demanding a dialogue in the team.

These preparatory discussions lead into the design work based on an **idea foundation**. This foundation is formed from the team's beliefs, experiences, basic design philosophy, principles (e.g. simplicity, clarity, integrity), and key values (e.g. sustainability, ease of operation, excitement). Sometimes design is based on a **fixed idea**, by which we mean a key good idea that forms the core of the project. Many companies have more or less explicit idea foundations, often hidden to the public or articulated as corporate image and brand.

Based on these ideas, we can see that the perception of the core ambition or necessity of a project, the **framing**, can change during development and might not be shared across an organization or team.

Example:
Development of energy-labelled pump. The Danish company Grundfos is one of the world's leading pump manufacturers. In 2005, they launched the Alpha Pro shown in Fig. 6.9. This was a pump for water circulation in heating systems in family homes. Its power consumption is 5 W giving it an energy rating of 'A', the first one on the market.

Fig. 6.9 The Alpha Pro pump from Grundfos with its energy label ranking: 'A'

The history of this project is interesting (Gish and Hansen 2013) because despite its 'newness', the core technology can be traced back to 1985. It was here that the motor division first identified the possibility of using a chip to control speed and energy consumption. This division's ambition to deliver innovation was, however, in conflict with management's cost reduction focus at the time. Similarly, in 1992 a pump with an integrated frequency converter was proposed to reduce energy consumption. Again this clashed with the company's focus on manufacturing and a belief that no one would pay for energy savings. However, a turning point came in the late 1990s when the company started using lifecycle analysis tools. These revealed that 99 % of the pump's electricity consumption was due to its running phase, opening the company's eyes to ecology.

Unfortunately, the importance of energy consumption was not yet recognized by the market. Throughout the 1990s the company tried different sales campaigns but with little result. The breakthrough came at the end of 1990s when the Grundfos CEO decided to try and ban low-efficiency circulators. Years of lobbying in the EU resulted (in the mid-2000s) in a voluntary EU energy label system. The Alpha Pro was then developed based on the technologies originally rejected to become the first product on the market with an 'A' label. It became a massive commercial success. This illustrates how a concept merges multiple possibilities, agendas, intentions, and past products all framed against the company's beliefs. What might seem a quick and successful development project might actually require a long path of preparation.

Focusing on 'ideals' is a key means for taking development beyond mediocre, setting references for ambitions related to needs, market, utilization of resources, and identification of abilities. Our ideals provide guiding lights.

The **perception of ideals** takes a project beyond the traditional or 'me too' mindset.

Companies report that new development projects often have a fuzzy start (Andreasen and Hein 1987), beginning life as skunk work and gradually attracting people and resources to the idea. This is in sharp contrast to correct progression in the early phases, justifying need, abilities, opportunity, technology etc. However, articulation of these early initiatives can be formalized via portfolio management of ideas, concepts, and projects.

6.5 Feed Chain: Task Perception and Formulation

Conceptualization starts with 'somebody's' intensions as illustrated in Fig. 6.10. In an architectural competition, the frames of the task are defined but the solution space is open. Similarly, in a company's product development the driver might be a business need that demands development effort but of what kind? How should the task perception and formulation emerge from such situations? The development task is closely linked to a web of prerequisites, e.g. situational factors like economy, resources etc., meaning the task will impact several dimensions. It is important to identify aspects that might disturb, support, or inspire the development. In certain situations, it may even be necessary to remove, change, or create new preconditions as discussed in Chap. 5.

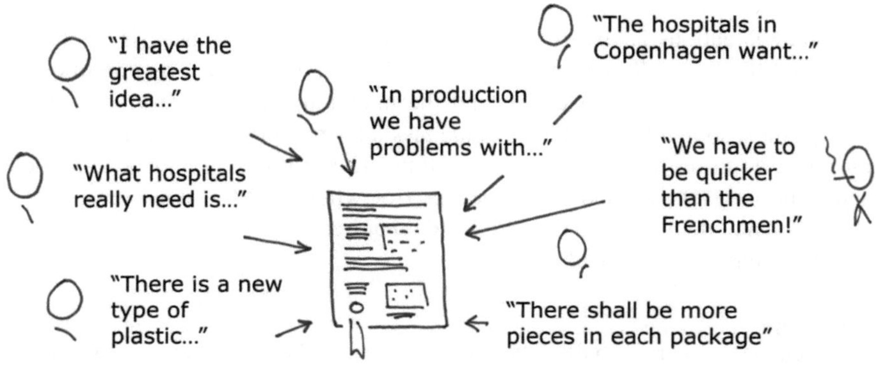

Fig. 6.10 'Somebody's' ideas go into the goal or task formulation document

Example:
Safety clamp and cutter. Knud Lykke Nielsen had the idea to create a combination clamp and cutter for severing the umbilical cord of newborn children, which became the product shown in Fig. 6.11. Traditionally, this is cut using scissors, creating a contamination risk for both child and nurse. Despite its value, the inventor needed support in the design activity and the transfer to manufacturing when realizing the idea. The task formulation should balance the transfer of the original idea and its empowerment, the link to hospital professionals, and the role of the consulting company. The Institute of Product Development provided a concept, design proposal, detailing, and support for the start up of production in Lykke Nielsen's sales company.

Fig. 6.11 Safety clamp and cutter, *courtesy* Institute for Product Development

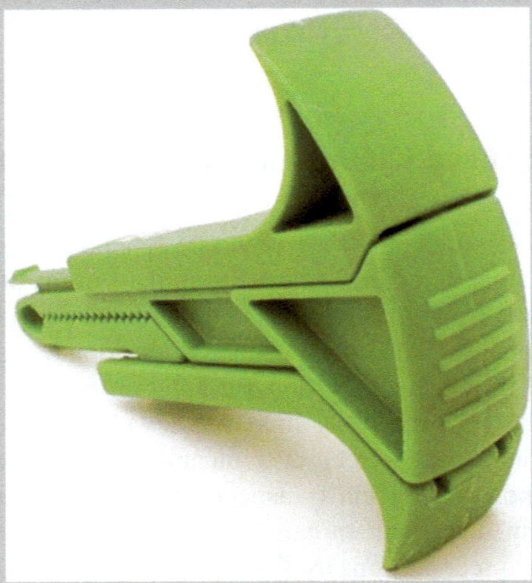

The many situations leading to new products are so diverse that it is impossible to try and list them all and naïve to summarize in a checklist. The core message is that the task solver must understand the intention, expected result, and task's prerequisites, so that they can control the task's formulation and fit with the conditions, e.g. duration, ambition, resources, time. The task has an understanding element: purpose, motive, and consequences (for whom), and a definition element: what to create, deliver, document, resource requirements, lifecycle focus areas, fit with the company, and launch plan.

Example:
Copenhagen market hall. The architect Hans Peter Hagens positioned himself at the head of a group of citizens aiming to change a messy, grey square in Copenhagen into a modern market hall. His ability to articulate

Fig. 6.12 *Left* early visualization of the market hall, www.arkitekturvaerkstedet v. architect Hans Peter Hagens. *Right* interior from the final market hall

and visualize their ideas became a key driving force, while a person with political flair and connections became the lever for the city's engagement in the project. A long period of development finally led to the high profile, hygienic, and environmentally conscious proposal shown in Fig. 6.12.
The proposal removed the existing traders (flower stands and a second hand market) and a messy parking area. Other traders that can afford the new (much higher) rent then took over. The project was a composed design task that changed the urban space, traffic conditions, trade pattern, square's environment, and life patterns. A question at the design stage was: will the square be attractive to the local people or tourists? Today the market has won a reputation as a new, exciting, and popular shopping and recreation area.

Our perception of task formulation can thus be articulated by the following mindset.

Task formulation is a mutual agreement between the design team and the client or manager. It articulates what is to be created, under what conditions, to be utilized in what way by the user and the company, and leading to what satisfaction and economic results.

It is vitally important to be aware of unfounded biases or random ideas built into the task formulation, and thus articulate correctly what is really wanted. The team should have enough degrees of freedom so that they can produce satisfying and successful results. The task often needs reformulation during development because of co-evolution or changing conditions.

6.5.1 Task Experience

At the start of a development project, a contract-like relationship is created between the management and the team that will actually undertake the task. This is often accompanied by a 'project start' seminar to clarify responsibilities and roles. Here, management or client meet the team, present the task, its mission and vision, go through the project together, and motivate the team. The idea is to jointly 'set the scene' and focus the project team on the task at hand. A key support for these activities is *scenario* creation, i.e. describing a composed picture of the future project in a time sequence, based on the participants' imagination, experiences, and ideas. Further, scenarios can be used to explore the new product, its production, application, recycling etc. Another type of scenario is to go through the project with all the actors from the different phases and discuss the challenges, cooperation, and coordination issues in a short, total simulation.

> There is no reason to let the project progression be dictated by what is designed. The team should be ahead of this progression in order to use plans and strategy effectively.

6.5.2 Team Identity and Role

A company realizes its role and identity through its mission, vision, and policies. This also includes logos and branding. The intent here is to focus goals such that synergy is created between the company's activities (development, production and sales) and that resources are optimally directed to a well-defined area of attack.

As with a company, the project team may also need to articulate its identity via a 'program'. Here, the focus falls on defining the team's task, situation interpretation, motivation, need for knowledge, idea approach, working style, goals, and quality criteria. The difference from a design brief is the team's analysis of its own abilities and what the team sees as central challenges. Example questions that a program answers are:

- **Who**? The identity of the company and the project. Interaction and dependencies of the client.
- **Why**? Status in the company's development activities, the motivation of the company and the team, limits and criteria for good solutions.
- **How**? The nature of the project, role distribution, project strategy, use of methods, phases, and milestones.
- **What** are we able to contribute? The team's abilities and knowledge in relation to the challenges of the task.

During conceptualization, when ideas are reasonably concrete, a *goal formulation* is normally created. This includes a *product specification* articulating desired functionality and properties, and a *business specification* articulating business expectations, sales, investments etc. (Andreasen and Hein 1987). Together these documents tell the team about the role and content of marketing and sales efforts, as well as the core competition. Finally, a *mission document* may be added to the goal formulation as proposed by, e.g. Ulrich and Eppinger (2004).

In industrial design, a *design brief* is also a common tool. The client creates this for or with the designer, although the designer may also formulate the document to articulate their task and vision. This document serves to complement a goal formulation as it articulates the vision and raison d'être of the proposed idea, while the goal formulation focuses on individual technical challenges (Keinonen and Takala 2006).

6.6 Feed Chain: Problem Formulation

Some authors make a distinction between problem and task, where problem relates to design situations and task to routine. There are a number of reasons for why the term 'problem' is used like this in the design literature. First, the historical perspective of the engineering designer as a problem solver, who finds the evil's root (i.e. the technical problem) and solves it (Glegg 1969; Wallace 1952; Harrisberger 1966). Second, the literature has been influenced by creativity research that focuses on similarities between design and problem solving (Osborne 1963; Gordon 1961; Guilford and Hoepfner 1971). Finally, design methodology has evolved from more general models of problem solving (Pahl and Beitz 2007; Roth 1994; Koller 1979). Together, these lead to a divided view of the word 'problem', sometimes used to start a work and sometimes to describe a need or root evil. Thus, we must answer whose problem is it that we speak about: the designer's, the product user's, or the inherent traits of design?

Cross (2008) states that design can be seen as the decomposition of an overall problem into sub-problems, see Fig. 8.2. When solutions are found to sub-problems, these sub-solutions can be combined into an overall solution. Cross highlights that the two patterns of problem and solution decomposition are very different. The importance of Cross' model is the separation of the two patterns and the necessary understanding of the mapping between them. Building on this dual perspective, we can substitute the word problem with task—used throughout the book. Many tasks relate to issues of, e.g. cost, reliability, noise, and environmental effects, which are distributed among a product's physical sub-systems. We prefer the word task and see a key challenge in the complex task identification and breakdown that is a precondition for synthesis and planning.

When a need is recognized or formulated and a company's management allocates resources, the design task is born. Asimow describes the design process as iterative problem-solving including sub-problems: "*In attending to the solution of a design problem there is uncovered a substratum of sub problems; the solution of the original problem is dependent on the solution of the sub problems*". Based on this, we can explore the link between problem and need. Asimow sees this as synthesizing a black box system, where "*the desired outputs of the system are derived from the effective needs of the customers. The language of the outputs should be more precise than that of the needs, and should reflect what the system does or provides in responses to the eliciting needs*". This describes a soft, non-technical need being translated into a sharper, more technical goal formulation. It is through this process that engineers recognize their 'problem', e.g. the control engineer sees control problems while the solid mechanics engineer sees strength problems. Unfortunately, the engineer's specific challenge does not point to which solutions might be best.

Many authors look to design strategies with a starting point deep in the problem—identifying the kernel of the task. The assumption here is that the technical solution is the core of the development and should be found before anything else is considered. However, we do not subscribe to this focus; conceptualization is not always of a technical nature and can include, e.g. service, environmental efforts, and use. For example, students involved in a Dutch project on flexible office layouts saw the core problem as making the furniture mobile. However, a small group of students found a different perspective, characterizing the problem as that of creating personal identity in the office space—leading to very different solutions (Restropo 2004).

Example:
Task kernel. There are a number of authors who advise that a 'product principle' should be articulated. Here a principle describes the kernel of a specific class of solutions, from which the design activity can be defined and progress. What principle is selected depends on the designer's professional background: a mechanical engineer may see a mechanism, an electrical engineer a control system, and an industrial designer the concept of man/machine interaction. In an experiment on designers' reasoning, this type of interpretative kernel selection was found to be dominant when following this 'product principle' type of strategy (Dylla 1991). Figure 6.13 shows some of the resulting principles, each mirroring the designer's interpretation of the task. The task was to design a support bracket for sensitive measurement equipment, requiring vertical and horizontal adjustment. In this situation, there does not seem to be a kernel that is fundamental to the task, rather its selection reflects the designers' interpretation of the task and their preferences.

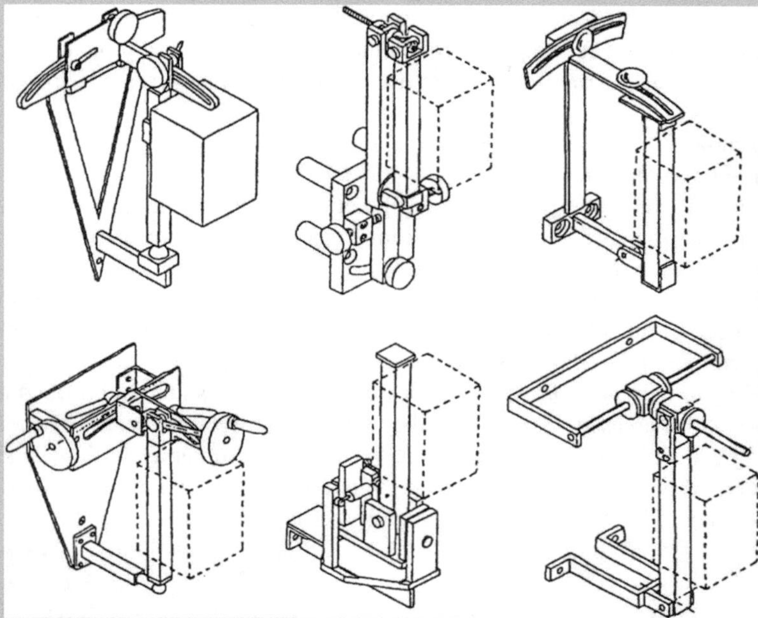

Fig. 6.13 Six concepts for an adjustable support (Dylla 1991)

Tasks typically contain intrinsic problems that reduce the quality/negate the solution unless they are solved. Building on this idea, Altschuller (1987) created a technique based on patents, reasoning that patents are solutions to intrinsic design conflicts. Thus, by analyzing a huge number of patents, he created a catalogue of frequent conflicts and a number of principles for their resolution. This is in line with our own views of problem formulation. To sum up our view on the 'problem of the problem', we formulate the following heuristics.

- Design starts with human needs, resulting in a task for the designer.
- Tasks should be broken down into subtasks, to allow for composed solutions to larger issues.
- It can be useful to abstract the core obstacle but the designer should always think in terms of need, problem, and task.

6.7 Feed Chain: Technology and Idea Elements

In Concept Synthesis concept alternatives can be created from new ideas or from the combination of known solutions. These known idea elements or technologies are often from other products, the wider company or research and development efforts. One of the key skills of a designer is to effectively build on these know elements. As such, this feed chain encompasses a company's past experiences, earlier products, R&D, and future plans. To maintain the integrity of a product range care must be taken in determining when to use known elements verses new innovations.

Example:
Surveillance camera
During the development of a new generation of surveillance cameras it was found that the cost of launching a range of different sizes was too high. An example camera is shown in Fig. 6.14.
Based on this cost issue it was decided to create a new drive system (motor and gear) compatible with all sizes variants, and maintaining the mechanical embodiment and visual design. The development allowed for a new housing, leading to new possibilities in aesthetic design, and resulted in both reduced costs and increased sales due to the new more attractive form. This points to a philosophy where a core idea is fixed in order to harvest rationalization effects. However, attempts to 'make things better' at the individual product level may kill the benefits.

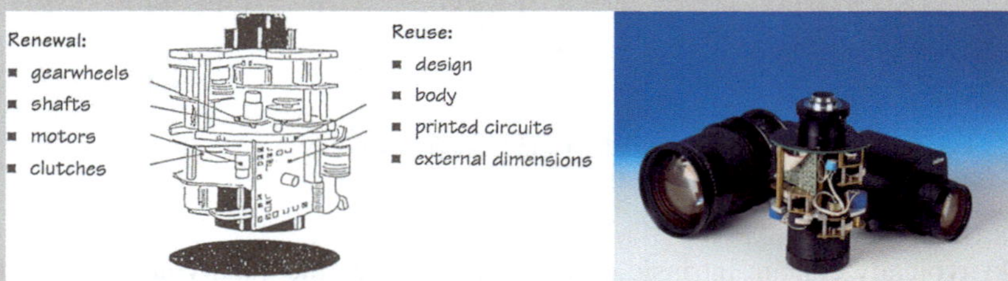

Fig. 6.14 *Left*, decisions regarding re-use and innovation in a planned development project. *Right*, the final products, *courtesy* Institute for Product Development

Re-use does not just relate to single parts but can influence variant and assortment decisions because the components should be used as widely as possible to deliver the greatest cost benefits. Further, re-use can be based on systematic competitor analysis leading to the identification of attractive solutions or components (Andreasen and Hein 1987). A product concept should consider re-use and carry over, based on path, assortment, risk, and cost. Because of this design can be seen as a balance between creating and searching for ideas or solutions. A concept is seen as a combination of sub-solutions and the new product is realized through a combination of known principles, existing components, and parts.

6.7.1 Research and Development

A special and very important feed element is created when a company arranges its development activities round the R&D of technological solutions. Here, the following forms are common:

- **Research**: using scientific investigations to identify matters of importance for products and technologies. When results are matured, they are transferred into the design activity.
- **Development**: arranging experiments and engineering efforts to bring competitive insight into the products.
- **External concept and product development**: external consulting companies are used to generate ideas, concepts, or product proposals that feed into the development process.
- **Buying patents, licenses or components**: the company purchases relevant technologies to support and shorten the development.

These development activities also relate to the clarification of other aspects, e.g. market, use, competition, and technology search. Although our treatment of this feed chain is short it has been and still is the main source of new product development. What we want to underline is the following heuristic:

> Technological initiation of new product development must be supported by proper clarification of the need and identification of relevant existing ideas that can be included.

6.8 Conclusion

Sections 6.3–6.7 have explored the five feed chains in our exploration model (Fig. 6.3), each of which support the dual activities:

- Searching for insight leading to new ideas, concepts, products, and new business.
- Searching for knowledge supporting the design activity and the argumentation for decisions.

Andreasen and Hein (1987) argue that investment in new product development is rewarded by clarification of: possibilities, solutions, arguments, insight etc., meaning that all activities in development are incrementally devoted to clarification. Thus we separate Exploration from Concept Synthesis because of the ‘global’ role of the search:

- The clarification of what to do based on the existing situation.
- The preparation of the design process by clarifying what it takes to perform it.
- The clarification and prediction of the product’s lifecycle for ensuring its fit for life.

> The starting point, core idea, or inspiration for development can be related to any one or combination of the **feed chains.**

New technology and 'physical ideas' are normally seen as the main feed chain for new development. However, it is important to balance the two idea dimensions: **the idea with** and **the idea in** the product, and thus trace ideas back to users' needs. Only in this way can the best possible argumentation for doing the right thing be established.

- Exploration should serve the full perspective of concept/product/business/use/life and if necessary be gradually expanded in accordance with the design progression.
- The design activity can expose decisive issues at any point in the process. Therefore, exploration cannot be seen as finished in this first step.

In Chap. 5, we introduced the Encapsulation Design Model that highlights the dilemma: how broad and how long should exploration be? From the breadth perspective, we suggest the gradual expansion of concern from concept design to business and life aspects. From the length perspective, we suggest that exploration should consider and support all phases of development. As such, exploration feeds into Concept Synthesis, which we will discuss in the next chapter.

References

Alger JRM, Hayes CV (1964) Creative synthesis in design. Prentice-Hall, Englewood Cliffs, NJ

Altschuller GS (1987) Creativity as an exact science: theory of the solution of inventive problems, 2 edn. Gordon and Breach, Philadelphia

Andreasen MM, Hein L (1987) Integrated product development. IFS Publications/Springer, Berlin

Andreasen MM, Tjalve E, Stahl H (1973) Konstruktionsprocessens faser (The phases of the engineering design process). Compendium Technical University of Denmark, Copenhagen

Asimow M (1962) Introduction to design. Prentice-Hall, Englewood Cliffs NJ

Beyer H, Holtzblatt K (1998) Contextual design—defining customer-centered systems. Morgan Kaufmann publishers, San Fransisco

Cross N (2008) Engineering design methods. Wiley, Chichester, GB

Dylla N (1991) Denk- und Handlungsabläufe beim Konstruiren. Carl Hanser Verlag, München

Dym CL, Little P (2000) Engineering design—a project based introduction. Wiley, New York

Gish L, Hansen CT (2013) A socio-technical analysis of work with ideas in NPD: an industrial case study. Research in engineering design, ISSN 0934-9839, 24:414-427, doi:10.1007/s00163-013-0159-z, 8 July 2013

Glegg G (1969) The design of design. Cambridge University press, Cambridge

Gordon WJJ (1961) Synectics: the development of creative capacity. Collier, New York

Guilford JP, Hoepfner R (1971) The analysis of intelligence. McGraw-Hill, New York

Harrisberger L (1966) Engineersmanship: a philosophy of design. Brooks/Cole, Belmont, CA

Keinonen K, Takala R (eds) (2006) Product concept design—a review of the conceptual design of products in industry. Springer, Berlin

Koen PA. (2004) The fuzzy front end for incremental, platform, and breakthrough products. In: PDMA Handbook of New Product Development, 2nd edn, pp 81–91

Koller R (1979) Konstruktionsmethode für Maschinen-, Geräte- und Apparatebau (design method for machine, device and apparatus construction). Springer, Berlin

Lehtonen T, Juuti T, Oja H, Suistoranta S, Pulkkinen A, Riitahuhta A (2011) A framework for developing viable design methodologies for industry. In Culley SJ et al (eds) 18th international conference on engineering design, impacting society through engineering design, design society

Markussen TH (1995) Et teoribaseret handlingsgrundlag for betjeningsdesign (A theory based action plan for interaction design). Ph.D. thesis, Technical University of Denmark, Copenhagen

Maslow A (1943) A theory of human emotion. Psychol Rev 10(4):370–396

Osborne AF (1963) Applied imagination—principles and procedures of thinking. Schribners, New York

Pahl G, Beitz W (2007) Engineering design: A systematic approach. Springer, London

Pugh S (1991) Total design—integrated methods for successful product engineering. Addison Wesley, Wokingham

Restropo J (2004) Information processing in design. Ph. D. thesis, Delft University of Technology, Netherlands

Roth K (1994) Konstruieren mit Konstruktionskatalogen (designing with design catalogues). Springer, Berlin

Thomas RJ (1993) New product development—managing and forecasting for strategic success. Wiley, Hoboken, NY, pp 50–57

Tjalve E (1979) A short course in industrial design. Newnes-Butterworths, London. (Facsimile edition 2003) Systematic design of industrial products. Institute of Product Development, DTU Copenhagen

Ulrich KT, Eppinger SD (2004) Product design and development, 3rd edn. McGraw-Hill/Irwin, New York

Wallace PJ (1952) The technique of design. Pittman, London

Chapter 7
Concept Synthesis

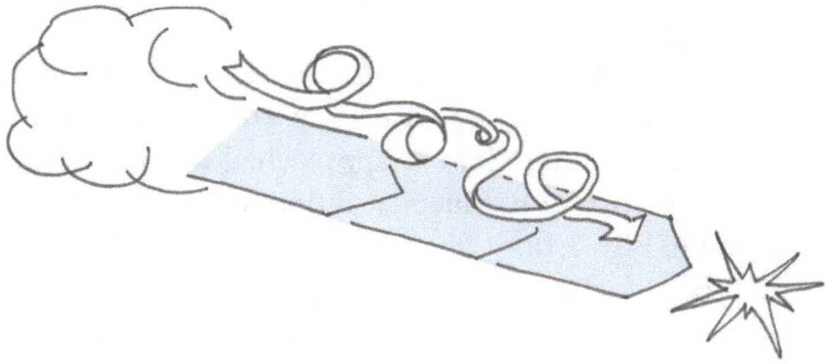

A concept both answers a need and intention and is a clarification of tractability in the realization. Thus, creating the final concept proposal demands the combination of ideas from multiple dimensions. We call this concept synthesis.

This chapter focuses on exploring the foundations for conceptualisation in theory, strategy, and models, including human creativity, visualisation, and combination. We subsequently bring these together in a procedural Concept Synthesis Model, which forms the basis for goal formulation, ideation, evaluation, and decision-making. Ultimately these lead to the selection of a final concept that becomes the starting point for Product Synthesis, discussed in Chap. 8.

7.1 Synthesis: From Dream to Proposal

Concept Synthesis is the process through which goals are made explicit, alternative concepts are established, and a final selection is made based on the requirements. Thus we define concept synthesis as:

Definition: Concept Synthesis is the phenomenon of creating a kernel of insight and ideas in the form of concepts. This provides the answer to need and intention, and is a proposal of the probable tractability and success in its development, realization, sales, and use.

The nature of Concept Synthesis brings together creativity and methodology i.e. we need to make creative leaps but also retain a systematic line of reasoning. This demands an understanding of the balance between the synthesis methodology and the need for explorative, opportunistic, experimental, and creative behaviour.

M.M. Andreasen et al., *Conceptual Design*, DOI 10.1007/978-3-319-19839-2_7

We will not focus on specific methods of creativity etc. as these are already well treated in many other works; instead we explore how they can best be deployed in reality.

Example:
Conceptualisation of the *Handpresso*. The Handpresso offers a portable means for making espresso style coffee using a manual pump (Fig. 7.1b). The inventor, Jul Nielsen, approached this project by varying the characteristics of existing products at different levels to produce new concepts, illustrated in Fig. 7.1a.
In order to achieve the final product it was necessary to combine ideas from several domains: the technical principals in the energy system, the aesthetic design, a new operational process, and manufacturing ideas. Further, it was also necessary to establish financing to support development, production, and sales.
The conceptualization requires several ideas to be combined: technical principle (e.g. the energy system), visual design, operation, manufacturing processes, and how to establish financing for development, production, and sales. Figure 7.1c shows the visual form of the different concepts. A condition for financial support is to ensure the ownership of the concept; (d) shows the patent drawing, which does not actually look like the final product but contains patent claims covering a broad variety of devices.

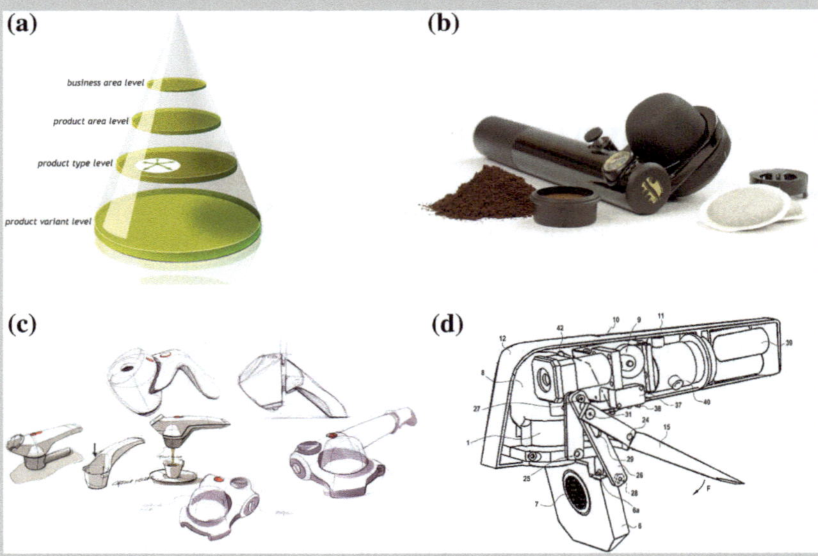

Fig. 7.1 Conceptualising the Handpresso **a** the inventor's approach of varying characteristics, **b** the final Handpresso product, **c** the conceptual proposals of form and use (Kind permission of Nielsen Innovation, France), **d** the illustration from the patent application

7.1.1 The Challenges of Concept Synthesis

A concept **answers** a need, intention, and product identity, as well as offering a **prediction** for tracing the success of the project. Successfully combining these elements requires much more than creative guesswork! Here synthesis describes the gradual concretisation of the product design from two angles:

- The **idea with the product**: decisions on need, user behaviour, market area, parameters of competition, and basic ideas for sales, branding, supply etc. These all lead to constraints that determine the solution space and thus help define the concept i.e. 'what' and 'who' questions.
- The **idea in the product**: decisions about product type, use, functions, mode of action, and other aspects like form, embodiment, operation, production processes etc. These all lead to a gradual determination of the product's structure i.e. 'how' questions.

The gradual bringing together of these two areas starts in the exploration activity and develops throughout the process. However, there are a number of significant challenges that must be addressed. First, the concept is a point of no return; once the final product concept is under synthesis it is no longer feasible to propose new core concepts without damaging the progression of the whole project. Further, it is not possible to solve all the details up front—making the concept selection even more difficult. Thus it is not enough to formulate an abstract, one-dimensional concept e.g. 'eggshell', because the following synthesis activities may reveal it to be flawed. The challenge is to articulate the key dimensions without over detailing. This ultimately means that the designer must embrace the fact that design decisions are about probability, not 'yes' or 'no' type answers. Thus many projects fail because the designers were not able to tie their decisions to the goal formulation, and ensure that the goal formulation actually reflects reality, not just their own imagination. Finally, the manifestation of these challenges changes from project to project, and across project types e.g. new product development and platform design.

Ultimately it is these challenges that give Concept Synthesis its unique importance in our Encapsulation Design Model. The concept is not just an idea, it must answer: What will satisfy the users' need? What should be developed? Is this feasible? And will the result be successful?

Creating concepts by articulating and combining ideas is almost totally reliant on human cognition. This is supported by three areas: *creativity*—dealing with the human element, *methods*—supporting the systematic requirements, and *visual thinking*—helping the designer bring these together.

> **Creativity** produces ideas, **methods** deliver the pieces and composition, and **visualization** make these tangible.

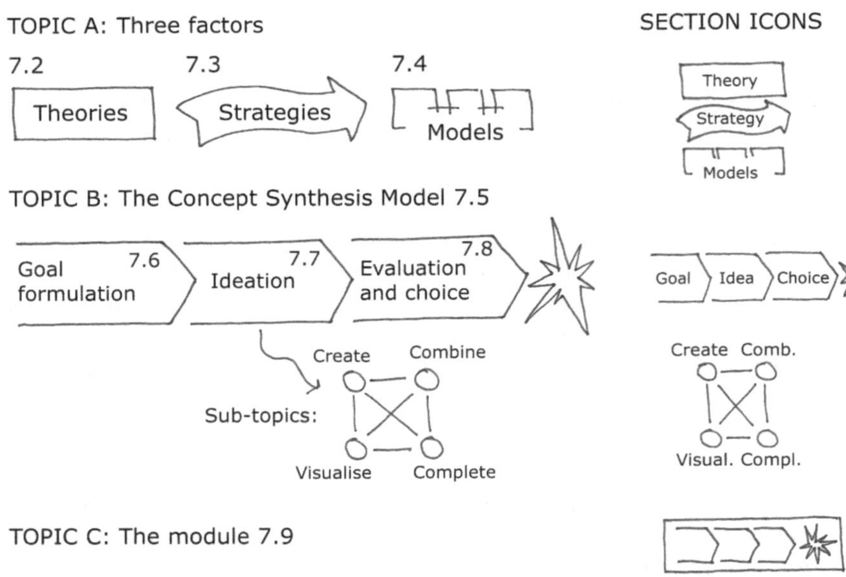

Fig. 7.2 The three-topic structure of this chapter and the icons used in the following for overview

The importance of concept synthesis for this book means that we will discuss the mentioned topics at length all in one chapter. As such, we have split the chapter by topic as shown in Fig. 7.2: (A) the three factors theories, strategies, and models, (B) the Concept Synthesis Model and the three phases of conceptualisation, and (C) reflecting on this module. Furthermore the topic 'ideation' is treated as four interrelated actions: create, combine, visualise, and complete. Figure 7.2 shows the icons used in this chapter to support the reader.

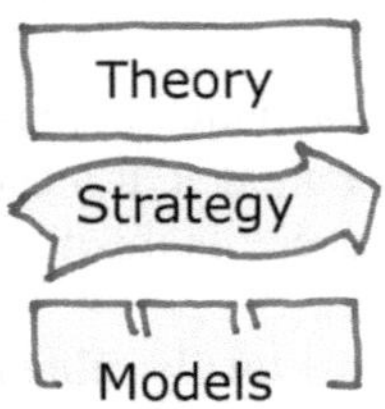

7.2 Theories on Ideation and Conceptualization

Conceptualization not only leads to the development of concepts but also to key design knowledge. This knowledge clarifies the arguments for a concept and lays down the rationale for why the selected concept should be successful. This builds on both **creative** and **systematic** thinking, which can both be supported by **visualisation**.

7.2.1 Creative Thinking

Creativity has two key meanings: the ability to create **unique solutions** and the ability to produce **many solutions**. This is illustrated in Fig. 7.3, which shows, on one hand the designers 'level' of creativity, and on the other the output from the creative process in terms of design ideas. Both of these are subjective and can only be realised with respect to a third part 'judge'.

Many researchers have evaluated design creativity in terms of problem solving. However, creativity is important throughout the whole design process, as well as in many other aspects of daily life e.g. jazz improvisation or artistic painting. Psychologists define creativity as the ability to combine known elements into something new. This definition contains the relative element 'new'; new to whom? In the moment of creation, it is difficult to know if something is really 'new'.

In the design context, new is relative to those solutions that already exist on the market. In order to create something new here a designer must understand the current solutions, their elements and their raison d'être. It is not sufficient to function creatively; you have to have 'some good pieces to move'. A second question that is relevant in the design context is: should the ideas be useful? We argue that in the first round the answer to this is *no* because over constraint can limit creativity.

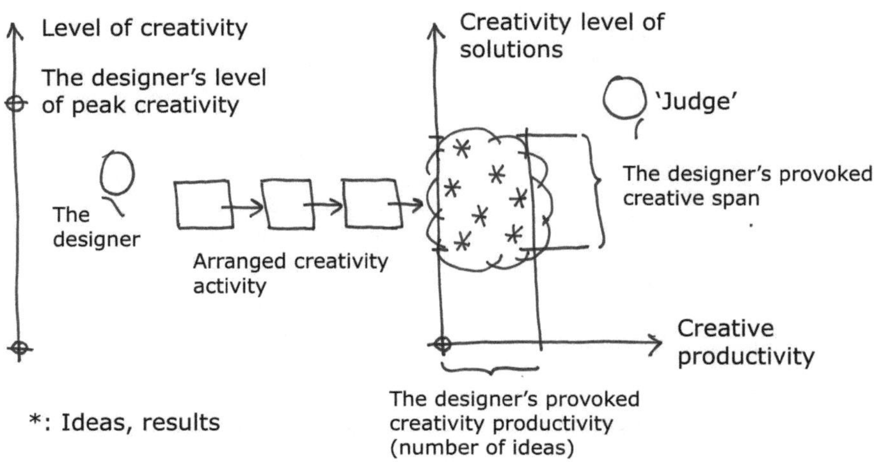

Fig. 7.3 Interpretations of creativity

However, the designers must know the basic needs to be addressed, and existing solutions, in order to understand what might be valuable for the user. Thus there is scope to explore wild ideas in these early stages before the focus turns to project timeline and pragmatic applicability.

Hatchuel and Weil (2003) link creativity and knowledge in C-K Theory to try to explain ideation. They distinguish two domains: concept (C) and knowledge (K). Here, concepts are proposals that have no logical status in the knowledge area i.e. they are things we know nothing about. Thus investigating these concepts leads to new knowledge and the empowerment of the concept. Ultimately, this development results in a new concept and new knowledge, which adds to existing knowledge, see Fig. 7.4.

Creativity also depends on **attitude**. It takes daring to create something new, believe in the creation, and work towards its realisation in the final solution. As such, we can learn from children who are creative simply because they have not yet recognised the barriers of culture, social norms, rules, traditions, and common sense or their own ego, rationality, and logic. Working to remove these barriers thus has many names e.g. thinking out of the box, lateral thinking or to be conscious of 'mental space'. When we work to free ourselves of these barriers we allow our subconscious to produce ideas without our conscious thought. This is called **incubation**, which has four phases (Shapiro 1980): *preparation,* where the designer recognizes and works on a problem i.e. "*…soaking themselves in the problem*". *Incubation*, where the subconscious works. This takes time and evidently is most productive when other things occupy us. *Illumination*, where the

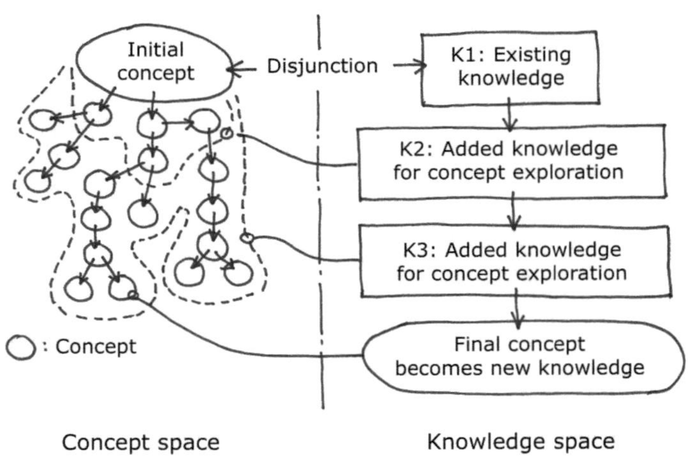

Fig. 7.4 Model of C-K Theory showing movements in the concept and knowledge domains

idea occurs as a moment of enlightenment. This is normally a short, critical period because the idea is easily lost. *Confirmation,* where the idea is developed and perhaps shows its potential. Based on this incubation can be trained by attention: *Do you need incubation time? How long, an hour, a day? In what situations do ideas occur?* A surprising fact is that ideas often occur in situations not devoted to creativity e.g. when walking in nature, while they are more difficult to force using creative techniques on demand.

Building on this combinatory perspective on creativity, one of the most productive ways of generating ideas is **association**:

- **Similarity**: linking the unknown with known things. Here we look for *direct analogies* i.e. similar geometry, function, sequence, visual impression, context, material etc. Alternatively, *symbolic analogies* focus on verbal similarities e.g. strong as a bear, fragile as glass.
- **Contiguity**: there is a mutual connection. Here we look for connections between known ideas, and then try and transfer these connections to the problem. These connections can be physical or abstract e.g. related in time.
- **Contrast**: linking solutions from the *opposite* problem to the current problem. Here we can use real opposites or simply imagine opposing requirements e.g. start a fire > extinguish a fire. These can then be used to generate ideas for our given problem e.g. put fuel in a fire extinguisher to start a fire.

As we have discussed design is not just a jigsaw waiting to be assembled. Instead it evolves through the resolution of the many issues encountered in producing the final solution, its elements and their interaction. Here, we are often confronted with **paradoxes** (Dorst and Hansen 2011; Hansen et al. 2009). A design paradox is a conflict between two possible, well-argued, interpretations of a design situation. Here, both interpretations seem valid, and the designer must balance them by assessing usability, cost or sustainability outcomes. As such, we can see that design is an almost continuous process of resolving paradoxes where the ideal is solutions that dissolve these conflicts, based on the values of the designer and the specific project.

Example:
Designing a Copenhagen Market Hall. First year students at the Technical University of Denmark were asked: How can we make the fruit market square in Copenhagen more attractive? The students were required to identify needs and propose solutions. Using a socio-technical approach the students identified a network of actors and collected the need statements shown in Table 7.1. The students were surprised to find that many of the statements were incompatible with each other, making a solution rather challenging.

Table 7.1 Important actor's perspectives on the market hall

Actor	Potential discourses	Need statement
Lord mayor	A landmark of Copenhagen	Being recognized as a visionary political leader of Copenhagen
Interest group	An integrated part of Copenhagen townscape	Do not obstruct the existing Copenhagen townscape
Food administration	Food hygiene level	Avoid contamination of food commodities
Fire brigade	Fire safety and rescue	Avoid obstacles, which hinder access for fire engines or evacuation of persons
Customers	Easy shopping, shopping must be an experience	A marketplace, which is worth visiting
Sales persons	Good display of fruit, protected against theft, shelter for sales persons and fruit	A good marketplace, which attracts many customers
Design team	We have to design an attractive market space	Being acknowledged for creating the right design

Dissolving design paradoxes demands **trade-off reasoning** i.e. to find a solution that is suitable for all conflicting criteria, rather than finding a completely alternative solution by separating the criteria. Solutions of this sort often emerge from 'constructive conflict' where different designers propose challenging solutions that force the design team to discuss trade-offs. This is linked to the fact that design is a learning process, driven by the designer's reflections on ideas and goal decisions (Bucciarelli 1984). This learning is highly dependant on the situation, described as **situatedness**. Thus when design paradoxes are encountered we have a number of ways of dealing with them (Hansen et al. 2009) e.g. reframing the goal formulation and task, deconstructing the assumptions behind need statements, looking for radical alternative ideas, or creating trade-offs. These are only possible once there is a good understanding of the problem and possible solutions; as such we explore trade-off reasoning further in Chap. 12.

7.2.2 Systematic Thinking

Systematic thinking brings order to our understanding of possible solutions by considering the whole solution space. This builds on the underlying nature of design, where we create things by combining their characteristics. A product is defined by a set of characteristics, thus varying these characteristics can lead to different products. Systematic thinking gives a structure for this combination and variation of characteristics.

Systematic thinking in conceptualization aims to promote awareness of and collect the pieces from which solutions can be composed.

Product classes can be defined by shared functions or mode of use. Thus these classes can be described in a systematic model. Figure 7.5 shows a systematic representation of various means of human powered transport (Andreasen 1984). Here we describe the product in terms of its key characteristics. It is from this type of systematic thinking that morphologies emerge (Zwicky 1948). Zwicky, originally demonstrated this by showing that all jet engines could be described by varying just 11 characteristics linked in a morphology, leading to 36,864 possible jet engine principals.

Morphology combines a set of characteristics and propositions for variants or *extensional characteristics*. These can then be combined to create new solutions, based on any number of different design perspectives, as illustrated in Fig. 7.6. This can also lead to the search for new principals, and thus use activities or functions based on variation of alternative principles, for example, using *catalogues of physical principles* as proposed by Koller (1979) and Roth (Roth 1994). However, a principle in itself is not a solution until it has an embodiment and an arrangement. As such, variation is key to many systematic approaches.

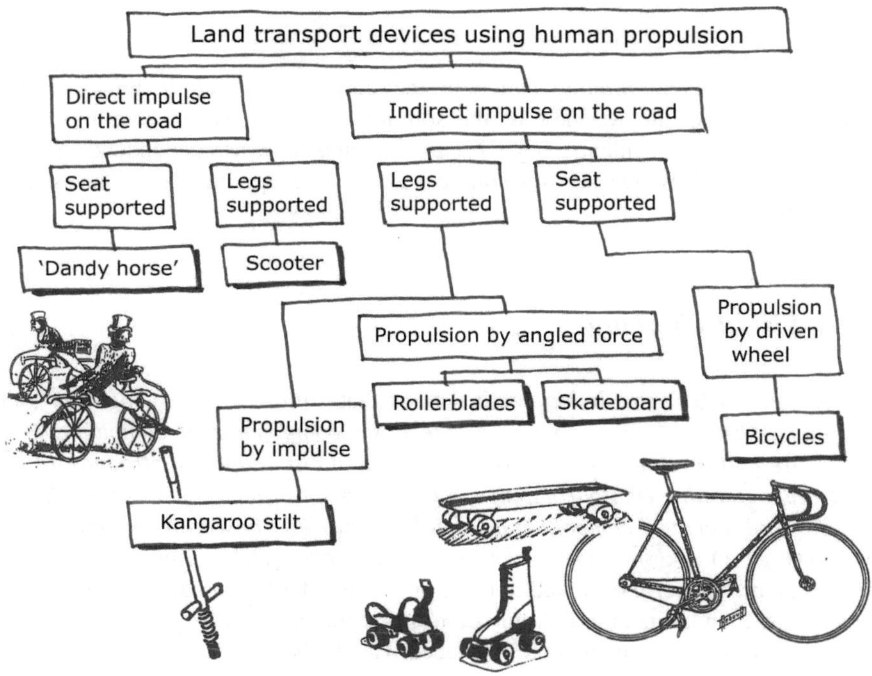

Fig. 7.5 A systematic representation of 'human powered land transport'

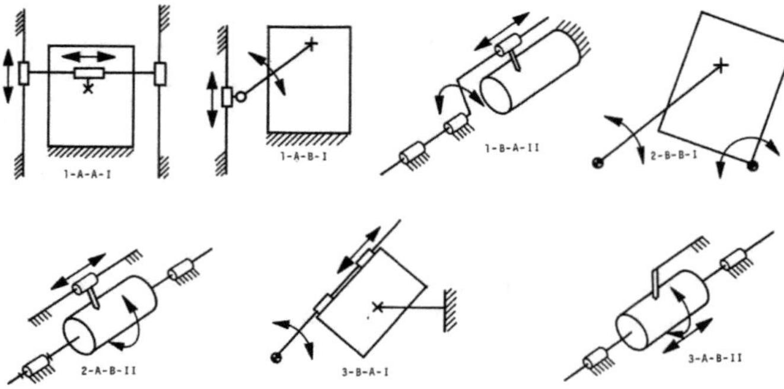

What moves?	1. movement	2. movement	Paper's form
1. 2 pen movements	A. Linear	A. Linear	I. Plane
2. Pen and paper	B. Circular	B. Circular	II. Cylindrical
3. 2 paper movements			

Fig. 7.6 Morphology with systematic variation of an idea for printing on paper, with some example solutions

Using systematic thinking can help the designer try and imagine 'all the solutions' and get an overview of the solution space. However, creating a systematic overview and identifying what to vary and how, are dependent on creative activities. Thus, we see systematic and creative approaches as necessarily complementary.

7.2.3 Visual Thinking

Henderson (1999) wrote in her seminal book that "*In the world of engineers and designers, sketches and drawing are the basic components of communication; words are built around them.... Visual representations shape the structure of the work and determine who participates in that work and what its final products will be. They are a central component of social organization based on collective ways of knowing*". Our knowledge as designers is described and codified in drawings. These can be free form or linked to more or less standardized symbols, techniques, and norms. Standardization not only concerns technical items or parts but also the processes by which drawings are produced. Hence drawing is often described as the **language of designers** (Tjalve and Andreasen 1975).

Drawings not only display information but also support the thinking process in their production and communication in their sharing. McKim (1972) highlights this cognitive aspect in his book "Visual Thinking", where there is an interaction between imagination, drawing, and interpretation of this drawing (Fig. 7.7). In this way details emerge from sketching that were not intentionally planned. Perhaps surprisingly we find that the hand and the brain support each other in a creative process. This is driven by our ability to imagine and develop ideas 'in the minds eye' by continuously sketching, reflecting, and modifying. Such imagined pictures are so important in design and externalizing them is seen as a basic characteristic of designers. "*I draw something. Even if it is 'potty', I draw it. The act of drawing seems to clarify my mind*", Cross (2008) quotes from an interview with a designer ['potty' is a British colloquialism for something that is a little bit crazy].

Tjalve emphasizes the importance of work sheets and collecting visualisations of alternative solutions as a means for what we call 'self communication' i.e. using visualisations to support ideation and reflection, see Figs. 7.8 and 7.9. This technique can be seen as 'graphical brainstorming'. In a similar way film directors conceptualise the scenes they will shoot and the flow between them in a continuous storyboard of sketches.

The multifaceted role of visualisations (Fig. 7.7) means they can make very useful boundary objects (Chap. 4) (Henderson 1999). They create a common understanding by making things explicit allowing people to communicate otherwise tacit knowledge. This is further supported by the fact that drawings are created interactively and can be dynamically shaped and redrawn.

Besides the situations noted above visualisations also play an important role in development, from idea sketches, to formal drawings describing a product's final embodiment and details for production. In this way visualisation brings together creative and systematic thinking to support the designer throughout the design process. The nature of sketching helps keep the design process flexible, sketchy, experimental and open, covering many possibilities until the final selection of the best concept. The main question is, what type of visualisation to use in a given design situation and for a given conceptualisation strategy.

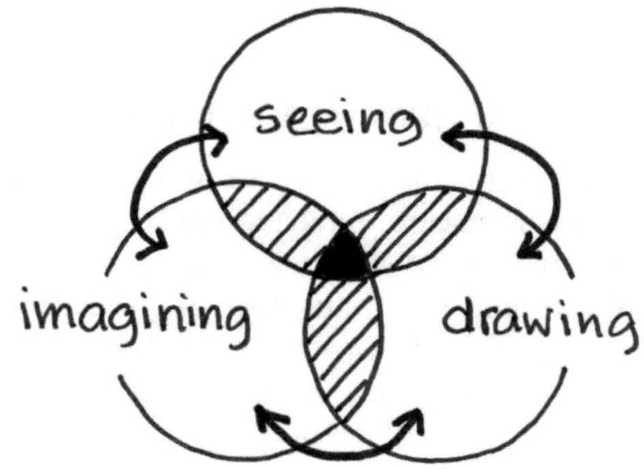

Fig. 7.7 Sketching in support of creativity

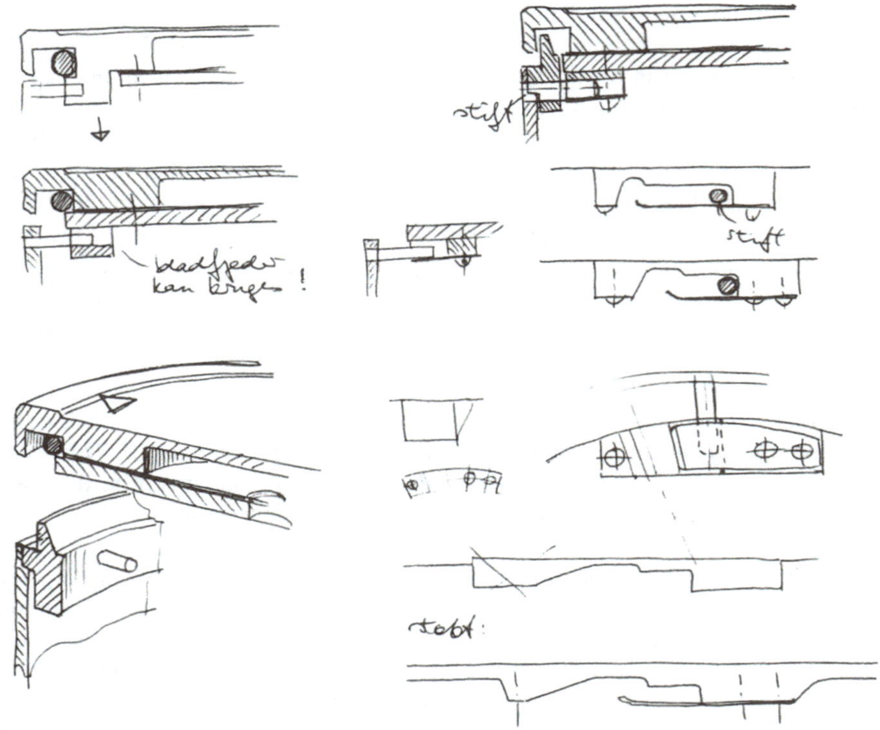

Fig. 7.8 Example of Tjalve's (1979) graphical ideation technique

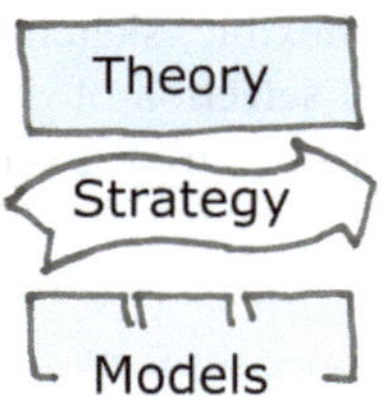

7.3 Strategies for Conceptualization

A concept strategy aims to provide the safest and simplest way to get to a fully detailed concept. Strategic thinking is necessary because of the many conditions influencing the designer, from optimisation of resource use to risk reduction. When properly formulated design strategy can have a significant positive impact on a project. However, this is dependent on the designer's ability to match the strategy to the task, the product, and the skills of the design team.

Fig. 7.9 Example of a designer's sketch for self-communication challenged by finding 'flexible concrete', courtesy Thomas Ulrik Christiansen

Although it may appear that the various models of design provide readymade strategies, in fact, very few really describe strategy or the rational behind the models strategy. One good example of a model with a strong, embedded strategy can be found in Ulrich and Eppinger's (2004) Five-step Concept Generation Method (Fig. 7.10). This combines multiple strategic elements: problem oriented search and solution creation, a systematic overview of the solution space, and reflection on the learning from concept generation.

A designer's selected strategy typically mirrors their interpretation of the main challenges and possible solutions. This gives rise to three main types of strategies: **cognitive**, **search**, and **systematic**.

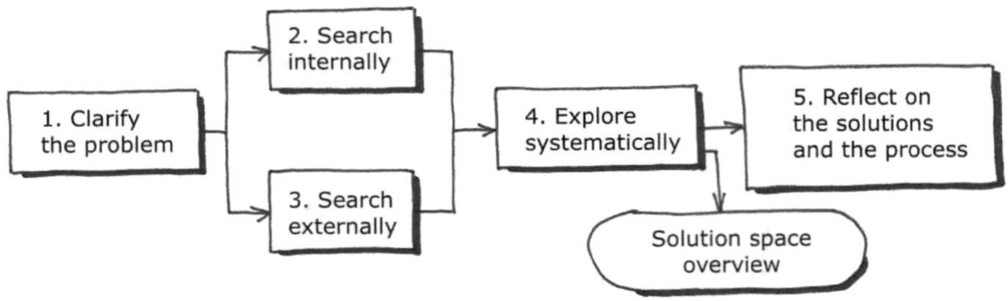

Fig. 7.10 A simplified five-step concept generation method, after Ulrich and Eppinger (2004)

7.3.1 Cognitive Strategies

Cognitive strategies relate to how we think, learn, and solve problems. This is in contrast to *activity strategies* that concern the activities and methods we use to tackle a situation. These activity strategies are closely related to staging as discussed in Chap. 4. Decomposing cognitive strategies, four basic types can be identified (Kruger and Cross 2006):

- **Problem driven**: the designer focuses on understanding and defining the problem.
- **Solution driven**: the designer leaves the problem only roughly defined and focuses on searching for and subsequently generating solutions.
- **Information driven**: the designer focuses on understanding and searching for information relevant to the design assignment, and uses this to point to further relevant information.
- **Knowledge driven**: the designer focuses on relating the current brief to past experiences and knowledge about similar problems. New aspects are then explored by gathering information.

These four types form the basis for a strategy space with more specific guidance on the different approaches (Fig. 7.11) (Hansen and Andreasen 2008).

In the context of the traditional design literature the problem driven approach is dominant. In Chap. 3 we introduced Kleinsmann's "Six Images" of design, characterized by the drivers: product, solution, value, business, experience, and vision. Each approach mirrors a designer's personality but also their cognitive strategy and, to a degree, the type of tasks they prefer. The scope and content of these cognitive strategies underlines the diversity of designers, as well as the need to carefully fit the strategy to the methods, tasks, problem, and design team.

7.3.2 Search Strategies

Search strategies are used when we believe the information or solution is already 'out there'. This uses broad-spectrum research to identify ideas and insights, which might trigger or enhance new product development (explored in Chap. 6).

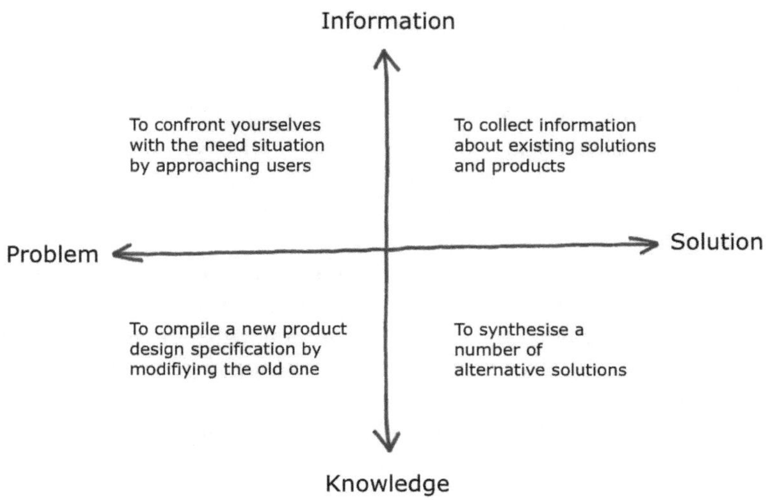

Fig. 7.11 Four strategic dimensions for attacking a problem

These search strategies typically follow one of the cognitive approaches outlined above: problem, solution, information or knowledge driven. Search strategies build on the idea that new technologies are formed from combinations of known technologies and require existing technologies as a precondition for their manufacture, distribution, service etc. (Arthur 2009). In this context the dominant search strategy is **starting from the known**—ideas do not come from nothing!

Example:
Combinatory development. Arthur (2009) is searching for the mechanisms behind technological innovation. It is not Darwin's mechanisms that produce radical new technology, but he seeks to understand how 'heredity' might work in this context i.e. how new technology is established through the combination of known technologies. He formulates an example: '*If you open up a jet engine (or aircraft gas turbine power plant, to give its professional name), you find components inside – compressors, turbines, combustion systems. If you open up other technologies that existed before, you find some of the same components. Inside electrical power generating systems of the early twenties century were turbines and combustion systems; inside industrial blower units of the same period were compressors. Technologies inherit parts from the technologies that preceded them, so putting such parts together – combining them – must have a great deal to do with how technologies come into being. This makes the abrupt appearance of radically novel technologies suddenly seem much less abrupt. Technologies somehow must come into being as fresh combinations of what already exists.*'

Search strategies can be articulated as different **modes of discovery**. These are often described generally as gradually funnelled divergent (ideation, synthesis) and convergent (evaluation, selection) activities. Here, the belief is that the remaining solutions are superior, like washing gold out of the sand. The Delft Design Guide (Boeijen et al. 2013) uses the label 'Discover' for a group of methods giving insight into the actual situation as a means for change and ideation including SWOT analysis, morphology, storyboarding, function analysis, and context mapping.

Search strategies are also closely related to **navigation** i.e. planning the synthesis and composition of the design such that feasible sub-solutions are selected and composition is productive. Typically this focuses on a goal-oriented exploration in the part of the solution space where there are potentially good solutions. As such we are not trying to map the whole solution space and focus on the area that might benefit us most, just like panning for gold! Here key strategies are the *Method of Controlled Convergence* (Pugh 1991), the *Design-Build-Test Cycle* (Wheelwright and Clark 1992), and *Set Based Design* (Ward et al. 1995). The most widely used is *Set Based Design,* which comes from Toyota's approach for automobile design. A set is a range of solutions or alternatives for a sub-system. Sets are explored in parallel and gradually narrowed concurrently, managing the interactions and interfaces between the partial solutions and the total product. The number of alternatives and the depth to which they are clarified is deeper than traditional variation approaches, however, the subsystems are 'frozen' early by a specification with no changes permitted later in the process. The power of this approach is reflected by the many versions that can be found in the design literature e.g. multi component and multi issues design: How can we most effectively develop components in parallel but also rapidly reduce the number of possibilities to a promising design? Set based design answers this by developing broadly but confronting solutions early in a kind of elimination race.

Example:
Toyota and Nippondenso. Toyota's set-based design included manufacturing concept, styling, manufacturing system design, product design, and the line of components (Ward et al. 1995). Figure 7.12 shows the development of an alternator program by Nippondenso. This eight-year project followed set-based approaches and intensive design modelling, here called prototyping. The first year's ideas and prototypes were combined, modified, and improved, and new ideas added. In the 4th year three different designs were considered with five prototypes each. In the 5th the final solution was selected as the basis for a product family of 700 alternators ready for multiple car models and clients. Goals were also set for radical breakthroughs e.g. the performance to weight ratio, leading to a 50 % weight reduction.

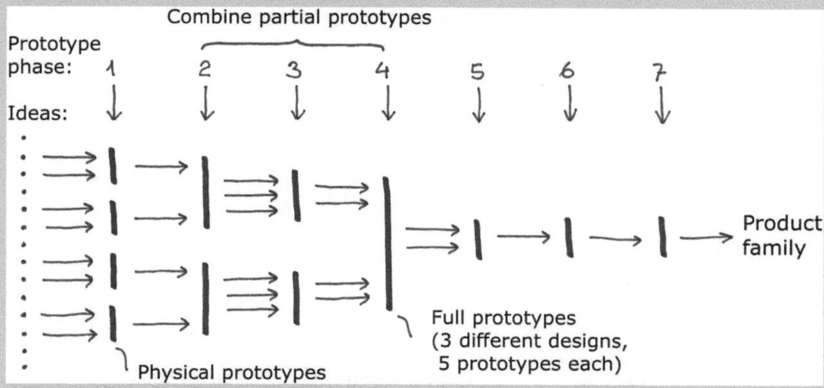

Fig. 7.12 Nippondenso's research and development process

7.3.3 Systematic Strategies

We introduced systematic thinking in Sect. 7.1, where solutions are identified by their characteristics that can be varied to cover the solution space. This provides the foundation for a structured overview of the solution space, where we can identify and fill gaps. This is in contrast to 'solution space blind' search strategies (called 'point based design'). One way of providing a structure for a problem is the arrangement of goals and means in a hierarchy. Here each element is both a gaol, when seen from below, and a means, when seen from above, as illustrated in Fig. 7.13. This can help determine at what level a problem should be addressed and what higher-level elements might contribute to this problem.

Alternatively, the systematic division of the solution space, Fig. 7.14, gives another structure. This example shows the principle options for protecting a welder from smoke inhalation. Generally, a solution space can be defined by intuitively creating a set of solutions with a wide spread. These can then be classified to find a pattern for dividing further expansion of the space and to help focus the design activity, Fig. 7.15.

Hubka's first law states that the *functions* and *means* of a product are related in a causal, hierarchical structure (explored further in Chap. 11). This can be visualised in the 'function/means-tree', illustrated in Fig. 7.16 based on Buur (1991). The tree can show not only a single product but also alternative means for certain solutions. Together, all the possible sub-solutions for all the sub-functions in a product form the solution space.

Design has a high degree of recursion i.e. patterns at a high level are repeated at lower levels. As such, many aspects of concept synthesis can also be applied to the sub-system, component or module level. It also means that strategies like set based design can evolve during the development activity.

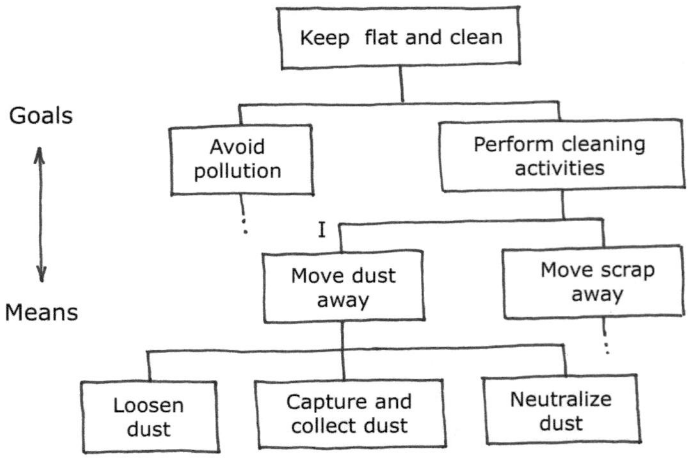

Fig. 7.13 A goals/means hierarchy for cleaning a house, intentionally made general to support the re-evaluation of vacuum cleaner solutions

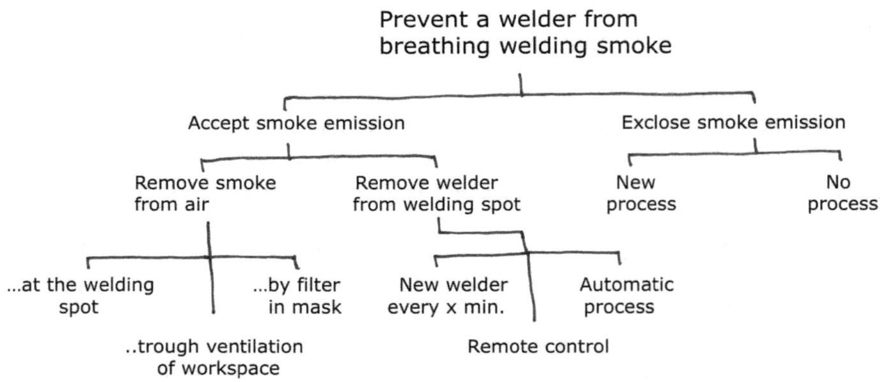

Fig. 7.14 Systematic division of the solution space

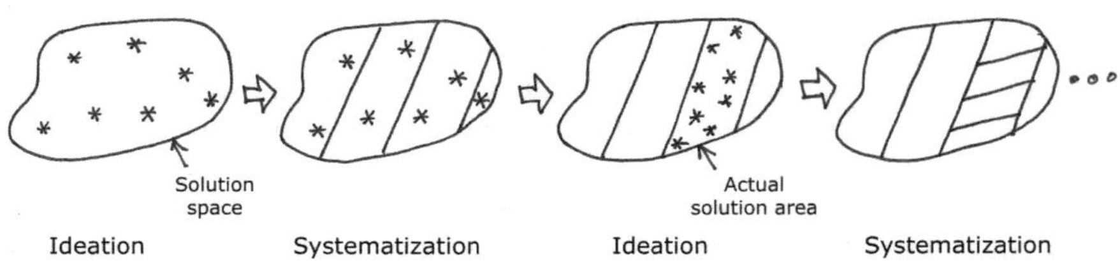

Fig. 7.15 Gradual identification of the solution space

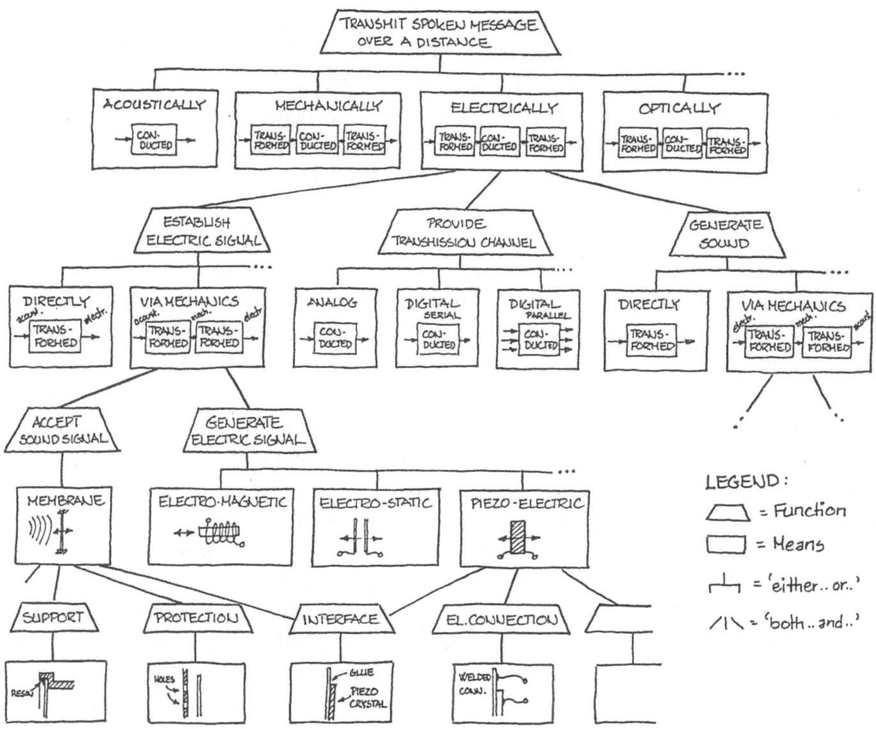

Fig. 7.16 An example function/means-tree for 'transmitting spoken messages over a distance'

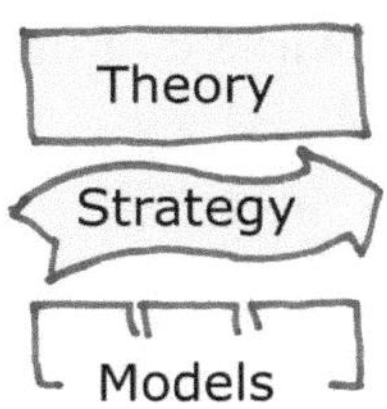

7.4 Models of Conceptual Design Activity

All of the strategies described above build on a foundational model of conceptualisation. There are many different proposals for this, which we will explore here, before we describe our own model in Sect. 7.5.

Conceptual design lacks a clear definition and identity; however, there are three consistent perspectives: **procedural** (identifying sequences of activities or tasks), **creative** (building on cognitive processes), and **strategic** (interpreting theoretical models of design). Design literature contains both descriptive models—trying to explain design activity in practice, and prescriptive models—setting out instructions to be followed when designing. Here, we examine a selection of these in order to lay the foundation for our own model, which seeks to bring together the disparate threads of conceptualisation into a more cohesive whole.

A basic description that applies to both problem solving and creation is the *TOTE-model* (Test, Operate, Test, and Exit), Fig. 7.17a. Here the idea is that solution proposals are formed by the designer and iteratively tested until they work. The *General Problem Solving* model, gives a more detailed procedure for this cycle, Fig. 7.17b. Here it is important to clarify the problem and 'good solution' criteria before seeking solutions (either creatively or through search). Further, multiple solutions should be created in order to select the 'best' solution through evaluation.

Problem solving models might seem like 'linear thinking' and an attempt to apply non-existent rationality to problem solving. However, the strength of these models is in their 'ought to' aspects. On the other hand their main weakness is the uncertain understanding the task before we have seen some solutions (co-evolution, see Chap. 5). Thus it can be difficult or misleading to follow a set of predefined criteria. This mirrors design's uncertain nature. Based on this Hubka and Eder (1976) added a set of *basic operations* to the General Problem Solving model: '*providing and preparing information*', '*verifying, checking*', '*communicating solutions*', and '*providing representation*' (specifying the solution so that it can be communicated and utilized). Further elaborating on this Lindemann (2007) proposed the *Munich Procedural Model* (German: Münchener Vorgehensmodell) as an elementary, situation adaptable model of behaviour or action, see Fig. 7.18. This links goal, task, ideation, selection, and consequence considerations. Hear a unique feature is the integration of use insight into the solution's properties to align goal, consequence evaluation, and selection. This model helps the designer to recognise the situation i.e. what aspects should be clarified? Where do I find new insight or hindering/non-clarified aspects?

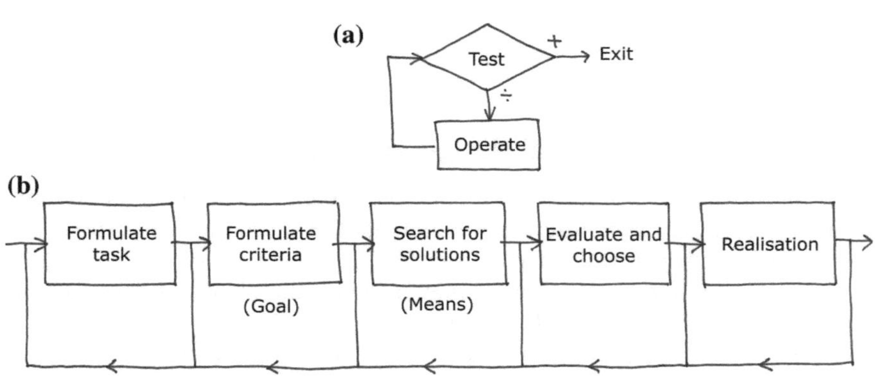

Fig. 7.17 Conceptualisation models **a** TOTE **b** general problem solving

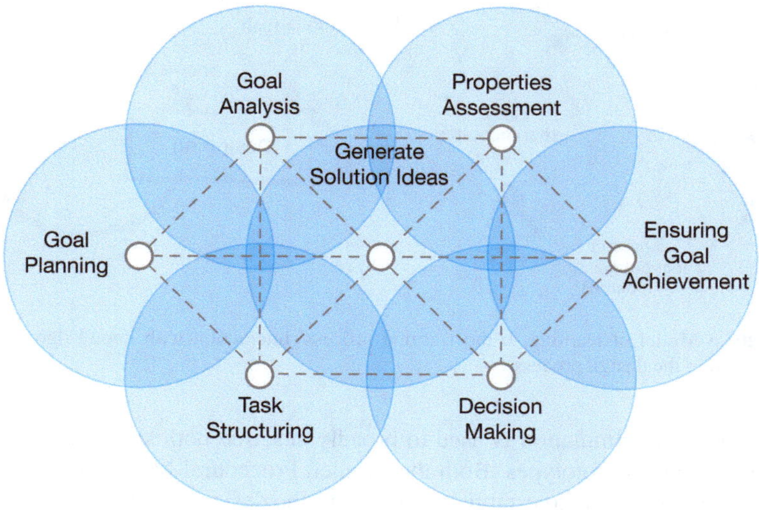

Fig. 7.18 The Munich procedural model for problem solving

The same philosophy of supporting the designer in recognising the current situation and navigating its resolution is found in the C-QuARK method of Ahmed and Wallace (2004). This very open method takes its name from the eight questions that it poses to the designer: reasons, optimisation, trade-offs, past designs, possible issues, validity of specification, limitations, and is the idea worth perusing. This focus on simple clarifying questions aids navigation in complex projects where other's contributions, existing solutions, past designers', and other designers' arguments all need to be synthesised. Both the Munich Procedural Model and the C-QuARK method focus on providing patterns for the designers thinking, neither model address design activity.

In the context of design activity the General Problem Solving Model can be applied by adding theoretical design elements e.g. Roozenburg and Eekels (1996) use 'form follows function' as the basis for their basic Design Cycle, shown in Fig. 7.19. Here a desired function (from need analysis etc.) forms the starting point for synthesis, leading to a 'provisional design'. Subsequently, the preconditions for selection are insights into the design's properties from prototyping and

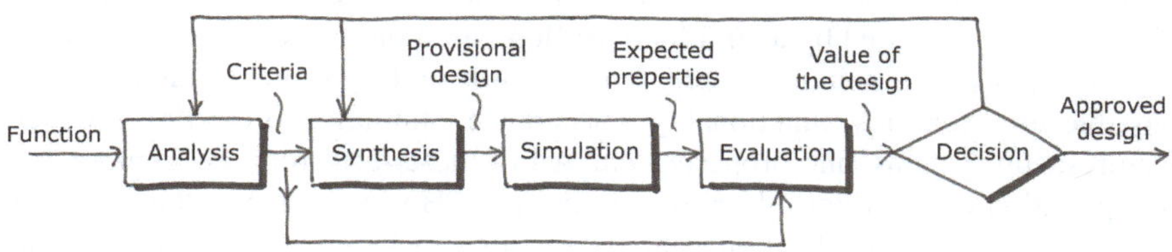

Fig. 7.19 The basic design cycle, after (Roozenburg and Eekels 1996)

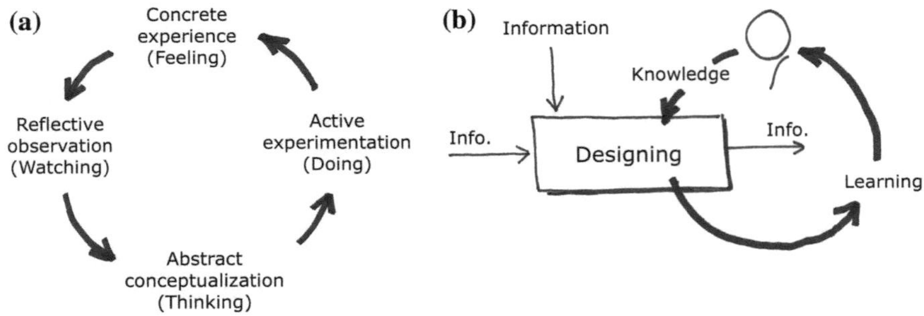

Fig. 7.20 **a** Model of learning, **b** the learning process' interaction with knowledge utilization and creation in the design process

simulation. Here simulation is used to broadly describe both simulation tools and physical drawings/prototypes. Both the Munich Procedural Model and C-QuARK build on similar efforts to establish insight into the design's properties as the basis for evaluation.

Underpinning all of these conceptualisation models is the concept of reflective learning (Kolb 1984). This is shown in Fig. 7.20a and explains how designers recognise and create insight and knowledge. Here design activity not only creates information but also knowledge. Thus the process becomes an interaction between problem solving and learning as in Fig. 7.20b.

The models introduced in this section both describe conceptual design activity in a project and form elementary operations occurring at different levels in the hierarchy of a composed design. This points to the need for a Concept Synthesis Model able to bring together these disparate elements.

7.5 The Concept Synthesis Model

The second major topic we cover in this chapter is our proposed conceptualization model and its use. Our starting point is the three elements discussed in the previous sections: theories, strategies, and models i.e. current explanations of conceptualization. Two key questions are what constitutes the model, and what insight or mindset the designer needs to be able to use it successfully.

The **model** dimension is articulated in three major steps, illustrated in Fig. 7.21. The synthesis is guided by a **goal formulation** based on the designers understanding of the possible solution space. This in turn builds on understanding about possible use scenarios, functionality, appearance, similarity to existing products, features, sub systems, and properties etc. All of these feed from the Exploration's interpretation of need, into the problem and task. However, due to the evolutionary nature of conceptualization goal statements can be created when something

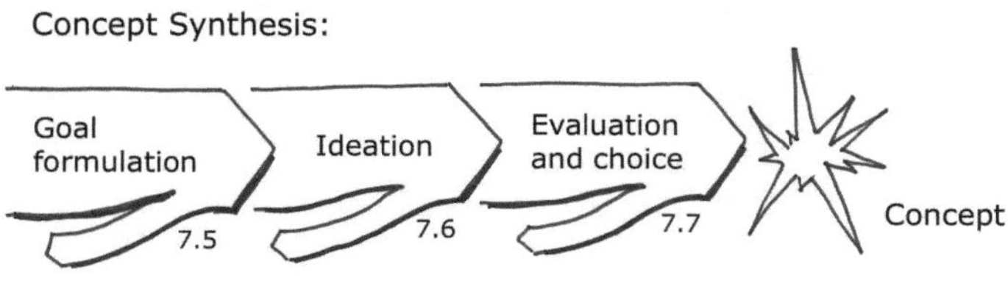

Fig. 7.21 The three steps in our Concept Synthesis model

interesting or important occurs throughout the process. The concepts themselves are then formed through **ideation** and externalised as sketched of modelled concept proposals. Finally, the synthesis elements are closed by the selection of the final concept(s) or as a sequence of **evaluation and choices**, gradually bringing the synthesis closer to the goal by iteratively refining solutions.

The **mindset** dimension is articulated as the knowledge, advice, methods, tools, and models used in the conceptualization. This has four major traits. First, the model uses the idea of a **solution space** delimited by the requirements of the solution. How this space is populated depends on the available knowledge, the team's perception of the task and their own limitations. Second, the concept is created **gradually** based on intention, ideas, and solution elements from the Exploration. Concepts are then **composed** by combining ideas from various domains including product, use, and issues. Third, we build on the basic **strategies** of *goal formulation before ideation*, and *concept selection based on evaluation of multiple proposals*. Other strategies may be added e.g. for strategic search, to best suite the design team and the specific staging. Finally, the model promotes **sketching** (and basic models) as a means of idea documentation, in order to retain the dynamic, creative freedom most conducive to successful conceptualisation.

Together these lead to the very basic articulation of our model shown in Fig. 7.21:

- **Goal formulation:** the formulation of requirements and criteria for the expected ideal solution.
- **Ideation:** the creation and combination of idea elements into concept proposals.
- **Evaluation and choice:** decision-making based on comparative evaluation and selection.

The model is shown with the ideal flow from one step to the next. However, we acknowledge that there are necessary feedback loops (illustrated by the feedback arrows).

Here we focus on concepts, rather than ideas, because concepts are **answers/** solutions that are sufficiently detailed and clarified that we can judge if they satisfy the need, intention, and task. Further, the concept is a **starting point**, forming the basis for decisions about the product development's more risky or costly activities. Thus the concept is at the heart of product development, linking all the activities described in our Encapsulation Design Model, see Fig. 7.22.

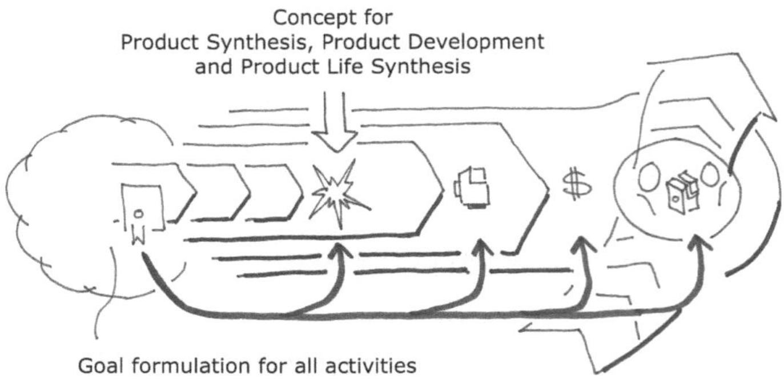

Fig. 7.22 The concept as the core of the Encapsulation model

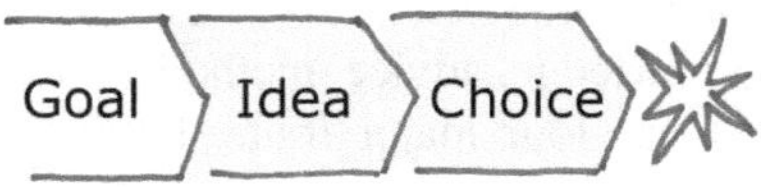

7.6 Concept Synthesis: Goal Formulation

The goal formulation has a major impact on the activities and results throughout the design process. This is illustrated repeated in industry where strong market insight and clear goal formulation are key success factors.

Literature reports about a line of studies on product's success and failure. Baxter (1995) analyzes data from Cooper (1993) and points out that strong market insight and goal formulation is a success factor. Baxter writes: "*Products, which are sharply and well defined in a design specification prior to development, were 3.3 times as likely to be successful as those that were not. The message – put lots of efforts into getting the product right at the start before beginning the design work*". Further, Hollins and Pugh (1990) write: "*In our research, one of the most surprising discoveries (and one of the most depressing) was the woeful inadequacy of the product design specification in companies*". They found that short, incomplete specifications based on poorly argued market research were all too common in practice. The importance of goal formulation, and its shocking lack in industry, is further illustrated by Foxley and Blessing (2000). Here, more than 400 UK companies were asked about their product development abilities. Although 44 % identified the need to improve customer focus, analysis of competitors and market, the majority could not articulate how they were competing! Similar results were found in Germany by Grabowsky and Geiger (1997) where again, customer orientation and market insight (in the form of effective goals) were the largest success factors, while most companies identified these as their weakest areas. The successful formulation and utilisation of goals builds on three key elements, illustrated in Fig. 7.23 (Andreasen and Wallace 2010):

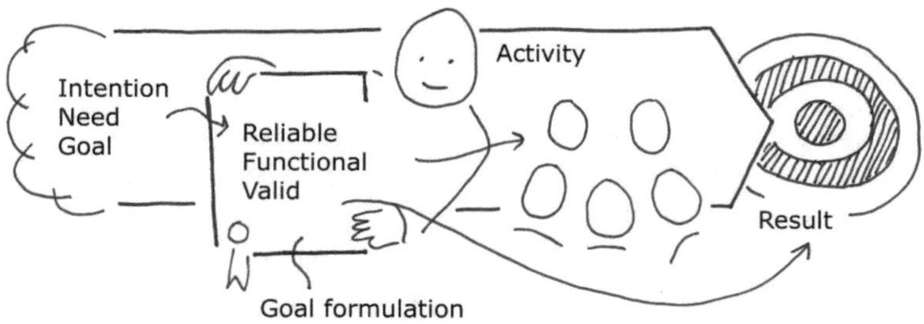

Fig. 7.23 Reliable, functional and valid goal formulation

- **Reliability**: the goals should accurately mirror intention, need, goal perception, and the idea behind the project. A key issue here is transforming speculative imaginings into concrete goals that can be agreed by the design team.
- **Functionality**: the goals should be clear and understandable, able to be transformed into tangible design tasks to be addressed.
- **Validity**: goals should be updated to reflect changes in need and context or because of co-evolution. Thus they form a moving target that balances flexibility with the underlying constraints associated with the project e.g. time or budget.

Successful goal formulations serve two main tasks: they allow management or clients to follow the projects progress, and they support team in arranging and directing the project. This leads to six key situations where the goal formulation is a critical document (Hansen and Andreasen 2004):

- **Communicating ideas**: intention, argumentation, motivation, mission, and expected results.
- **Managing** the synthesis activity: directing exploration of the solution space and identification of attractive solutions.
- **Evaluation** of design alternatives: helping to decide what should be kept in the progression.
- **Navigating** between solutions and activities: helping to find a possible solution within the constraints of resources and time.
- **Decision making** at milestones: helping decide the continuation of the project and fostering agreement between management and the design team.
- **Breakdown** of the goal formulation: helping identify discreet sub goals and specifying their associated sub systems.

Mapping these conceptual elements to the real world, the tasks linked to the goal formulation are illustrated in Fig. 7.24. As this mapping is complex, and the role of goal formulation diverse, it is essential to carefully check if the goal formulation is working as expected in the actual project.

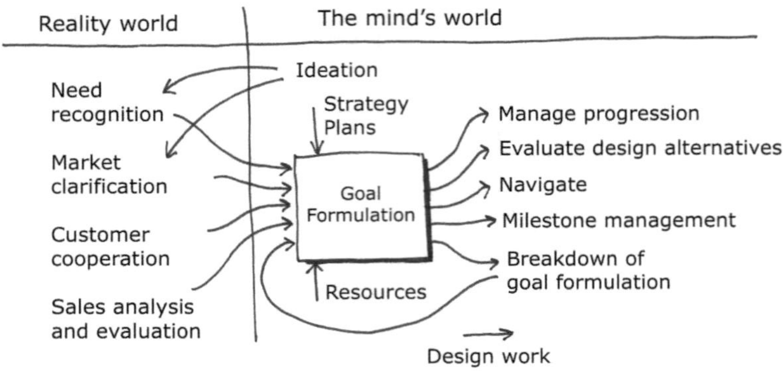

Fig. 7.24 Requirements of the goal formulation's functionality (Hansen and Andreasen 2004)

To create an effective goal formulation we need to transform the users' needs and the life cycle actor's interests into entities 'to be designed'. These entities should also be coupled with success criteria, again linked to the users and actors.

7.6.1 User Needs and Actor's Interests

When we dream about owing a new product, be it a car or a kitchen knife, we usually express our desire as wishes for the product and its characteristics e.g. a big engine or crafted blade. These paint a patchwork picture of the ideal solution and reflect the specific personality of a user e.g. one car buyer might only want enough space for their golf clubs, while another might want to go camping for a week. These need to be synthesised into a concrete unambiguous ideal solution that describes the whole product—not just bits. Thus when we create an ideal solution we bring together many statements and wishes, while also remembering the key needs that must be addressed—no one will buy a car that doesn't move, no mater how much storage it has!

Satisfying the user means balancing the dream product with an available solution that meets the demands of the use activities. However, the user is just one of many actors involved in a product's life cycle. Thus the goal formulation must include criteria for 'goodness' reflecting all the involved actors, not just the user. This links to what we call universal virtues, illustrated in Fig. 7.25. Here, we find that in each life phase these universal virtues provide criteria for how we can relate the goal formulation to the actor's interests via e.g. cost, efficiency or quality.

The life cycle actor's interests can be articulated as values related to the activity's performance but also to the development and utilization of equipment, see Fig. 7.26, where the product's packaging properties radically increase transport efficiency. Here the product's transport properties are determined by the interaction between product, packaging, and transport means. Focusing on this, *Chair number 14* by Thonet can pack 36 units into just one cubic meter (Gleiniger 1998).

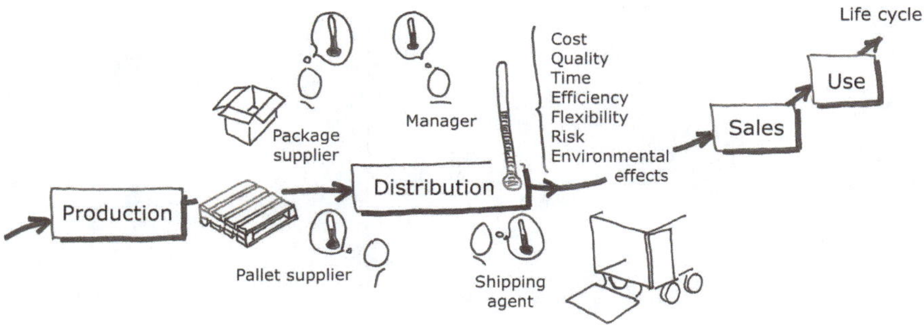

Fig. 7.25 The product's life cycle and the universal virtues

Fig. 7.26 Example of a product's design fitting with a life phase system: the chair and an old photo shoving 36 chairs in a glass box for illustration, *courtesy* Thonet GmbH

Many different entities, like the transport box, appear during the life cycle and all need to be reflected in the goal formulation. This can range from spare parts to actual necessary systems. This is particularly important from an environmental perspective. For example, the materials required to produce, package, protect, and transport a new car weight ten times more than the car itself. In response to this example, it is now mandatory for car manufacturers to be able to scrap and recycle much of their product. Figure 7.27 shows recycling initiatives.

Beside the design entities introduced in Chap. 2 there are entities related to the arena in which the company's products are established, sold, and used. These include establishing a network, identifying a value chain, and supplying services, accessories and spare parts. Many products exist with the express purpose of driving sales of consumable elements. For example, games consoles are sold at a loss in order to move larger quantities of the profitable games; similarly printers often generate more income via ink refills than sales of the basic print unit.

Fig. 7.27 Demolition scrap sorting and household sorting system

Ultimately the model in Fig. 7.28 means that as the design entities in a project are gradually clarified the goal formulation must be updated. Further, as the project is detailed it may also be necessary to work out sub documents for components, especially if they are to be outsourced. In this context, the goal formulation focuses on use, functionality, and properties—defining the performance of the design entities without necessarily constraining the entities themselves. In order to achieve this level of cohesive, encompassing goal formulation we must build in value statements tied to tangible elements in the concept (Hansen and Andreasen 2007).

- Formulate value statements. Describe 'the good product' from the actors' perspective.
- Relate value statements to the product's functions. Describe what the product should do. A product with high-performance, precision functionality is normally better.
- While designing, consider the requirements from multiple perspectives. Statements will be more or less ambiguous in their formulation, thus interpretation can lead to surprising solutions.

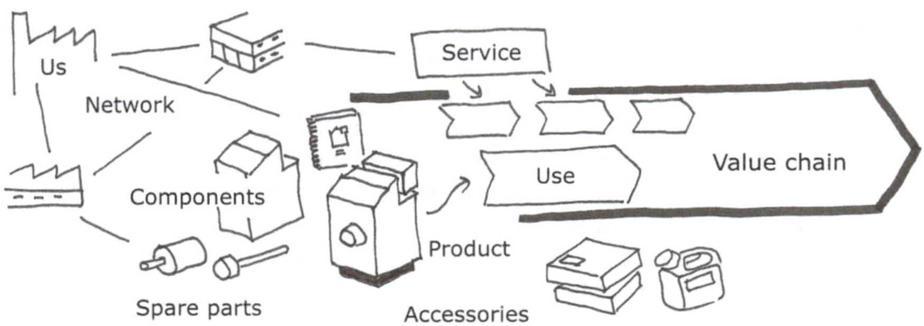

Fig. 7.28 Examples of design entities that can either be incorporated in conceptualisation or separated into independent design tasks

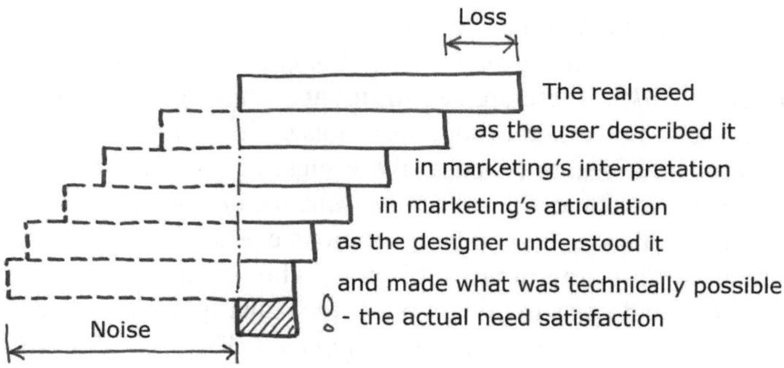

Fig. 7.29 Noise and loss distort the goal formulation

These tasks can be supported by the simple exercise of creating an internal 'sales brochure' linking the product description to emotional points: What the user can do with the product, supported by a story about a characteristic user's experiences, feelings related to use etc. This is particularly relevant in a business-to-business setting where user communication is usual neglected.

As highlighted in the third heuristic above interpretation is critical to the realisation of the goal formulation (Jul Nielsen 1990), see Fig. 7.29. The complex information, from Exploration, must be formulated into an idea and intention for the project leading to the desired goal.

7.6.2 Articulating Goodness and Value

A goal formulation can be articulated as statements about what the entities should do and their desired properties. Statements about properties include the required functions and functional properties, product properties, and relational properties concerning a product's interface with the user, life phase systems etc. Typically we distinguish requirements and criteria as in Fig. 7.30.

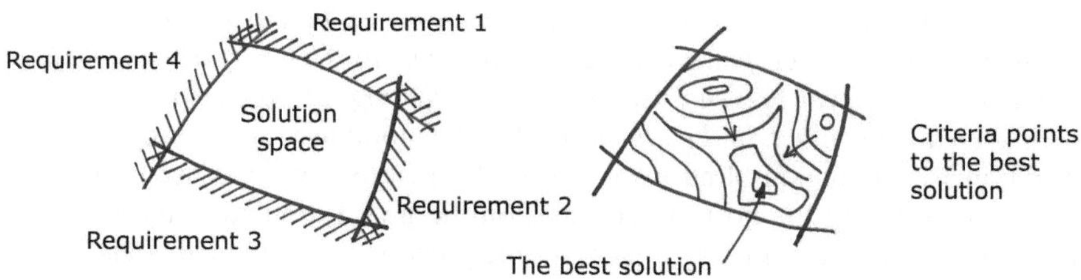

Fig. 7.30 Illustration of solution space requirements and criteria

Example goal formulation statements for a production machine are: *noise level lower than 30 dB, shall be able to fit through a standard door, use hydraulic power,* and *Yield 10 kg*. These are **requirements**. They describe both characteristics (size, hydraulic power) and properties (noise, yield), formulated as absolutes. In order to validate a design proposal these must be satisfied. Some other statements for the same machine are: *easy to operate, waterproof,* and *attractive*. These are **criteria**. They allow us to compare solutions e.g. assessing ease of operation. Here, the first and last criteria are subjective, while 'waterproof' can be defined in technical terms and measured via a prototype. Finally, we also encounter statements such as: *show the number of parts completed*. These are **desired functions**. We discuss how these can be articulated and built into a product in more detail in Chap. 13. However, it is relevant here to understand that when we synthesise a product we bring these elements together, reasoning about functions and properties in parallel.

7.6.3 Formulation and Use

In practice a company defines a team's goals and responsibilities in a set of documents. Of importance here is the *product goal specification*. In the design literature there are many proposals for how this can be formulated and used (Roozenburg and Eekels 1996; Dym and Little 2000; Ulrich and Eppinger 2004). This core document is usually supported by the *design brief*—clarifying mission, vision, and ideas, and the *business specification*—clarifying market strategy and economical goals (Andreasen and Hein 1987). Although these might seem straightforward they pose a double-edged sword. First, the more statements we include the greater the difficulty in articulation and communication, shifting the focus from useful progress to documentation and formalisation (Almefeldt et al. 2003). Second, the more we work on the goal formulation the more confidence we place in it, rather than checking the actual results in reality.

Experience from practice suggests that the ideal approach is to create the goal formulation solution neutral, but to set out the tasks based on what you know. It is difficult for designers (or anyone else) to imagine goals and tasks without taking a starting point in known solutions. This leads to the following quote from Wallace and Andreasen: "*A goal statement may never be correct, can never be complete, can never be final, can never represent the views of all stakeholders and actors, can never prevent incorrect and conflicting interpretations, can never resolve creative conflict, and it should not!*". The moral here is that we should not let the goal formulation blind us to the need, intention, etc. We must inevitably move forward with the project and build on the goal formulation through ideation. This provides contrasting drivers for the process, the analytical goals and the creative ideation.

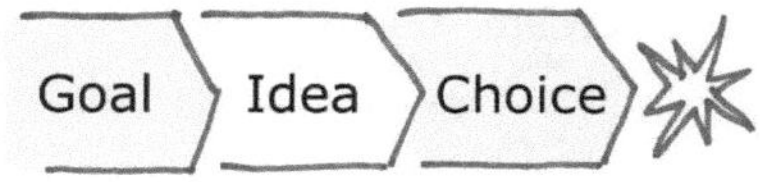

7.7 Concept Synthesis: Ideation

In this section we focus on transforming ideas into concepts, where they are ready for evaluation and choice. This transformation builds on four elements: **create** (generate ideas), **combine** (ideas are combined into concept proposals), **visualise** (illustrate and document the concept's key traits), and **complete** (link concepts to the most relevant activities and life phase conditions). Figure 7.31 shows Concept Synthesis together with the four conceptualization activities. This is a reminder that although we will discuss these activities sequentially, they are in fact interconnected, with clarifications in one dimension leading to changes in the other dimensions.

Ideation is extremely open, no approach is out of scope and there are no clear start or end points. This leads to key challenges, illustrated as questions: 'I have a reasonably good solution, should I continue to search for better solutions? I have a lot of ideas already, how should I proceed? I want to generate a convincing set of alternatives, what should I do?'

Our concept synthesis model and the four activities in Fig. 7.21 thus provide a safety net supporting the designer's personal approach to ideation. Together the four activities help ensure that ideas are articulated and captured in a useful way for the design process.

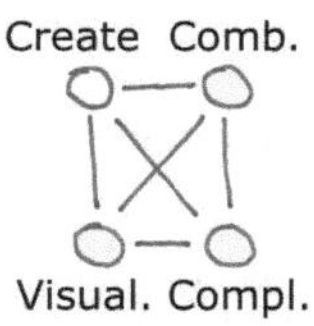

7.7.1 *Create*

We cannot expect ideation to be a single planned and focused activity. Ideas emerge during Exploration, while goals are formulated, when found ideas are scrutinized etc. Moreover, ideas continue to be generated throughout the project, and might even challenge or seem better than the chosen concept. Thus a key element in creativity, beyond strategy and methods, is actually **capturing ideas**, making them tangible and combining them into concepts. One way to start this process is to look at the characteristics of established, successful businesses (Jul Nielsen 1990). New business can then be found by more or less systematic variation of

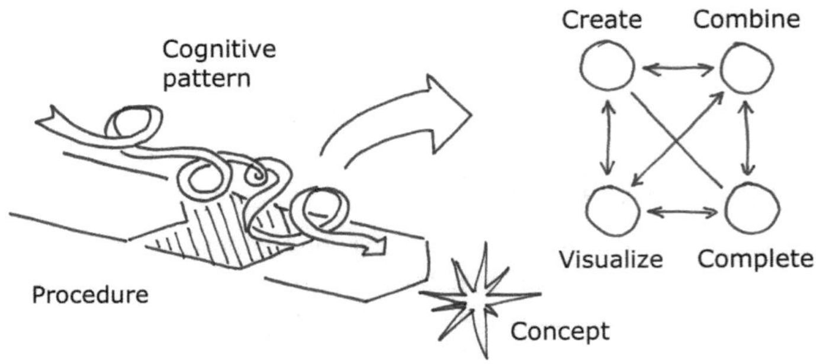

Fig. 7.31 The four-element transformation of ideas into concepts

the relevant characteristics. Here core characteristics can be found in the different issues related to products, for example:

- **Usability**: ideas that make use easier, more efficient or enjoyable may lead to other innovations (Hede Markussen 1995).
- **Environmental characteristics**: reducing a product's total environmental impact or influencing users behaviour to be more environmentally friendly can both yield new innovation (McAloone 2000).
- **Service**: introducing service elements can attract buyers' attention, change behaviour, and increase value perception (Tan et al. 2008).
- **Business**: many companies launch generic products, not daring to do things differently. Exploring new areas of technology, product functionality, visual design, sales mode or delivery type etc. can provide a starting point for variation and innovation (Jul Nielsen 1990).
- **Business strategy**: new ways of operating as a business can spark innovation, and innovation can point to new ways of operation (Kirkegård 1988).

These do not cover all the possible areas of innovation, but they do highlight the need to explore the various perspectives that can be used to raise questions and thus produce ideas. Figure 7.32 gives an example of a systematic mapping of different issues to be explored in ideation. There are many other sources of inspiration including: nature, sport, logistics, games, science fiction, play, and so on. Similarly, prototyping, working with products, and experimentation provide more tangible sources for ideas. All of these can be facilitated with a range of methods.

The simplest creative method is to *create dialogue*. Its staging is simple: one person seeks help from another in solving a task. The helper's role is to formulate questions, to provoke new perspectives, and to help develop new lines of through by 'bouncing' ideas back and forth. The helper is key, as this does not work as a solitary activity, Fig. 7.33.

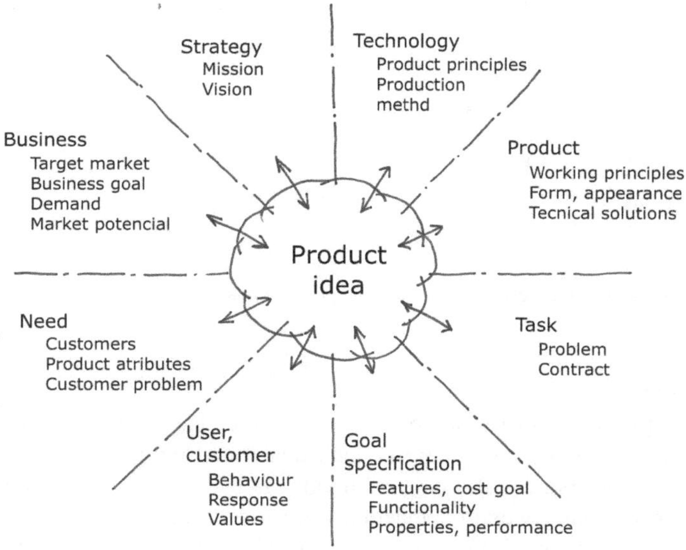

Fig. 7.32 One systematic mapping of issues related to ideation (Hansen and Andreasen 2005)

Fig. 7.33 The dialogue method and its pitiful substitute

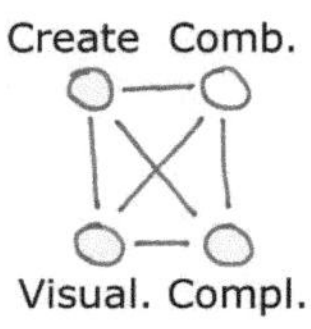

7.8 Combine

Chapter 2 described the composed nature of concepts, while Chap. 6 showed how exploration provides the various elements to be combined. Thus the composition of a concept reflects the various dimensions of the product e.g. realisation, use, and life. This links to the 'ideas in the product' and 'ideas with the product'. In any given situation previous products can be used to help clarify these dimensions—simplifying composition as illustrated in Fig. 7.34.

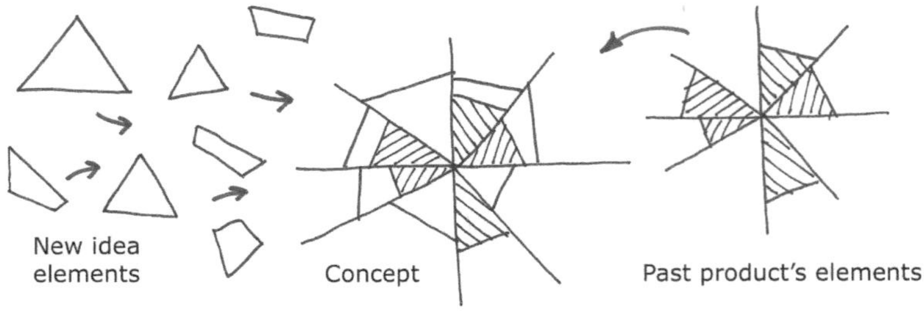

Fig. 7.34 An idealised representation of a concept's composition

Example:
Combinatory design of a hemodialysis machine. The reliability and operation of hemodialysis equipment is such that it can be operated by patients at home. This reduces the cost of treatment by over 50 % for hospitals. However, current generation machines are not designed for the home. To address this user workshops were organised to pinpoint key issues, gain insight into the functionality required for home use, and identify the criteria for a good home machine. This was further supported by field studies in hospitals and in patients' homes, focusing on the nurses and patients' operation of the machine. These resulted in insights regarding economy, stigmatization, aesthetics, safety, and hygiene.
Using these insights four concepts were created based on a building block approach, combining components from existing machines and arranging them in new layouts focused on home use. Finally, a concept proposal was produced that was well received by patients, and could be easily produced based on past components as shown in Fig. 7.35. The project was performed in cooperation with the Association of Kidney Patients in Denmark. Courtesy Anna Heuschkel and Laura Fokdal.

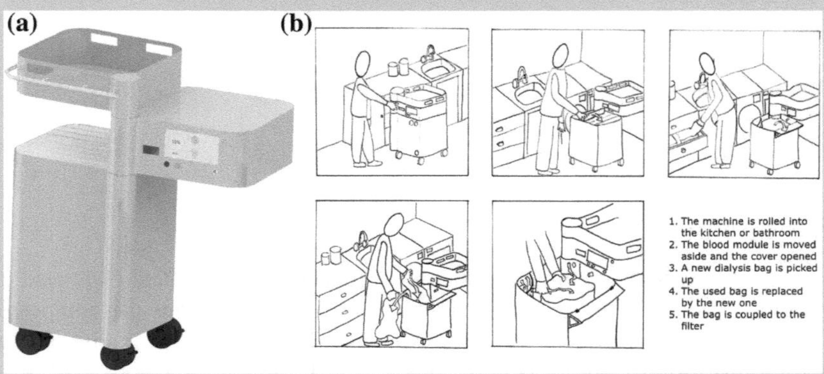

Fig. 7.35 A CAD rendering and use scenario illustration for the home dialysis machine

The process of searching for and maturing ideas is often pictured as the narrowing funnel shown in Fig. 7.35a. As the funnel reduces ideas are rejected, combined, and developed until a concept is selected. This is sometimes called the 'death curve' (shown in Fig. 7.35b) where ideas are rejected until only one remains. Both of these illustrations highlight the importance of combining ideas and the principals of set based design (Andreasen and Hein 1987).

Figure 7.36a shows the two perspectives on combination: A *post mortem* view where it is relatively easy to identify if there are sufficient good ideas, and a *proactive* view where ensuring a flow of potential ideas is an open question. In this proactive sense a good strategy is to explore as many areas as possible as illustrated in Fig. 7.32b.

Example:
Colostomy pouch. Colostomy pouches are a relatively simple product, however, maintaining a dominant market position demands expertise in areas including plastics, gluing, packaging, distribution, disease development and treatment, and relation with the public health infrastructure. In the context of conceptualization this highlights the significant challenges faced by designers trying to combine ideas from these very different domains.

7.8.1 Organizing the Combination

The difficulty highlighted above raises the question, who should be responsible for combination? There is no easy answer to this question, although there are a number of different approaches that can help. These aim to better align the various areas of innovation: product creation, marketing, sales, and production. Further, they often demand new organisational structures to work effectively. Below we highlight five strategies for supporting combination and strategic alignment.

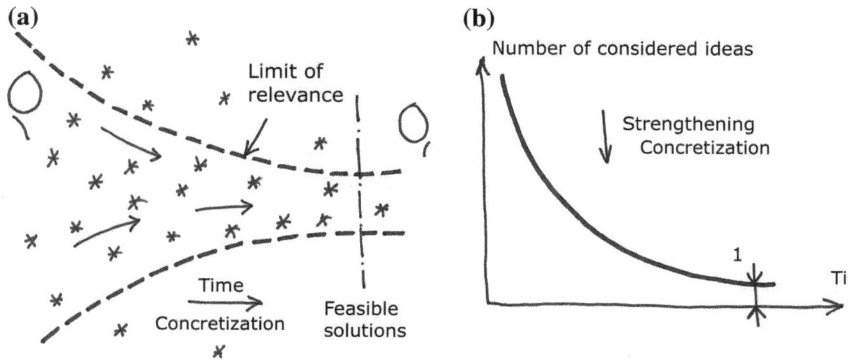

Fig. 7.36 Two illustrations of idea flow leading to a single, usable idea

1. **Composing a product committee**: this brings together function managers from finance, marketing, distribution, sales, production, and development to manage the establishment and utilization of new product development.
2. **Composing concept proposals**: concepts are mapped out and clarified with respect the various relevant functions in order to mature the idea and identify possible issues.
3. **Composing a project team**: a specific team is made responsible for concept creation—usually with a 'quirky' name such as 'idea land'. This team can access any expert in the company and is empowered to involve employees from any functional unit.
4. **Company theme days**: informing about new production technology, network partners, lead customers etc. is introduced to the whole company to encourage employees to suggest ideas from any unit.
5. **Business-responsible team leader**: similar to the project team, a single leader is appointed who is responsible for ensuring alignment between the business needs and ideation across the functional areas.

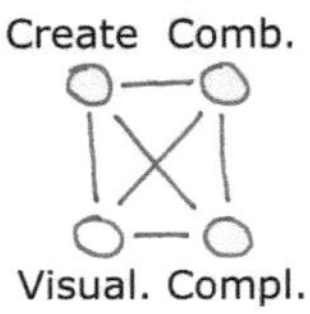

7.8.2 Visualise

One of the key tools for helping bring together the various actors needed in combination is visualisation. In particular visualisation supports actors from different backgrounds developing a shared understanding of a concept. There are three general visualisation rules in this context.

1. They should illustrate as many different concept **perspectives** as possible (Fig. 7.22) in order to foster a broad understanding of the concept on display.
2. They should use different mutually supportive **modes** of communication e.g. text, figures, data. This should make the visualisations both self-explanatory, and accessible to actors from different backgrounds.
3. When alternative concept proposals are presented, they should be visualised in the same style, using the same techniques, and showing the same elements. This means all the concepts should be **detailed** to the same level in order to allow for a more balanced comparison.

It is generally true that well illustrated ideas tend to be more attractive, particularly when presented to management teams who might be unfamiliar with the details of the project. This introduces two risks in visualisation. First, care must be taken to ensure the viewers of a concept illustration understand what they are seeing. For example, it is quite possible to make a visualisation photorealistic, which can lead issues where clients get a false sense of certainty and do not understand that there are many questions left unanswered in the concept, see Figs. 7.35 and 7.37. When we illustrate a concept it does not need to comply with the laws of physics, as such the transition to final product can produce quite radical changes. Second, where concepts are visualised at different levels of detail it is common for viewers to select based on the quality of the visualisation rather than the quality of the idea. This is particularly true of sketches where much is open to the interpretation and imagination of the viewer.

Example:
Conceptualisation of a bicycle helmet.

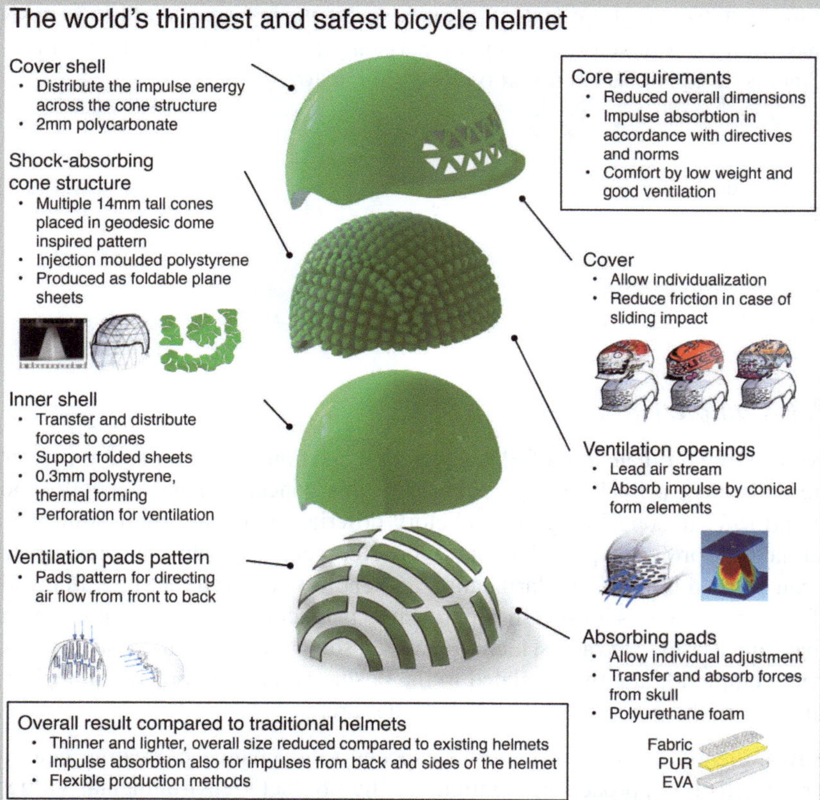

Fig. 7.37 Example visualisation of a bicycle helmet concept proposal

During a student project a new bicycle helmet was proposed. First, the main challenge was identified based on interviews, questionnaires, observations, and workshops. This resulted in the goal to make a more elegant and comfortable helmet suitable for the social bicycling scene in Denmark. Here a key trade-off was the size—closely tied to the perception of elegance—and the protection offered by the helmet. Several concepts were composed: the use of conical elements for impulse absorption, the production of absorption cones on plane sheets to be folded to the tree dimensional form, and customisable foam inserts allowing helmets to be adapted to the user. Finite element analysis was used to assess the protection characteristics of each concept, which were prototyped using vacuum formed absorption cones. The final concept offered the thinnest helmet on the market and superior safety performance. The final stage in this process has been bringing the product to market. Here the concepts were presented as illustrated in Fig. 7.37. This highlights the key elements of the concept and how analysis has been used in selection and comparison with existing designs. Courtesy Anders Sund Nielsen and Jakob Filippson Parslov.

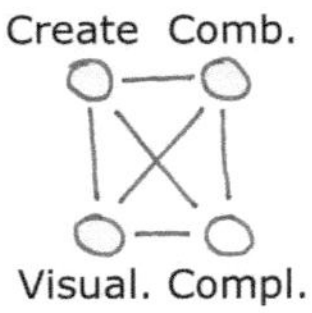

7.8.3 Complete

Obviously the ultimate aim of idea generation and concept combination is to complete and select a final concept. Preferably this concept should be both innovative and safe, unappealingly contradictory criteria. In seeking innovation we can often neglect project aspects leading to development risk; while in seeking safety we can focus too much on clarification and endless research that kills innovation. Unfortunately there is no specific approach to balancing these and it must be based on careful case-by-case assessment. However, some areas that are easily addressed during completion are those related to the business, and not directly related to the realisation of the product. Some general rules are:

- **Sales**: identifying new or otherwise special sales channels; creating new sales or advertising approaches; highlighting the product's environmental virtues in special marketing efforts, design elements etc.; creating a 'history' related to the product or incorporating it into a larger narrative.
- **Users**: focusing on new user or buyer segments; tailoring announcements to specific user segments; empowering the product's graphical or iconic identity.

- **Distribution**: identifying ways to simplify, combine or enhance the distribution; developing new distribution or logistical approaches.

A particularly effective approach, when used well, is the idea of incorporating the product or the brand into a larger narrative. This can be used to drive a nostalgic, romantic imagination of the brand's ethics, quality consciousness or other desirable attributes. Alternatively it can be used to identify the brand as an 'underdog' fighting the system and thus more worthy of support. It is worth noting that if such stories are found to be false or are poorly executed the customer backlash can be significant—use with care!

Example:
Product story. The Danish company Lego was established in 1930 by Ole Kirk Christiansen, who manufactured wooden toys. The name was made from the Danish 'leg godt' meaning play good. Later the local priest told him that the Latin meaning is 'I collect' or 'I construct'. In 1947 wood was substituted by plastic, and in 1958 the connection concept for current Lego pieces was patented. The toy range has gradually developed from basic pieces to functional pieces, thematic toys, and computerized elements. Today Lego figures are related to a wide range of popular media series and movies (Fig. 7.38).
The Lego brand is so strong that no story need be added. The reputation for creative play is supported by research, while customers create webpages or form clubs and other open source networks. However, Lego has still worked hard to develop a narrative around its ethical standards. Here they have focused on maintaining quality and fighting inferior imitations. They have also spent a significant effort in communicating many of their design decisions and product choices to the public in order to support their narrative as an open and ethical company.

Fig. 7.38 **a** Children playing with Lego. **b** A serious use of Lego for conceptualization, *courtesy* Georg Kronborg Christensen

With the completion phase finished the ideation phase of concept synthesis ends, leading directly into evaluation and choice of the produced concepts. In reality there is usually some overlap between these phases as evaluation and ideation are both continuous processes, and there is often a more gradual selection and development of the final concept.

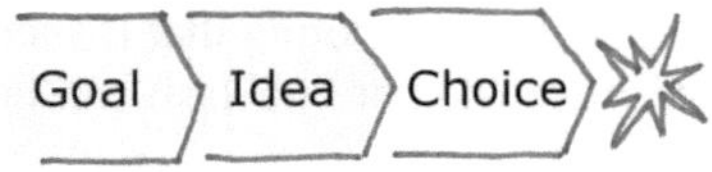

7.9 Concept Synthesis: Evaluation and Choice

The main concern in evaluation and choice is the quality of the decision, which can only be indirectly influenced by methods. The concept decision is the point of no return in terms of both fulfilling the need, intention, and task, and ensuring a good estimation of tractability, probability of business success, and possible negative effects. This relates to the five core dimensions to be satisfied by the goal formulation during decision-making: evaluation, validation, navigation, concern of totality, and making the decision. Evaluation methods tend to focus on how well the goal formulation is satisfied by the actual design. In terms of validation we recommend a broad perspective following five heuristics (Hansen and Andreasen 2004):

- It is the designer's responsibility to consider the full product lifecycle and identify critical issues or effects that would be caused by the concept under consideration.
- Where identified, these critical effects should be described and avoided—if possible—such that the concept has the best possibility for demonstrating fit for life under all conditions.
- The concept's utility should be fully described and enhanced where possible to give the concept the best possible fit for purpose. Note, this does not guarantee a competitive edge.
- Legal considerations should be documented. Legislation and standards etc. often demand systematic testing, which can be significant for some concepts (Alexander and Clarkson 2000).
- Similarly, validation should be correct, complete, and clearly documented in compliance with any legal or systematic requirements (Alexander and Clarkson 2000).

Evaluation and choice are inherently uncertainty activities. In particular we encounter the following paradox: lack of data leads to uncertainty and the desire to further clarify the concept, however, it is not possible to move forward with conceptualisation if every concept must be fully concrete before a decision is possible. We cannot bring all our concepts to market just to see which is the best in the real world! Identifying the balance between our desire for more clarification, and making an early decision is not simple and relies heavily on experience. Thus involving more experienced designers and other relevant actors in evaluation and review is essential. The aim is to strike a balance between the natural optimism of younger staff and the natural caution and pragmatism of more experienced designers.

An example of where this balance can fail can be taken from our experience with a Danish design firm. The designers typically presented very detailed, embodied product proposals to the management team, leading to complaints such as "I know what I get, but not what I could have had". Here, the balance has gone to far towards concretisation with the design team losing sight of the conceptual aspects that would have allowed them to identify alternatives.

In general companies make evaluations by formulating a list of criteria by which to compare alternatives. However, design deviates from these types of methods due to its dynamic progression. We do not stop the design activity to perform externalized, systematic presentation of alternatives, unless the situation is decisive e.g. a review meeting. Typically decisions in the design context grow organically from the iterative refinement of concepts. Further, this evolving characteristic means that different decision dimensions become relevant at different points, requiring a more flexible evaluation approach. This allows designers to make tentative decisions that can be later verified.

Example:
Tentative decisions. Mortensen and Tichem (1994) analysed integration decisions at Bang and Olufsen. Here they found that a key goal for the new product was a clean, 'silver' surface finish (see Fig. 7.39). Three solutions were developed with aluminium machining appearing the most promising. Thus a tentative decision was made to follow this production solution and then later verify it based on experiments. This allowed them to further develop the concept while experiments were carried out. The ultimate result was that the decision was validated and changed from tentative to final. Thus the team was able to start purchasing of the machines, tools, and materials in parallel to the continued development.

Fig. 7.39 CD-drawer on a B&O system, Beosound 9000, courtesy Bang and Olufsen A/S

The importance of tentative decisions is further highlighted by Badke-Schaub and Gehrlicher (2003), who found that of 40 decisions recorded in a project 21 were made in the project group, 13 involved management, and 11 involved external partners. Less than half of these decisions were planned beforehand i.e. there was no agenda and preparation. Thus it is critical that designers consider alternatives. Nutt (1999) found a 14 % increase in project success rate when companies actively worked to consider and incorporate alternatives. Further, formal evaluation methods are seldom used in industry, with only 15 % using them routinely in one study (Janhager et al. 2002). Thus a significant responsibility is with the designer! We recommend the following heuristics.

- Make decisions visible. Make alternatives explicit with detailed visualisations, and record the key criteria, assumptions, and arguments.
- Stay focused on the most important criteria: "Why do we believe that the customer will buy our product?" Identify the criteria that point to this question.
- Visualize the expected product lifecycle (via e.g. the multi board technique, see Chap. 10) to help identify consequences and service possibilities.

7.9.1 Making the Evaluation

Design concepts are ambiguous in their nature and thus difficult to evaluate. Evaluation and choice can be based on guesswork, gut feeling or intuition. However, it is normally better to use a rational method or at least an open procedure (Cross 2008). Here the starting point is always the goal formulation, or part of it. This can then be confronted with the concept.

- Does the proposal satisfy the requirements? If not is it actually a "non-solution"?
- Criteria from the goal formulation can be used to compare alternatives, leading to a ranking.
- One traceable evaluation approach is to link each criterion to a scale representing goodness, and then to sum the scores.
- Key criteria can then be recognised by weighting the scores before they are summed.
- Criteria related to different stakeholders should be treated separately as they cannot be summed i.e. a benefit for one doesn't necessarily balance out problems for another.
- Despite the apparent 'rationality' of using a scoring system it is important to also apply common sense—no system is free of bias!

Many authors recommend using measurable criteria by setting a metric and/or measurement method. This technique can be valuable for judging detailed component alternatives or choosing between components from suppliers. However, value aspects are always subjective as they are related to user perceptions. A metric may lead to a technically good product, but cannot guide us to products with competitive advantage, value, and utility.

7.9.2 Evaluation Criteria

It is important to underline that the criteria for concept evaluation and choice must be found in the goal formulation. This is important because we often see that 'special criteria' are invented for final selection (fitted to the intended winner?), or arbitrary lists of advantages and disadvantages are applied.

Example:
Telephone standard. An architectural competition was launched in 1980 to create a new standard for public telephones. The winner became a sculptural standard in stainless steel (Fig. 7.40). This was widely considered to have won because the proposing architect was relatively famous at the time. Key to the competition requirements was that the standard should be accessible to those with disabilities, including wheelchair users. In this context many claimed that the winning design was in fact a non-solution, causing significant public backlash. The key here is to always respect the goal formulation, inventing new criteria to favour a particular person or concept is unprofessional design practice.

Fig. 7.40 The winning standard for public phones called 'the question mark' from 1980

There are stakeholders related to many criteria. For example, management aim to reduce risk, while the design team try to ensure tractability of the project technically and economically; or the project and its marketing should generate profit, while the user desires utility and good value. How to treat such conflicts? One way is the use of multiple successive levels of evaluation criteria. Here the first level might focus on the products raison d'être e.g. possible, profitable, desirable, and sensible (Blessing 2014). A new product might be key to the survival of the company if it is actually profitable. The dimension 'sensible' is more difficult to describe. From one perspective it is that the product has a raison d'être for both user and producer. From another perspective, the product should be sustainable and respect legislative, ethical, and moral criteria. A set of superior criteria is economy, tractability, and attractiveness, as in Fig. 7.41.

Once these high level criteria have been considered the goal formulation can be used for a more careful evaluation of need, value, and competitive advantage.

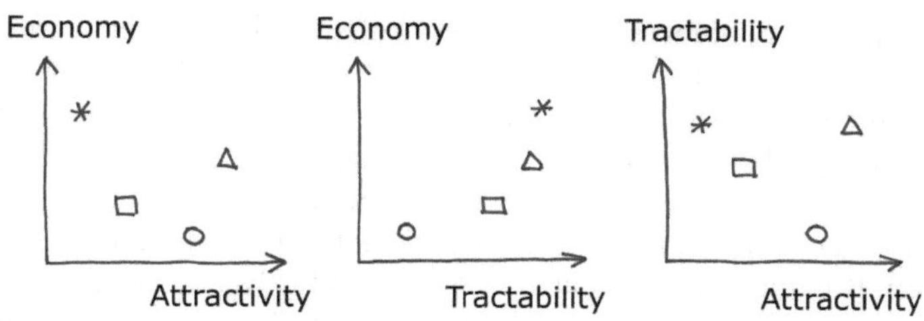

Fig. 7.41 Visualization of a decision with four alternatives and the criteria: **attractiveness** for the customer, **economy** for the company, and **tractability** for the design team

- Remember that a goal formulation can't be fully correct, complete, and unambiguous.
- Check if user preferences have changed since the goal formulation was made.
- Identify the criteria key to competitive advantage and ensure they are considered.
- Aim for impartiality when evaluating design alternatives.
- Aim for the same level of detail describing the same aspects for each proposal.

Evaluation will reveal both strengths and weaknesses in individual concepts. As such, a good strategy is to combine concepts. Remember that the task is to find the *best concept*, not necessarily the perfect or optimal concept.

7.9.3 Organizing Concept Decisions

Arranging formal, methodical decision-making processes will empower the decision-making but are not sufficient in isolation. Reliable decision makers use both logic and intuition (Patton 2003). Concept decisions should be based on input from different functional areas e.g. market, economy, and production. Kihlander et al. (2008) identify five factors influencing concept decisions, which say a lot about the context and culture for decision-making.

1. **Project and goal**: the project specification and goal formulation are the basis for choice but can reflect specific requirements regarding e.g. features that are seen as sales arguments. There may also be cross-organizational links to other projects on re-use, resource sharing, coordinating launches etc.
2. **Procedures and support**: the various organizational levels and functional areas related to a project can paralyze a decision making in meetings via lack of clear leadership or responsibility and unwieldy documentation. Care should be taken to efficiently manage those involved.

3. **Individuals' competences and motives**: technical insight concerning products and competition is important but a products' multidisciplinary nature makes it difficult for specialists to work outside their area. As such, it is important to focus on the goals and market needs. Experience is important but can be based on out-dated insights or be fixated on a particular favourite area. Thus leaders' must take responsibility for decisions and also ensure that procedures are followed. It is important to note that if a company has bonus arrangements, or other incentives, these can easily twist concept decisions towards these systems rather than the real goals or needs.
4. **Teamwork and company culture**: unwritten rules, forces in the informal organization, and power struggles regarding whose ideas and areas should be supported often come to the fore in concept choice situations. Thus procedures and methods can be used to try and defuse political decision-making by focusing on the goal formulation.
5. **External aspects**: a company's robustness relies on its ability to operate over a long period, and maintain its resources in the face of changing market and economic conditions. Systematic methods are key to ensuring these long-term factors are considered in concept evaluation.

7.10 The Module 'Concept Synthesis'

We distinguish Concept Synthesis as a separate module because it can be seen as an independent activity performed by professional 'concept makers' or 'idea makers'. This feeds into Product Synthesis, Product Development, and ultimately Product Planning, as in Fig. 5.9, and illustrated in more detail here (Fig. 7.42).

Together with the Exploration module concept synthesis leads to a definitive concept. Thus the concept synthesis module is a significant decision milestone in both Product Synthesis and Product Development.

The six dimensions shown in Table 7.2 illustrate the forward-looking aspect of conceptualization (Belliveau et al. 2002). These all deal with future situations—ensuring product tractability, risk reduction, and positive economic outcome. These provide a checklist for things to be considered in a concept proposal. They are critical to success or failure but it is a matter of judgement how they are related or should be weighted in a given situation. This underlines the fact that a concept is much more than a sketch or rendering, it is a comprehensive clarification and documentation of the Exploration and Concept Synthesis activities.

7.10.1 The Milestone Meeting

Dym and Little (2000) highlight the many stakeholders associated with concept choice. This might be because of special knowledge, because they will be influenced by the consequences of the choice, or because they have some

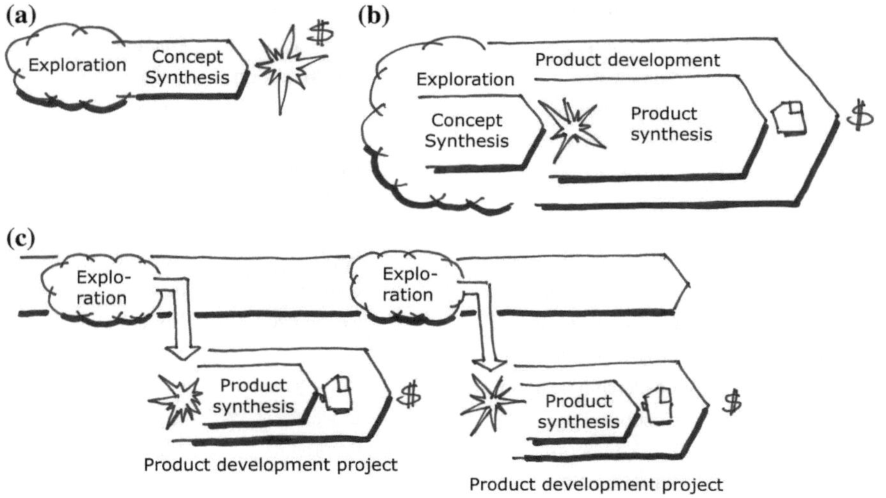

Fig. 7.42 Concept synthesis as a module in **a** conceptualization **b** product planning **c** product development

responsibilities related to the decision. Thus each part brings different levels of interest and input. Empirical research on milestone behaviour shows significant deviations from what is seen as good decision-making. Christensen and Varnes (2006) report that few decisions taken at milestone meetings build on rational methods, instead these meetings simply become a 'rubber stamp' for decisions which have already been made, or a forum for political argument. Thus there is no general practice for a milestone meeting.

This 'rubber stamp' approach is actually in line with recommendations from Andreasen and Hein (1987): management should not evaluate and choose, but sanction the proposed choice. Here, the management's task is to judge the quality of the analysis and synthesis with respect to strategy, resources, competences, risk, and business creation (Jensen and Andreasen 2010). Together these make milestone activities an ambiguous, multifaceted phenomenon.

7.10.2 Reflecting on Concept Synthesis

It is critical that projects and activities like Concept Synthesis be finalized by reflection. We must learn from our own experiences if we are to improve! Dym and Little (2000) write: "*The second kick of a horse has no real educational value*". Reflections lead to better practice based on:

Table 7.2 Example of evaluation criteria for choice of concepts (Belliveau et al. 2002)

Factors	Specific issues	Attractive	Unattractive
Market	Market size	>$100 million	<$10 million
	Market growth	>20 %	<5 %
	Market drivers	Satisfy all	Meets at least one
	Market access	Existing business	Needed
	Potential market share	>20 %	<5 %
Competency	Business infrastructure	In place	Needed
	Customer familiarity	Current base	Few
	Core competency	Recognized	None
Competitive issues	Proprietary position	Yes	No
	Leadership position	#1 by year 5	No lead
	Cost position	Lowest	Highest
	Key competitive advantage	Proprietary	None
	Sustainability of position	High	Low
Time factors	Time to sales	<2 years	>5 years
	Full commercialization	<5 years	>5 years
	Competitive time advantage	>2 years	<1 year
	Operating at break-even	<3 years	>5 years
Technology	Technology availability	In place	Needed
	Technology readiness	Proven	Discovery still needed
	Technology skill base (people and time)	Available	Needed
Financial	After-tax operating income	>12 %	<8 %
	Maximum cash hole	<$20 million	>$50 million
	Revenue stream	>1 product line	1 product
	Business potential	>$100 million	<$20 million

1. **Review of the project's goal**: was the goal formulation a good interpretation of the situation and our abilities or were there substantial adjustments? Did we actually end up with a different project and product from what we expected?
2. **Review of the process**: was the project goal oriented or chaotic? Would the same process make sense for the next project? Did the process fit the task? Did we waste time on methods we did not understand or properly integrate? Were there too many blind roads?
3. **Review of plan and budget**: did our planning help? What could be improved for next time? We authors have seen companies with 17 years of experiences not being able to keep plans.
4. **Review of result creation**: effectiveness and efficiency are two strong criteria for evaluating the results. Were the results unique and influential? Were our efforts efficient?

The aim of reflection is not to point to scapegoats, but help to make steps in improving support and community of practice i.e. the team's way of working. Other mechanisms like peer review, mentoring, and coaching give individual feedback, but this is secondary to the team's performance.

7.11 Conclusion

The chapter Concept Synthesis deals with the unique challenge of creating a concept; in its simplest form: finding something yet unknown. In Topic A we discussed theories, strategies, and models of conceptualization before Topic B introduced our three step Concept Synthesis model, which emphasizes the methodical aspects of conceptualization. Finally, we sought to formalise our discussions of evaluation and choice with respect to the concept: it must answer the need, intention, and task; and provide a convincing starting point for the concretization, production, and launch of the product. The importance of formalized evaluation is also highlighted by our reflections on the problems of company politics and conflict.

The following three chapters describe the continuation of the design process and its dependency on Exploration and Concept Synthesis. The Russian Doll metaphor tells us that Product Synthesis, Product Development, and Product Life Synthesis all build on these early activities.

References

Ahmed S, Wallace KM (2004) Identifying and supporting the knowledge needs of novice designers within the aerospace industry. J Eng Des 15(5):475–492. http://www.tandfonline.com

Alexander KL, Clarkson PJ (2000) Good design practice for medical devices and equipment, part I: a review of current literature. J Med Eng Technol 24(1):5–13

Almefeldt L, Andersson F, Nielson P, Malmquist J (2003) Exploring requirements management in the automotive industry. In: Folkeson A et al (eds) Research for practice proceedings of 13th international conference on engineering design, Design Society, Stockholm

Andreasen MM (1984) Systematiske idemetoder (Systematic idea methods). Institute of Engineering design, Technical University of Denmark, Copenhagen

Andreasen MM, Hein L (1987) Integrated product development, IFS (Publications)/Springer, Berlin. Facsimile edition (2000) Institute of Product Development, Technical University of Denmark, Copenhagen

Andreasen MM, Wallace K (2010) Private communications

Arthur WB (2009) The nature of technology—what it is and how it evolves. Penguin Press, London

Badke-Schaub P, Gerlicher A (2003) Patterns of decisions in design: leaps, loops, cycles, sequences and Meta processes. In: Proceedings of ICED'03, Stockholm, Sweden

Baxter M (1995) Product design—practical methods for the systematic development of new products. Chapmann & Hall, London

Belliveau P, Griffin A, Somermeyer S (eds) (2002) The PDMA toolbook for new product development. Wiley, New York

Blessing L (2014) Teaching material from SSEDR Summer School on Engineering Design Research (unpublished)
van Boeijen A, Daalhuisen J, Zijlstra J, van der Schoor R (2013) Delft design guide. BIS Publishers, Amsterdam
Bucciarelli LL (1984) Reflective practice in engineering design. Desi Stud 5(Is 3):185–190
Buur J (1991) Apparatkonstruktion (Mechatronic design). Institute for Engineering Design and Control, Technical University of Denmark
Christiansen JK, Varnes CJ (2006) From models to practice: decision making at portfolio meetings. Int J Qual Reliab Manage 25 (1):87–102
Cooper RG (1993) Winning at new products—accelerating the process from idea to launch, 2nd edn. Addison-Wesley Publishing Company, Boston
Cross N (2008) Engineering design methods. Wiley, Chichester
Dorst K, Hansen CT (2011) Modelling Paradoxes in novice and expert design. In: Proceedings of ICED'11, Copenhagen
Dym CL, Little P (2000) Engineering design—a project based Introduction. Wiley, New York
Foxley DM, Blessing L (2000) Closing report—partnership for profitable product improvement. The Royal Academy of Engineering, London
Gleiniger A (1998) The Chair No. 14 by Michael Thonet. Verlag form, Frankfurt am Main
Grabowsky H, Geiger K (eds) (1997) Neue Wege zur Produktentwicklung (New directions for product development). Raabe, Berlin
Hansen CT, Andreasen MM (2004) A mapping of design decision making. In: Marjanovic D (ed) Proceedings of DESIGN 2004. TU Zagreb, Dubrovnik
Hansen CT, Andreasen MM (2005) On the content of an product idea. In: Samuel A et al (eds) Proceedings of 15th international conference on engineering design ICED 05, Design Society, pp 1–14
Hansen CT, Andreasen MM (2007) Specifications in early conceptual design work. In: Proceedings of 16th international conference ICED'07, Design Society
Hansen CT, Andreasen MM (2008) On the content of design problems. In: Proceedings of NordDesign, Tallinn University of Technology, Estonia, pp 110–119
Hansen CT, Dorst K, Andreasen MM (2009) Problem formulation as a discursive design activity. In: Proceedings of ICED'09, Stanford, USA
Hatchuel A, Weil B (2003) A new approach of innovative design: an introduction to C-K theory. In: Proceedings of ICED'08, Stockholm, Sweden, pp 109–124
Hede Markussen T (1995) Et teoribaseret handlingsgrundlag for betjeningsdesign (A theory based action plan for interaction design). PhD-thesis, Technical University of Denmark
Henderson K (1999) On line and paper. Visual representation, visual culture, and computer graphics in design engineering. The MIT press, Cambridge
Hollins B, Pugh S (1990) Successful product design—what to do and when. Butterworth-Heinemann, London
Hubka V, Eder EW (1976) Theorie der Konstruktionsprozesse (Theory of the design process). Springer, Berlin
Janhager J, Persson S, and Warell A (2002) Survey on product development methods, design competencies, and communication in Swedish Industry. In: Proceedings of the 4th international symposium on tools and methods of competitive engineering, Wuhan
Jensen TE, Andreasen MM (2010) Design methods in practice—beyond the 'systematic approach' of Pahl and Beitz. In: Marjanovitc D et al (eds) DS 60 proceedings OF DESIGN 2010, Dubrovnik, Croatia, pp 21–28
Jul Nielsen H (1990) Systematisk produktsøgning (Systematic search for new products). PhD-thesis, Technical University of Denmark
Kihlander I, Janhager J, Ritzén S (2008) Dependencies in concept decisions in complex product development. In: Proceedings of DESIGN 2008, Dubrovnik, Croatia, pp 1159–1166
Kirkegård L (1988) Fastlæggelse af udviklingsopgaven (Determination of the product development task). Institute of Product Development, Technical University of Denmark, Copenhagen

Kolb DA (1984) Experiential learning; experience as the source of learning and development. Prentice-hall, Englewood Cliffs

Koller R (1979) Konstruktionsmethode für Maschinen-, Geräte- und Apparatebau (Design method for machine, device and apparatus construction). Springer, Berlin

Kruger C, Cross N (2006) Solution driven versus problem driven design: strategies and outcomes. Des Stud 27(5):527–548

Lindemann U (2007) Methodische Entwicklung technischer Produkte (Methodical development of technical products), 2nd edn. Springer, Berlin

McAloone TC (2000) Industrial application of environmental conscious design. Professional Engineering Publishing, London, UK

McKim RH (1972) Experiences in visual thinking. PWS Publishers, Boston

Mortensen NH, Tichem M (1994) Observations about decision making in design—a case study. In: Proceedings of WDK workshop evaluation and decision making in design (EVAD), Technical University of Denmark, pp 1–6

Nutt PC (1999) Surprising but true: half of decisions in organizations fail. Acad Manage Prospect 13(4):75–90

Patton JR (2003) Intuition in decisions. Manage Decis 41(Is 10):989–996

Pugh S (1991) Total design—integrated methods for successful product engineering. Addison Wesley, Wokingham

Roozenburg NFM, Eekels J (1996) Product design: fundamentals and methods. Wiley, Chichester

Roth K (1994) Konstruiren mit Konstruktionskatalogen (Designing with design catalogues). Springer, Berlin

Shapiro HI (1980) Cranes and derricks. McGraw-Hill, New York

Tan AR, Andreasen MM, Matzen D (2008) Conceptualization of Product/Service-Systems through structural characteristics. In: Marjanovic D (ed) Proceedings of international design conference DESIGN 2008, Dubrovnik, Croatia

Tjalve E (1979) A short course in industrial design. Newnes-Butterworths, London. Facsimile edition (2003) Systematic design of industrial products. Institute of Product Development, Technical University of Denmark, Copenhagen

Tjalve E, Andreasen MM (1975) Zeichnen als Konstruktionswerkzeug (Drawing as design tool). In: Konstruktion nr. 2, pp 41–47

Ulrich KT, Eppinger SD (2004) Product design and development, 3rd edn. McGraw-Hill/Irwin, New York

Ward A, Liker JK, Christiano JJ, Sobek DK (1995) The second Toyota paradox: how delaying decisions can make better cars faster. Sloan Manage Rev (Spring) 1995:43–61

Wheelwright S, Clark K (1992) Revolutionizing product development: quantum leaps in speed, efficiency and quality. The Free Press, New York

Zwicky F (1948) The morphological method of analysis and construction. Courant Anniv Vol. Wiley Interscience, New York

Chapter 8
Product Synthesis

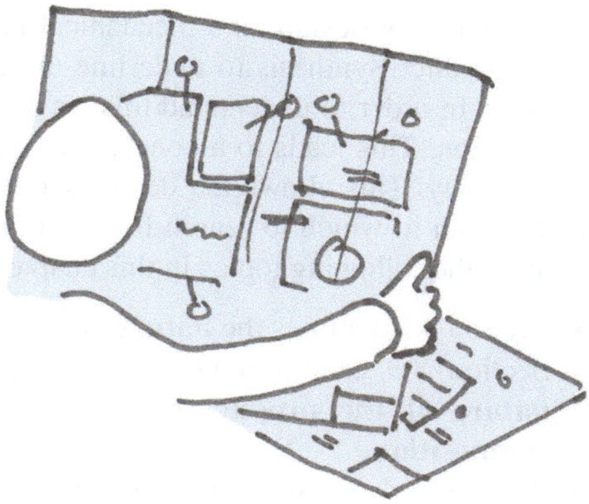

The Product Synthesis activity, the third module in our Encapsulation Design Model, leads to the definition of how the product will be materialized, i.e. the composition of the design and its parts. The starting point is a selected concept, and the results are a full specification of the product. Parallel to this product synthesis also justifies the realization. The product related activities and the functions of product and activities need special care, with a focus on the search for proper solutions. Materialization means the definition of producible parts and their assembly, normally called embodiment design.

The core concept in this chapter is Domain Theory, which explains three interacting domains: activity, organ, and part. This theory serves as a backbone for explaining products' nature, the nature of their use, and the synthesis of these. This chapter explores how conceptualization is found in the details and how the concept synthesis actually leads into Product Synthesis activity.

8.1 Understanding Products' Nature and Synthesis

We use the sequence *Task/Concept/Design/Business/Use and Life* as the backbone for our model. The element **design** is the topic of this chapter. Following traditional terminology, however, we will talk about product and product synthesis, even though the real output is a specification of the product. Actual products only appear in the Product Development activity as a result of manufacture and sales,

M.M. Andreasen et al., *Conceptual Design*, DOI 10.1007/978-3-319-19839-2_8

see Chap. 9. Before we come to that, we must ask what is a design and what does Product Synthesis mean?

Product Synthesis is core to all types of design practice including, architectural design, food design, textile design, software design, and so on. Although we focus on engineering design, and mechanical products in particular, we use the general label Product Synthesis to underline the generic nature of the discussions in this chapter. In order to understand this we use Domain Theory (Andreasen 1980) as a foundation. This leads to a concept for 'spelling' the structure of the product and use activity. This allows the designer (and the reader) to decompose the degrees of freedom in which to search for solutions and compose the design. As such, we address the following topics in this chapter:

- Section 8.2 explores the **nature of Product Synthesis**.
- Section 8.3 examines how Product Synthesis merges knowledge about the **nature of the product** and about how people reason about a **product's composition**.
- Sections 8.4–8.7 show how Domain Theory can be used to **'spell' the product and use activity** in three different **domains: activity, organ, and part**.
- Section 8.8 looks at how Product Synthesis is achieved by **progressing through these domains**.
- Section 8.9 examines how this gradual concretization and definition of the design can be captured in a **final product model** using Domain Theory.

8.2 The Nature of Product Synthesis

Product Synthesis is traditionally called engineering design (German: *Konstruieren*) but offers substantial insight for all areas of design application as noted above. Although many of the examples used here focus on products our treatment of Product Synthesis is such that translation into their other areas is feasible for the reader.

Historically, design was separated from production and sales resulting in almost all design starting with a goal statement (from management) and ending with a set of drawings (handed over to production). This is similar to our framing of Product Synthesis. However, as explained in our Encapsulation Design Model, Product Synthesis flows from Exploration and Concept Synthesis, and is part of the wider Product Development organization including marketing, sales, distribution, and service. Creating a concept can be seen as the core of new product development but Product Synthesis is the unavoidable craftsmanship needed to realize these conceptual dreams.

Product Synthesis is generally complex due to the numerous layers of sub-functions and the various interactions needing to be solved in the embodiment of the product. This leads to multiple part design and interface clarifications. Figure 8.1 illustrates the complexity of even a seemingly simple product, a bicycle. At the

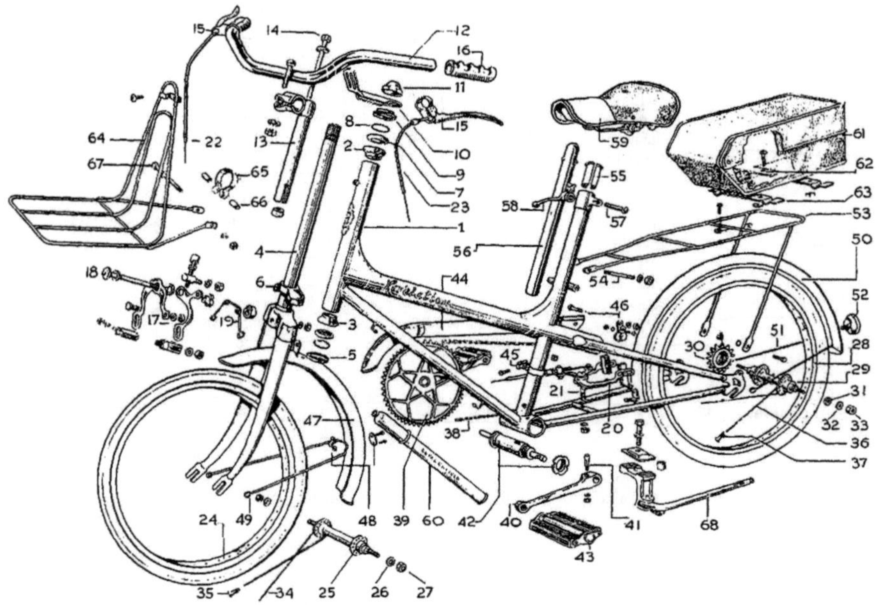

Fig. 8.1 The result of product synthesis is a complete determination of the design's parts, here illustrated as an exploded view of a 'Royal Enfield Revelation' from the 1960s

end of Product Synthesis all these loose ends will be tied up with an unambiguous definition of the product's composition, ready for production.

Even when using powerful CAD and product data management (PDM) tools the complexity of product synthesis results in a number of commonly experienced problems. First, **rework**, corrections, and changes often result from failures in cross-disciplinary communication or vague goal statements. These continuous changes result in significant difficulty in **managing change** propagation. For example incrementally new products aim for 10–20 % changes from the previous version, but often end up with 70 % or more. In this changing context remaining focused is a difficult proposition and results in the loss of the **designers' intent,** i.e. the traceability of experiences and decisions. Here rework and incremental design are almost equivalent to start from scratch. Finally, these issues combine to reveal the **limitations of CAD**. In this context CAD's geometry focus means it is of little help when reasoning about, e.g. functions, interactions, and activities. As such, product synthesis must be addressed through design tools.

Our focus can be illustrated using Cross (2008) model which highlights the layers of synthesis shown in Fig. 8.2. The internal layer defines eight stages in the design process, which are similar to concept synthesis. The core traits are goal formulation, search for solutions, and selection based on criteria. The outer layer is the symmetrical problem/solution model that concerns the composition of the solution. We can substitute 'problem' with 'function' in Cross' model without

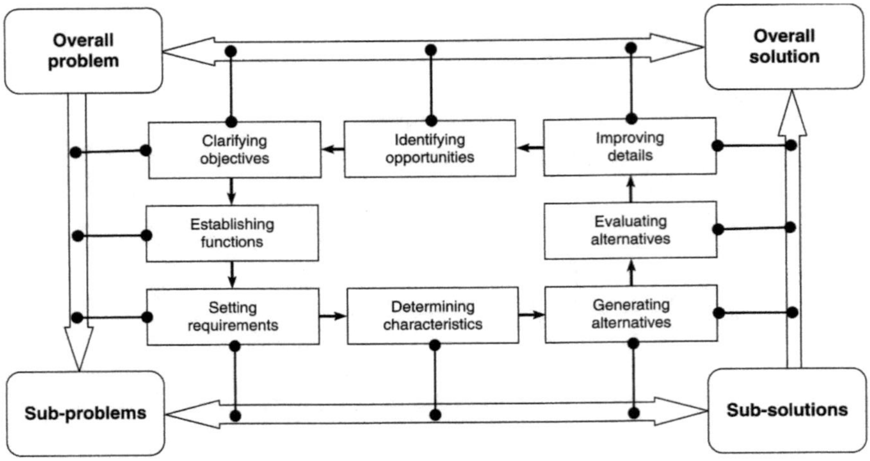

Fig. 8.2 The eight design process stages in the problem/solution model (Cross 2008)

changing its validity. As such, we focus on this outer layer, using sub-functions and their solutions as the key elements to be addressed when creating an integrated product.

Product synthesis is a design activity determined both from the nature of the artifact being designed and the designer's cognitive abilities. We focus on the artefact dimension in what we call 'spelling the product'.

8.3 Differences and Identities in Synthesis

Until now, we have discussed conceptualization generally, i.e. independent of discipline or the product's nature. However, here we narrow the scope to physical products, which are realized in industrial settings. This is a very broad field where each product area has its own synthesis theory, or at least practice. For example mechanical engineering works with machine elements and parts, electronics with circuits and components, food with recipes and ingredients, and so on. Each has their own language, way of reasoning, and rules.

The lack of a bridging, general synthesis theory is seen as a weakness of design methodology and is visible in the problems of multidisciplinary design. How far can we go in articulating a common ground? Which core aspects are common across products? There are commonalities related to the activities in which a product is used, the functions it realizes, and its material manifestation. This brings us to Domain Theory, which uses function as a bridging concept in order to help develop a shared understanding of product realization, relevant across product areas. It corresponds with our view that good design support is obtained by creating a common articulation of the design entities, i.e. being able to create, model, and talk about the design in a cohesive way—what we call the 'spelling'.

Our advice is to focus on approaches that primarily support the overview and progression of the design, and secondarily management and coordination. How specific designs in the shoe industry, in health care, in the food industry, etc. should be articulated requires special insight and is left to the reader. What we supply is the basic pattern of reasoning.

8.3.1 A General Foundation: Systems Theory

Before we explore Domain Theory, we first need to briefly discuss Systems Theory. This is an approach to the analysis and synthesis of complex artefacts, preferably composed of discrete, well-defined elements. Its development started in the 1950s to help deal with complex dynamic systems, such as strategic defence or airport control. Figure 8.3 illustrates the core elements of Systems Theory. Here a system is modelled as an object with respect to its composite **elements** and their **relations**. Together these form a **structure** with certain **behaviours**. The definition of elements and relations, and level of resolution is left up to the observer. The basic rules for system modelling are that elements should be of the same kind (e.g. not mixing devices and activities), and that relations should also be consistent (e.g. not mixing flow and time relations).

In daily language system has a broad meaning but in Systems Engineering (Blanchard and Fabrychy 1998; Ashby 1968; Hubka 1973) a system describes a model of a given object and its elements as outlined above. As such, Systems Theory forms a foundation that is used in many design theories (Pahl and Beitz 2007; Gero 1990; Suh 1990; Cross 2008) where a product's components are seen as elements. However, one thing Systems Theory does not describe is how to proceed in synthesizing a design, instead it offers analytical insight. In order to

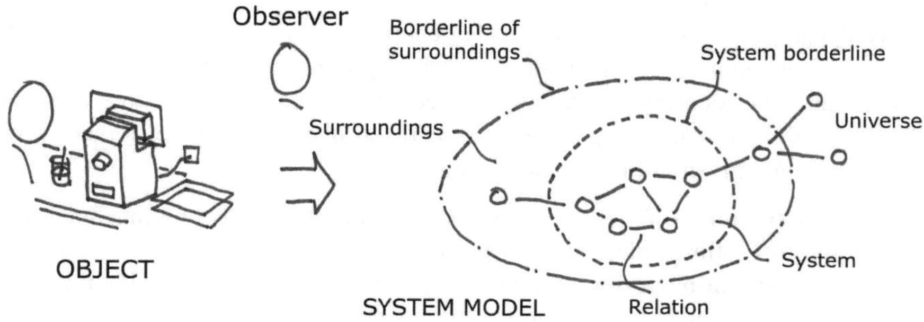

Fig. 8.3 The basic concepts of systems theory

address this synthesis aspect we look to Domain Theory, which utilizes Systems Theory in order to articulate its core concepts. We define a system as follows:

Definition: A **system** is a model of an object (a real or conceived product or activity) based on a certain viewpoint, which defines the elements of the system and their relations. A system carries **structure,** i.e. the elements and their relations (arrangement, architecture) and **behaviour,** i.e. the system's response to a stimulus depending on stimuli, structure, and state.

The game rules for system modelling are: elements should be of the same kind (not mixing, e.g. devices and activities), and the same with relations (not mixing, e.g. flow, 'kind of', and time relations). The power of Systems Engineering is that the basic features can be combined with technical theories and models (control theory, physics modelling, flow theories, etc.) to create more powerful models. Its strength is in analytical applications, not in synthesis. In the following, we show how Domain Theory utilizes Systems Engineering to articulate its basic concepts and add the nature of artefacts to Systems Theory.

8.4 Domain Theory and the Three Ways of Spelling

Domain Theory is a development from an earlier theory originally formulated by Hubka in 1973 (Hubka and Eder 1988). In 1980 Andreasen transformed Hubka's theory into a "model based design theory"—Domain Theory. This provides a comprehensive set of concepts that allow the designer to model and synthesize products. These concepts are similar to those originally proposed by Hubka but have been developed to better support educational and practical application (Andreasen 1980). Domain Theory uses three domains (activities, organs, and parts) to cohesively articulate the nature of activities and products, providing a synthesis approach. It is a theory teaching us to "spell a product in different ways" using the different domains, illustrated in Fig. 8.4.

By departing from Systems Theory's general view where 'everything' can be seen as an element, Domain Theory prescribes three views: one focusing on the activity, and two focusing on the product (technical system). We propose a vocabulary of different 'languages' associated with each of these three domains. We illustrate each with an example of a drilling machine.

- The **activity domain** focuses on how the product is used (drilling machine: how it is used for drilling holes), including the product's lifecycle.
- The **organ domain** focuses on what the product is able to do, i.e. how the functional elements interact to create the effects necessary for the activity (drilling machine: motor, transmission, and clutch transmit force, momentum, and rotation to the drill).

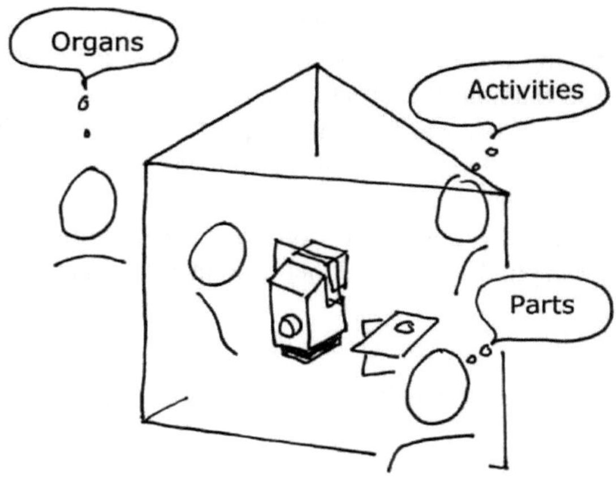

Fig. 8.4 Domain theory's three perspectives

- The **part domain** focuses on the products' parts and their interfaces. This is realized in manufacturing and brought together in the assembly process (drilling machine: the housing of the drill carries the components and creates a hand grip).

How are these domains related? Basically, each is an independent explanation of the product: **how it is used, how it functions, and how it is built.** This is illustrated in Fig. 8.5 with respect to a classic Moka Pot. Here we can explain the pot via each domain: *activities,* e.g. to prepare the Moka Pot and put it on the stove;

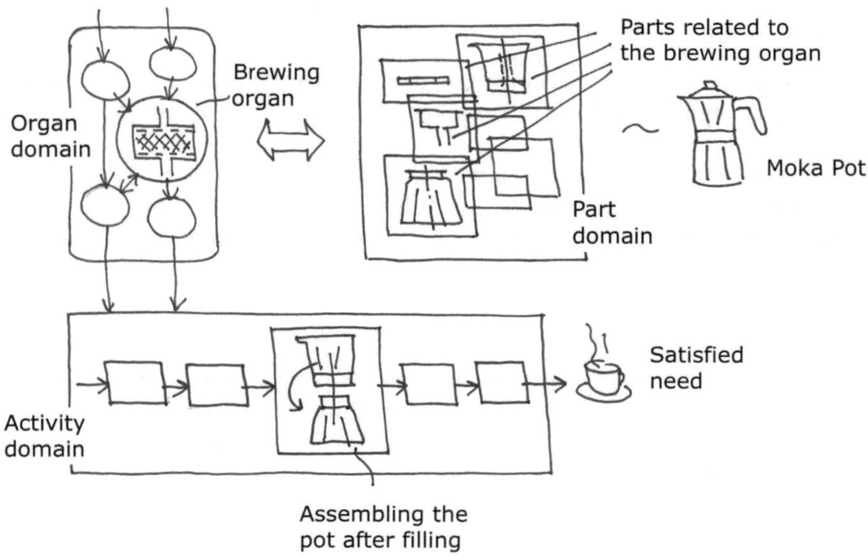

Fig. 8.5 Symbolic illustration of the relationships between domains

organs, e.g. the brewing organ, and *parts,* e.g. the threaded connector linking the boiling and serving pots. We can now consider the links between the domains. For example the organ structure (the product) should deliver the effects necessary for the activity, and the part structure should realize the organs as a viable product. These relationships underpin Domain Theory and its role in Product Synthesis.

In the following sections, we will examine each domain, their contribution to conceptualization, and their role in synthesis.

8.5 The Activity Domain: How the Product Is Used

The activity domain describes how the designer imagines or observes a product being used in practice. Our main interests are what the product can be used for and what the product can do. This close relationship between product and use activity is articulated in the definition:

> **Definition: Product** is any kind of materialized and executable artefact, i.e. able to carry behaviour and properties in order to realize functions and be deployed in a **use activity**.

The link between product and use activity can be decomposed with respect to a number of operators. These were originally proposed by Hubka and Eder (1988) but have been adapted here for clarity. The four main operators are outlined here and illustrated in Fig. 8.6.

- **Products** are tools or machines and deliver effects in the form of material, energy, information or biological entities interacting with the operands (the subject of the operation).
- **Humans** that operator directly or indirectly through the product.
- **Methods** are the work patterns, knowledge, information, and data necessary to control the product or humans' action. These can be seen as a procedure that directs the activity.
- **Active surroundings** are the environmental influences (gravity, water, wind, etc.), as well as given systems, e.g. energy supply, waste removal, cooling air.

All activities are based on natural phenomenon, e.g. the cutting of paper with scissors or brewing beer based upon yeast culture. Utilization of natural phenomena is also found within the products themselves and is core to their operation, e.g. electromagnetism in a drilling machine's motor. In the following example, we explore how these elements are combined in a real product.

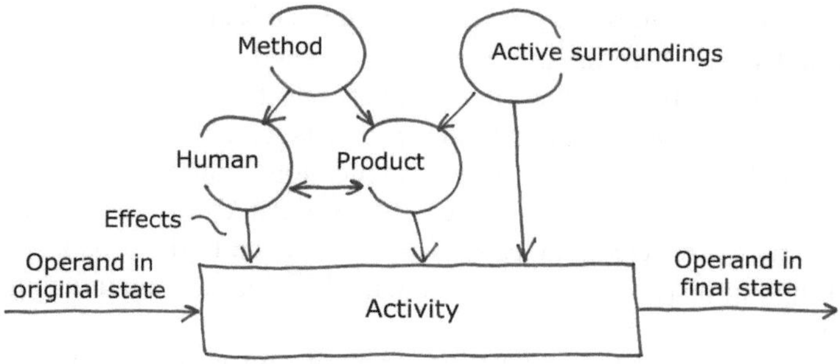

Fig. 8.6 The fundamental interaction between product and activity explained by operators and operands

Example:
A grill as a technical system. When a grill is used to prepare food the grill can be considered as a technical system where the operands are the food. The operator controls the process, as well as preparing the grill (positioning, lighting the charcoal) and closing it after use (disposal of ashes, cleaning, storing). The charcoal provides the energy input, while the grill transmits heat and supports the food. Further, environmental preconditions for grilling are available, oxygen, gravity, and air movement. Figure 8.7 shows a model of the activities related to preparing and finishing the overall grill activity including the food preparation.

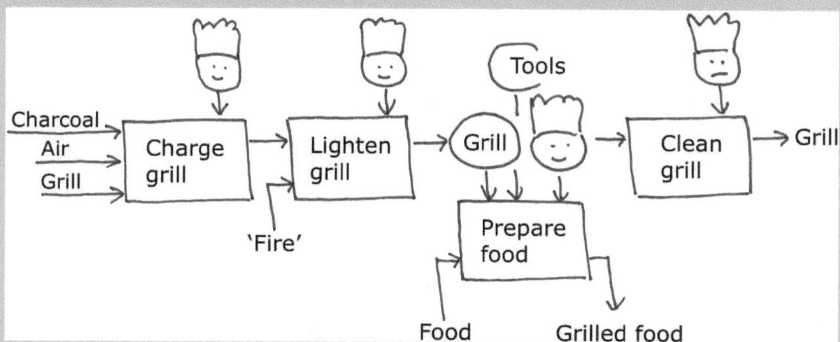

Fig. 8.7 Activity model of grilling. The operators 'method' and 'active surroundings' are not shown

The grill example highlights the two main patterns of activity. The first is the core **operation activity**, where the product is the operand and the food is grilled. The second is the **use activity** where the product, i.e. the grill itself, is the operand being influenced by the human and other products, e.g. cleaning devices. The use activity normally contains many single applications that together constitute the product's service period. As such, this can be formulated with respect to the lifecycle: The **product lifecycle** includes all activities related to the product from manufacture through to disposal. These are explained in Fig. 8.8, again using the Moka Pot example.

In an activity there are four different types of **operands** that are transformed: **material,** e.g. liquids or gases, **energy,** e.g. electricity or inertia, **information,** e.g. control data or speech, and **biological objects,** e.g. active bacteria or humans. Output from the activity is described by the operands in their final state, see Fig. 8.4, and constitute the outcome of the products use. For example a final state operand of the grill is grilled food, or brewed coffee in the Moka Pot. The value and utility of the product is connected to these outputs. The satisfaction of needs is established when the product is applied, not when it is sold.

The operation activity is related to a product's use. The product, together with the user, leads to the result of the operation activity, which (hopefully) satisfies the initially unsatisfied need.

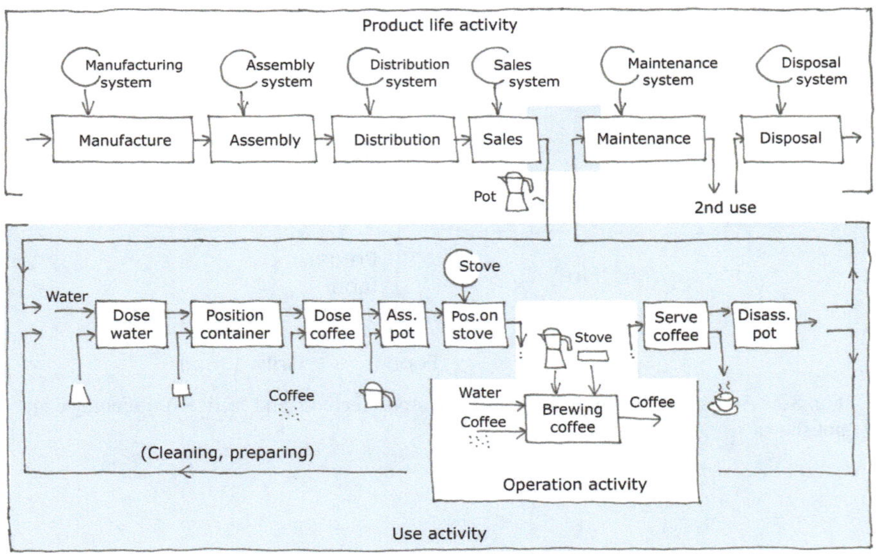

Fig. 8.8 A product lifecycle for a Moka Pot

8.5.1 The Activity System

The activities related to a product can be seen as a **system of activities**. The elements of this system are single activities and the relations are the operands. In the examples above we find simple linear, material-oriented systems. These are derived from a product's **mode of action** based on its key principles. In the same way an activity's underlying principles create its **mode of operation**. A drill's mode of action is the transformation of electrical energy into rotation of the clutch; its mode of operation is drilling via the rotating bore, which creates the desired holes. Here, technology describes the physical way in which this is achieved.

There exist an enormous number of technologies like harvesting-, food-, telecommunication- and manufacturing-, all characterized by a great number of machines, products, and tools. A technology is the sum or interaction between a product, the activity, and its result. For example one role of riveting technology is the use of riveting machines to insert rivets when assembling metal plates in the aircraft industry. However, the designer may see the connecting rivets themselves as a technology. Figure 8.9 shows some simple examples of technologies described in our terminology.

8.5.2 Designing an Activity

Here we can look back on a classic question: which came first, the chicken or the egg or perhaps in our context, cycling or the bicycle? This leads us to the idea of dual synthesis, which in some situations is obvious: a can opener design starts with the activity of use. However, a chair designer rarely starts by asking, what is sitting? Fundamentally, when we aim to produce something that is a radically new way of satisfying a recognized need we must first understand what ultimately satisfies that need. In doing this we must answer the following questions:

- What shall be transformed into what final state? What is the operand and what should its final state be, e.g. opening a can.

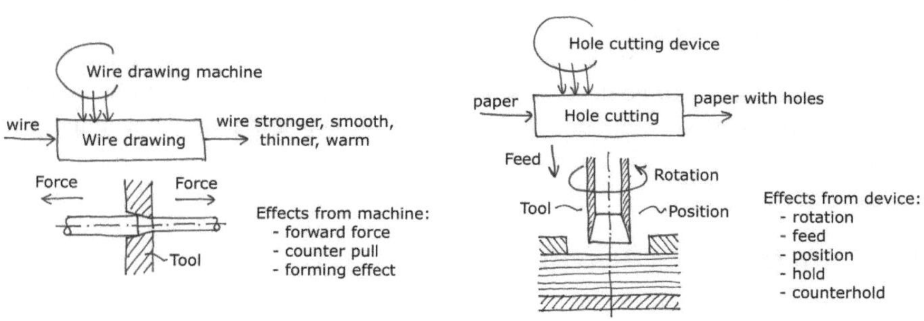

Fig. 8.9 The mode of operation for wire drawing and hole cutting activities

- What natural phenomenon, principle or mode of operation can be used to create this transformation? For example a traditional can opener uses a cutting knife that proceeds along the can's rim.
- What environmental conditions are necessary for the proposed mode of operation? For example an electrical power supply.

These raise the more philosophical question: can an activity be designed independently from the product? Typically, the relationship between a product and its associated activity is fundamental. For example the bicycle and cycling are innately linked, cycling alone is hard to imagine as an independent invention. As such, it typically requires more than one idea or design in order to invent a new product: one concerning its use activity and one concerning its functionality. For example the Segway builds on a novel use activity in its steering, coupled with a new application of existing technologies in order to provide its motor and stability functions.

A general approach to finding new solutions is to systematically modify the characteristics of known solutions, leading to the following heuristics:

New concepts related to activity may be found by varying how the technical activity is characterized:

- The mode of use or mode of life.
- The nature of the operand or alternative operands.
- The sequence of activities.
- The allocation of tasks between the user and the product.
- The type of effects from the operators.

New designs often build on known technologies, frequently with the existence of production technologies as a precondition (Arthur 2009). As such, one strategy when searching for new ideas is to focus on understanding the underlying phenomena and subsequently find means to solve, or bypass, inherent technical problems. For example consider the problem of generating continuous suction when vacuuming. The apparently inherent problem of the waste bag becoming clogged was bypassed by using bagless technology. An alternative strategy is to search for novel applications or use activities for existing or emerging technologies. For example laser cutters combine well-established technologies to give a new output. In either case ideas can be developed that idea require experiments to be clarified, and both strategies can yield concrete current solutions and more conceptual principals to be incorporated in totally new designs.

8.5.3 Models in the Activity Domain

The synthesis of product and use activity is often overlooked in the creation of new products simply because an underlying assumption is that the use shall automatically be derived from the way a product works. This not only results in the potential poor performance of the design but also reduces the possibilities for identifying original ideas in the use dimension. As such, we recommend explicitly modelling the use activity as a support for the creative process, focusing on the following criteria:

- The temporal and/or physical **sequence of activities** leading to the transformation of the operands.
- The **state change** of the operand. In particular, what the input operands should be and what steps are required in their state changes.
- The **mode of use** or natural phenomenon to be utilized in the transformation and its arrangement in space and time.
- The **allocation** of tasks between the product(s) and the operator(s).

There is a rich palette of modelling types able to support the designer in this synthesis activity ranging from less formal sketches, computer animations or verbal presentation to more formal models, such as use cases and Function Analysis Diagrams. Here, pictures are easily understood, informal text is instructive, and formal models may be sharp but can be difficult to relate to reality. Together with spelling the activity these serve a key role in creating activity concepts.

8.5.4 Spelling an Activity

When creating a new concept there are several elements in the activity domain that demand clarification. This provides a core for the concept and can provide inspiration for its solution. The following heuristics provide guidance for using the activity domain in practice. These elements are all potential starting points for conceptualization. Considering these activity elements ensures that an idea can address what the product will actually do as well as its overall function.

The **human need** is satisfied by the output from the use activity. Therefore, the design and explicit articulation of the use activity are critically important.
When **new technologies** are needed to realize the use activity the search for, or design of, these new technologies can form the core conceptualization activity.
The **role allocation** between the product and the user must link to the use activity. As such, new concepts can be created through novel allocation.
When the use activity is expanded to include the whole **lifecycle** the designer must also conceptualize the lifecycle elements, e.g. service, upgrade, reuse, and disposal.

8.6 The Organ Domain: How the Product Functions

As illustrated in Fig. 8.5 two of the domains in Domain Theory are internal considerations related to the product's functional units—the organs and its physical elements—the parts. To clarify, the organ domain describes the functions in a product, while the part domain describes the parts and their assemblies. In order to identify organs a different type of reasoning and abstraction is required from what we might use to identify simple parts. Returning to the bicycle as a common frame of reference we can explore this reasoning further.

Example:
A bicycle's organs. Figure 8.10 shows the basic structure of a bicycle and below we outline one possible proposal for the bicycle's organs:

Rolling organs: This organ provides the ability to move with low resistance via the wheels and their interaction with the road.
Force take-up organ: This organ transfers the forces from the user through the pedal arrangement and the crankshaft to the transmission organ.
Transmission organ: This organ consists of two chain wheels, the chain, and the stiff connection between their bearings in the frame. The front chain wheel shares the crank bearing with the pedal arrangement, while the rear chain wheel shares the rear wheel bearing.
Drive organ: This organ transforms rotation energy to a force on the road, driving the bicycle forward. It consists of a driven wheel and output via the tire surface in contact with the road.

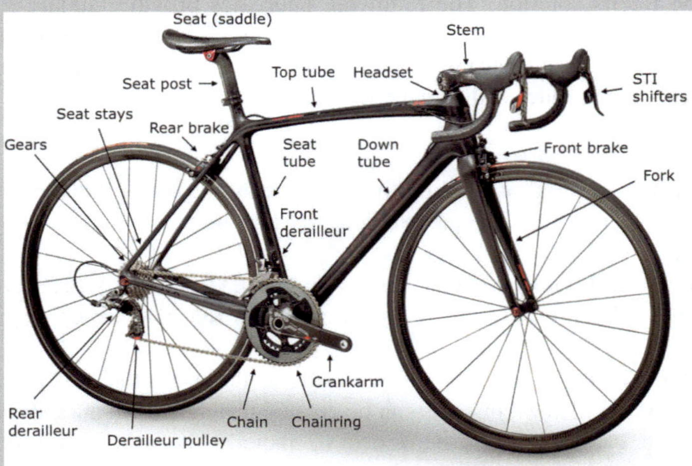

Fig. 8.10 Bicycle terminology, courtesy Daniel Barreneche

Steering organ: This organ directs the bicycle through the front wheel, head tube axis, and handles. Its output is a force on the road and a counterforce on the front wheel. A precondition is force take-up from the road and force transfer through the headset containing a bearing system.
Structural organ: This organ supports the user and transmits their weight through the frame and wheels to the earth. This also positions and supports all the other organs.

From this example we can draw some general points:

- Singular organs necessarily interact with other organs to achieve their functionality as well as contribute to the overall functionality.
- Entities can serve more than one organ, e.g. the rear wheel contributes to the rolling organ and the drive organ.
- Individual parts do not follow the organ composition. The individual parts shown in Fig. 8.10 do not offer a full explanation of the bicycle.

When identifying functions we ask: what are entities doing? And answer with action words, such as drive, force, transfer, etc. In contrast, during synthesis we start with a function and ask: what could do this? In this way we can use functional statements as a basis for finding solutions, i.e. organs. As such, we can define an organ as follows.

Definition: An **organ** is a system element of a product (when we see the product as a system from the function perspective). An organ is characterized by its function and mode of action, i.e. what it does and how it works.

In order to identify organs we must take a dynamic perspective on the product, examining what happens over time, i.e. what state transitions occur (see Chap. 12 for more). As organs are based on natural phenomena, the organ interacts with the surroundings when external effects (stimuli) act on it. Some of these effects can then be utilized as functions.

Organs are the functional building blocks of a product. They are the solutions that the designer constructs such that their interactions lead to the desired overall function. These interactions are normally chains of effects. The composition of organs is called an organ structure.

Definition: From a functional perspective a product is a **system of organs**. As such, a product's **organ structure** is defined by the organs (its elements) and their interaction (its relations).

Using this logic we can re-examine the bicycle example from earlier. Here, the interactions between the organs can be easily identified in the example: rotational force from the feet is transmitted through the transmission organ to the bicycle's drive organ and then to a force on the road, driving the bicycle forward. The organs' functionality depends on these interactions.

In order to realize these functions organs build on natural phenomena, such as physical, chemical or biological effects. Thus, the kernel of product synthesis is in the identification and arrangement of such phenomena (via the organs) into a whole that is able to realize the necessary functions.

Example:
A 'corkscrew's' mode of action. Wine bottle corks can be removed by penetrating the cork with a needle and pumping air into the bottle until the pressure is so high that the cork pops out. The product is similar to a small bicycle pump. Its main functions are to create air pressure and give access to the air space under the cork. These are realized by the pump and the hollow needle respectively. Another concept for a 'corkscrew' is to press a thin steel plate between the bottle's neck and the cork. This changes the friction and when the plate is pulled out the cork follows. The physical effects utilized in the two products are thus pressure increase and changing the coefficient of friction.

8.6.1 Models in the Organ Domain

In Fig. 8.2 we saw Cross' illustration of the problematic mapping from sub-functions to sub-systems, i.e. organs. Composition models in the organ domain thus describe both the **organs** and their **interactions**. There is again a range of approaches to modelling this system; however, they should typically aim to clarify the **mode of action**. Ultimately, considerations from the organ domain are linked to progress in the parts domain and it is thus typical for these to progress in parallel. Although organs are fundamental to product synthesis, they are much more difficult to immediately imagine in comparison to parts. In an effort to help mitigate this we will present a number of different organ models in order to help the reader get used to thinking in terms of organs.

One possibility is using **function language**. Here, instead of focusing on the organ as a concrete solution we focus on the organ's functions. Functions are the useful effect delivered by the organ. As such, we define the product via a system of function labels, e.g. create force, conduct current, create closure, and position needle. Figure 8.11 shows one such model, where a product's organ composition is articulated via functions.

Example:
Function labels. Bang and Olufsen sponsored project aimed to design a multi-store CD player (Hede Markussen 1990). Discs to be played are brought into the device (a), clamped (b), and transferred to the player (c). Alternatively, discs can be loaded without being immediately played. Here they continue from the clamping (c) into a circular store where the disc is secured (d). The functions related to these activities are described in the spatial arrangement illustrated in Fig. 8.11. Here, a, b, and c each represent an organ.

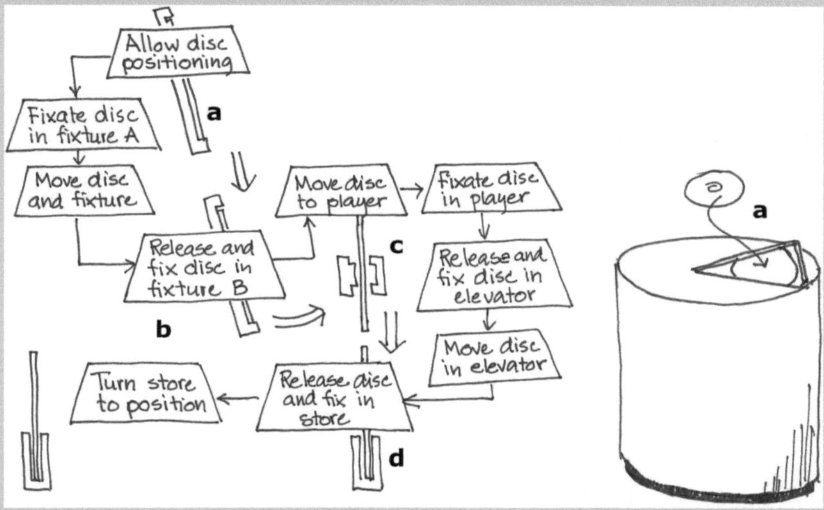

Fig. 8.11 An example visualization of a product's organs using function labels

An organ's **mode of action** (German: *Wirkungsweise*) can be articulated using simple symbols to describe a product diagrammatically. This technique is typically used for diagrams in, for example electronic or pneumatics circuits and train systems. Figure 8.12 describes the functions of a Tea maker in symbolic language.

Another way of modelling the organ domain is illustrated in Fig. 8.13. This shows the layout of an actuator with the parts and their relations. On the left is the technical layout drawing and on the right is the same assembly represented as a symbolic mechanism. This shows three arms that transmit force and move relative to each other about a central pivot. In this way the symbolic mechanism representation can be used to describe the organ (Courtesy Troels Petersen and Jens Peter Poulsen).

In electrical engineering there is a standard language for components (compositions of parts) that allows organ structures to be modelled at a generic level (circuit solutions) and at a quantitative level. In mechanical engineering it is more difficult to use similar standard representations because solutions are often superimposed over each other such that components serve multiple purposes, e.g. the frame in a bicycle.

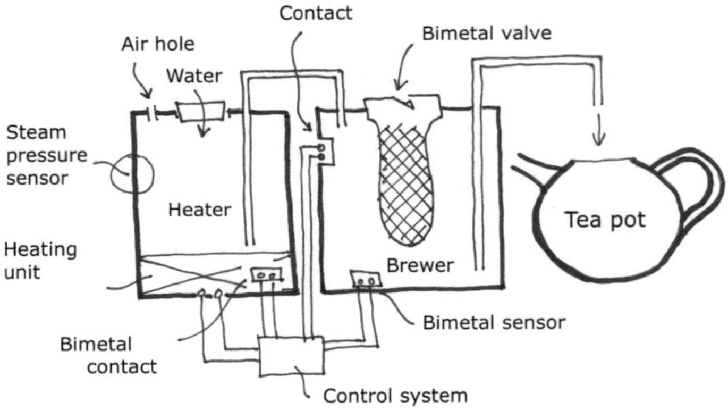

Fig. 8.12 Symbolic representation of the organs and their interaction in a tea maker

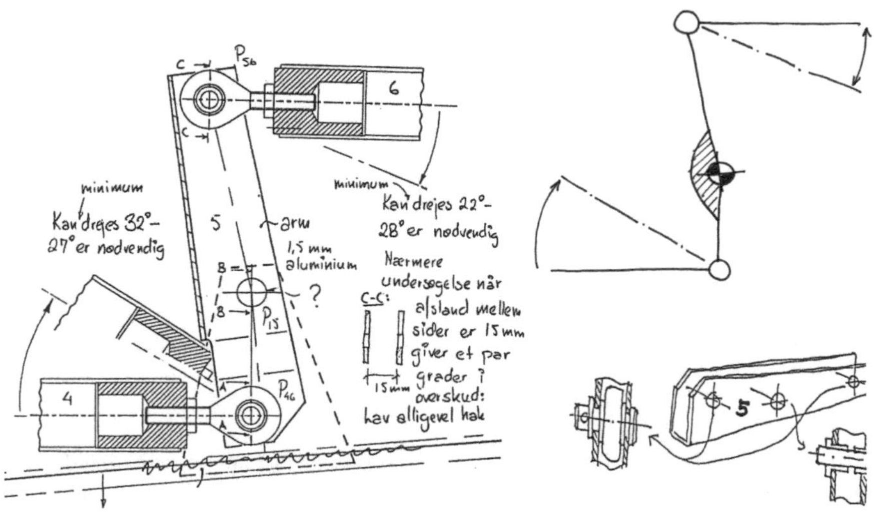

Fig. 8.13 Reading an organ structure from a part structure drawing

8.6.2 The Organ Domain as Concept Inspiration

Any technical idea actually describes organs by finding the mode of action based on natural phenomena. Therefore, many of the approaches mentioned in Chap. 7 (Concept Synthesis) can also be applied in organ synthesis. Organ thinking includes both of those organs hidden in the interior of a product and those that interact with the environment, e.g. interacting with the user or tool organs like a robot's gripper. When we articulate or 'spell' a product concept several organ heuristics should be considered.

- One starting point for synthesis is in the search for appropriate organs that **deliver** the **effects** demanded by the use activity.
- Articulating functions in a **solution neutral form** forces us to consider alternatives.
- **Articulating functions** focuses the search for solutions by providing abstractions that lead to organ classes, e.g. motor or operating organs.
- Organs should be **composed** with respect to the sub-tasks and sub-functions of the product (Fig. 8.2).

Ultimately, the model of the organ structure should provide sufficient detail that it can be handed to another designer who could reasonably be expected to complete the detailing with minimal further input. As such, organs need to be described in detail where their characteristics are specified with respect to the principle of the organ. Tjalve (1979) illustrates the transition to part structure in the example below, showing the qualitative and early quantitative considerations.

Example:
Tilting mirror. In an optical device, e.g. a periscope, a mechanism is needed such that an operator's input movement tilts a mirror in order to change the angle of the reflected light. Figure 8.14 shows alternative principles (a), a selected principle (b) and its variations (c), and the variant that leads to the illustrated part structure (d). The transmission mechanism is one single organ. The final picture shows how this is attached in the device, as well as its input lever.

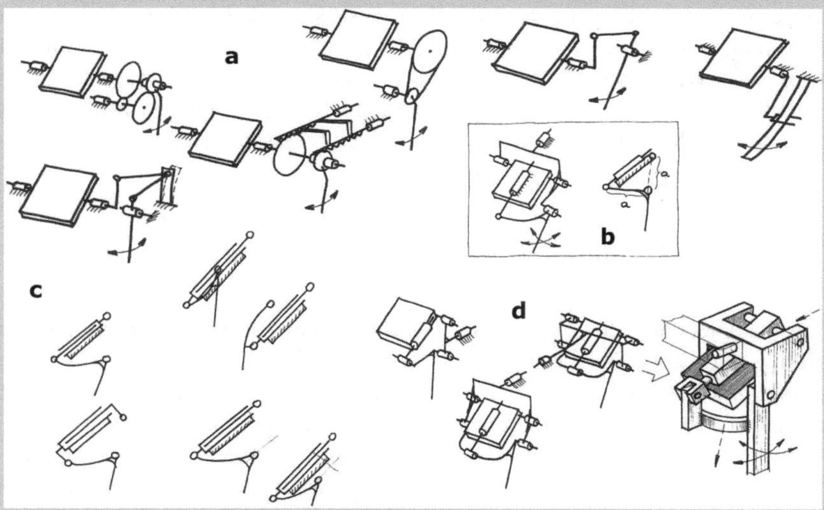

Fig. 8.14 Tilting mechanism's organ design and materialization illustrated by Tjalve (1979)

8.6.3 Spelling the Organs

One of the key reasons for the three domains of Domain Theory is to allow the designer to identify and utilize different design perspectives or degrees of freedom. The following heuristics articulate the degrees of freedom in the organ domain.

> The organs define a product's functions and properties. Therefore, it is important to understand their physics and **mode of action**.
> Understanding an organs **operating principle** allows for assessment of alternatives, as well as providing a foundation for materialization.
> The organ structure describes how the effects necessary for the activity are achieved. However, do not aim at completeness; **move to the part domain** when concretization is needed.
> Although some designers prefer to design based on part reasoning we recommend that at least a rough organ structure be described. This is because in the part structures' complexity the **function aspects of organs are easily lost**.
> The organ domain provides a **common ground** between different disciplines. Using function reasoning allows actors from all disciplines to contribute to the product synthesis discussion.

We see reasoning about functions and organs as a necessary precondition for a deeper understanding of how a product functions and thus how to generate alternatives, to reuse or 'steal' solutions, and to empower their performance. In order to complete this understanding organs and their functions must be materialized as parts. The part domain, explored in the next section, addresses this materialization.

8.7 The Part Domain: How the Product Is Materialized

Many designers feel themselves most comfortable working in the part domain. Working with specific parts gives a tangible basis for reasoning resulting in sketches and concepts that focus on the part structure. The limitation of this approach is that the parts do not necessarily explain the functionality and interaction of the organs. In this sense the part orientated approach does not support the search for alternative solutions because it is always limited to a search for substitute parts rather than entirely new approaches or phenomena. In the mirror example above (Fig. 8.14) consideration of organs allows us to search for entirely new ways to transmit the light, which could include video or other alternative means. In contrast considering only the parts might allow us to refine the current assembly e.g. replacing the mirror with a new material but does not allow us to re-examine the fundamental function of the mechanism. As such, the part and organ domains should be used in complement to each other throughout Product Synthesis.

As with organs, parts' behaviour is based on natural phenomena, e.g. stiffness or conductivity. In this way specific parts can help achieve the desired effects, e.g. piezoelectrical crystals can be used to create electric current. These characteristics link the parts to the organ domain. Here, the organ domain explains behaviour and functionality, while the part domain explains how this is 'built', i.e. the embodiment of the parts and their relationships in the assembly. This leads to the following definition of a part.

Definition: A **part** is a system element of a product (when we see the product from the embodiment perspective). A part is characterized by its physical properties, e.g. form, material, dimensions, and surface qualities.

In order to identify parts we need to consider how a product will be created via individual produced and assembled parts. This can be derived from the organs (which are themselves determined by the desired functions). As such, the part structure is primarily dictated by the organs but also takes into account how the product will be manufactured. This composition of parts, or part structure, is also referred to as architecture in the US. For clarity, we use the following definition.

Definition: From an embodiment viewpoint a product is a **system of parts**. This **part structure** consists of parts, seen as elements, and their assembly interfacing, seen as relationships.

This is illustrated in the example below.

Example:
Toy torch. The toy torch in Fig. 8.15 can be viewed from both the organ and part perspectives. In image (a) we see the organ for transmitting the finger forces (the housing also belongs to this organ) and the organ for creating rotation. In image (b) we see the finger grip part and how its details contribute to the organs: for returning, for end stop, and for locking. As such, this one part integrates the two main organs as well as contributing to other organs through its additional features.

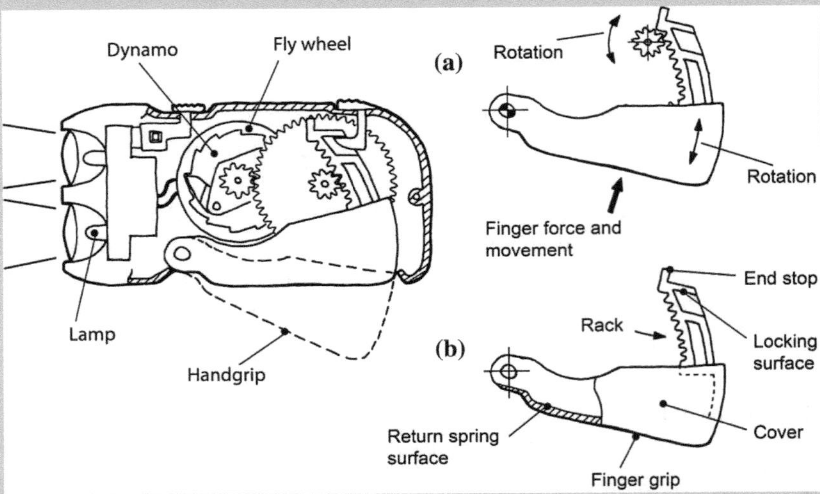

Fig. 8.15 A hand-driven toy torch: **a** Two organs integrated through the part **b** The part used for additional organs: end stop, locking, and returning

A key challenge for designers thus lies in successfully merging conditions from the organ domain with those from the part domain to make an integrated part structure. In this context each part can have multiple modes of action, contributing to multiple organs, and have many of its physical characteristics determined by production and assembly.

Example:
Electric lighter. An electric lighter based on a piezoelectric crystal creates an electrical spark to ignite the gas when a force is applied to the crystal. As such, the crystal is a single part, which requires another part to act on it in order to generate the spark, as well as a circuit leading the current to where the spark is delivered. Here, the crystal is a component specific to the lighter, while the housing and fuel delivery parts are generic to the whole range of lighters produced by the company. This is how the 'create current' organ works.

Certain parts and compositions of parts are found identically in many products and therefore available as off-the-shelf products or standard components (machine elements). Compositions may be defined from production point of view as subassemblies or from development point of view as modules in product family architectures. Some parts are serving specialized tasks, for instance, where the part is active as a tool in an activity: The sewing machine's needle, a scissors' two sharp edged parts cutting paper, the propeller giving propulsion to a boat, etc. Such parts normally serve or realize organs operating directly on the activity's operands.

In particular, parts and sub-assemblies are often designed such that they can be used in multiple different products to form either off-the-shelf products or parts of larger product family architectures. Where parts are specialized for a particular product they often realize those organs operating directly on the activity's operands as illustrated below.

It is important to try and articulate an organ's mode of action, as well as precisely define a part's characteristics. From the organ perspective we might ask in which direction the lighter actuator (see above example) must transfer force and how it moves with regard to the desired degrees of freedom. From the part perspective we might ask what form should the actuator take to fit in the operator's hand and interface with the rest of the lighter. These two perspectives are merged in the following part characteristics developed from Hubka (1973):

- **Form** which can be modelled or interpreted as geometry
- **Material**, which determine properties like elasticity, strength, conductivity, etc.
- **Surface quality**, which determine properties as wear, smoothness and reflection.
- **Dimension** in the meaning size and measure; slightly unsystematically can **tolerances** be seen as dimension also.
- **State** like tension, magnetic, warm, etc. necessary for its task in the organs.

In order to complete the part structure it is also necessary to take into account the relationships between the parts, e.g. positioning, fixing, and movability (rotation, sliding). This is generally achieved via the addition of specific **interface** surfaces, e.g. threads, grooves, and chamfers.

Traditionally, the output from this product synthesis process is a set of drawings specifying the parts characteristics as well as how they are assembled. These use a standardized language for projections, cross sections, symbols, etc. However, these only give a partial explanation of the design: we cannot see the liquid in a pump, the current in a circuit, the actions of the user or the use activity when applying the product. As such, we need to more closely consider the overall synthesis of the part structure.

8.7.1 Conceptualization and Models in the Part Domain

When developing the final product it is important to remember that most models or representations of the design are evolving and do not generally describe the final model that will be brought to market. This is important because design is recursive, i.e. we find the same pattern of activity on all levels of the design's resolution into main functions, sub-functions, and so on. However, all these levels relate to the organs and parts. As such, the designer may use the same concept synthesis and product synthesis operations again and again. Thus, these evolving models allow the designer to flexibly bring together the three domains in an iterative evolution of the final product. This is exemplified by the fact that the visual elements are often the last to be considered. For example, the overall **visual impression** of the product emerges late in Tjalve's development of the Tea machine seen in Fig. 8.17. This stems from the need for an overall understanding of the major parts before aesthetic form can be defined. This means that in incremental design it is often possible to start with a visual design, into which the organs and parts are forced to fit. This is also related to the use of **pattern thinking** in the design to create similarities in a family of products, giving a visual identity and style to support a brand. Similarly, **modular design** aims to create a varied product family by combining common modules. This can give a high level of diversity whilst retaining visual and functional commonality between products.

Alternatively, developing new concepts for parts can create different product solutions. Here, the manufacture of the product incurs significant cost and is therefore worth optimizing were possible. A second alternative is **structural variation**, which changes the spatial position of organs and parts, and varies their interfaces. This can again be undertaken from different perspectives, such as optimizing the product structure or reducing space and material use. These can all be realized using **building method thinking** to support the choice of production method. If we consider, for example gearboxes, door hinges, bicycles, and wristwatches, we find that there is a limited set of preferred alternatives. As such, it can be highly beneficial to start with a known building structure, instead of trying to be innovative in such a constrained environment.

8.7.2 Spelling the Parts

The part domain concerns the product's realization: how can we produce and assemble parts so that the overall functionality is achieved? Conceptualization proceeds from abstract considerations about need, use, functions, and organs to parts, but it is also possible to reason from given components. The argument for the longer reasoning path via functions and organ domains is that it allows the possibility of innovation and utilizes all the design degrees of freedom. Further, it ensures a more systematic fit between need, task, and solution. However, in order to be successful it is necessary to remember the parallel need for clarification activities in conjunction with the synthesis of the design. The many choices made in embodiment are based on argumentation that needs documentation in order to support a companies' future development, as well as for legal reasons.

> The composition of clarification activities is just as complex as the product's exploded view. However, it is not obviously visible and thus easily forgotten.

Our treatment of the part domain may give the impression that synthesis is just completing a jigsaw puzzle with a mix of new and reused pieces. It looks like this because we have not discussed the need to create the product's goodness as part of the synthesis. This goodness comes from attractive functionality, fulfilment of the required properties, and a balance between quality, cost, manufacturability, and sustainability.

8.8 Product Synthesis: A Three-Domain Progression

Proposals for how to progress with Product Synthesis are rare outside of concretization and detailing of the part structure. Here, the conceptual core can come from many different areas, e.g. new business ideas, a novel way of using an existing product or new features of organs. However, whichever path is chosen there are certain characteristic patterns that describe the progression of the design activity. One major example of this is that the significance of decisions follows two defined progressions. First, the significance rapidly decreases from the early stages where quality of need satisfaction and business cases are considered. Second, the detailed decisions become more important towards the end of the process with the increasing importance of quality and, e.g. manufacture of the product. These two curves describe the importance of keeping focus on both the need satisfaction and business matters, and the details of realization (Andreasen and Hein 1987), illustrated in Fig. 8.16a.

A second characteristic is the start and finish of design. Here, the end is given by a fully specified part structure, ready to go to production. In contrast, the start can be found in any step, although the causality chain must be respected. For example the use activity's result satisfies the need > realized via the use activity > facilitated by the product's effects and functions > delivered by the products organ structure > and materialized in the part structure. We call this a causality chain because each level's goal is defined by the level above, as illustrated in Fig. 8.16b.

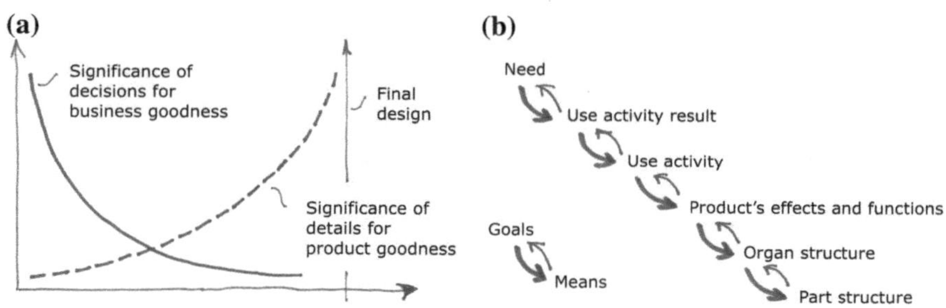

Fig. 8.16 Two characteristic patterns of product synthesis: **a** the importance of decisions **b** the causal chain

This causality chain is not articulated in the literature except by Hubka's General Procedural Model of the design process (1976) and implicitly by Tjalve (1979). However, Tjalve does give an exemplary articulation of product synthesis beginning with problem analysis. Tjalve takes four further steps: function thinking, search for solutions, defining the basic structure, and the quantitative structure. These steps are illustrated in Fig. 8.17. **Basic structure** describes organs and

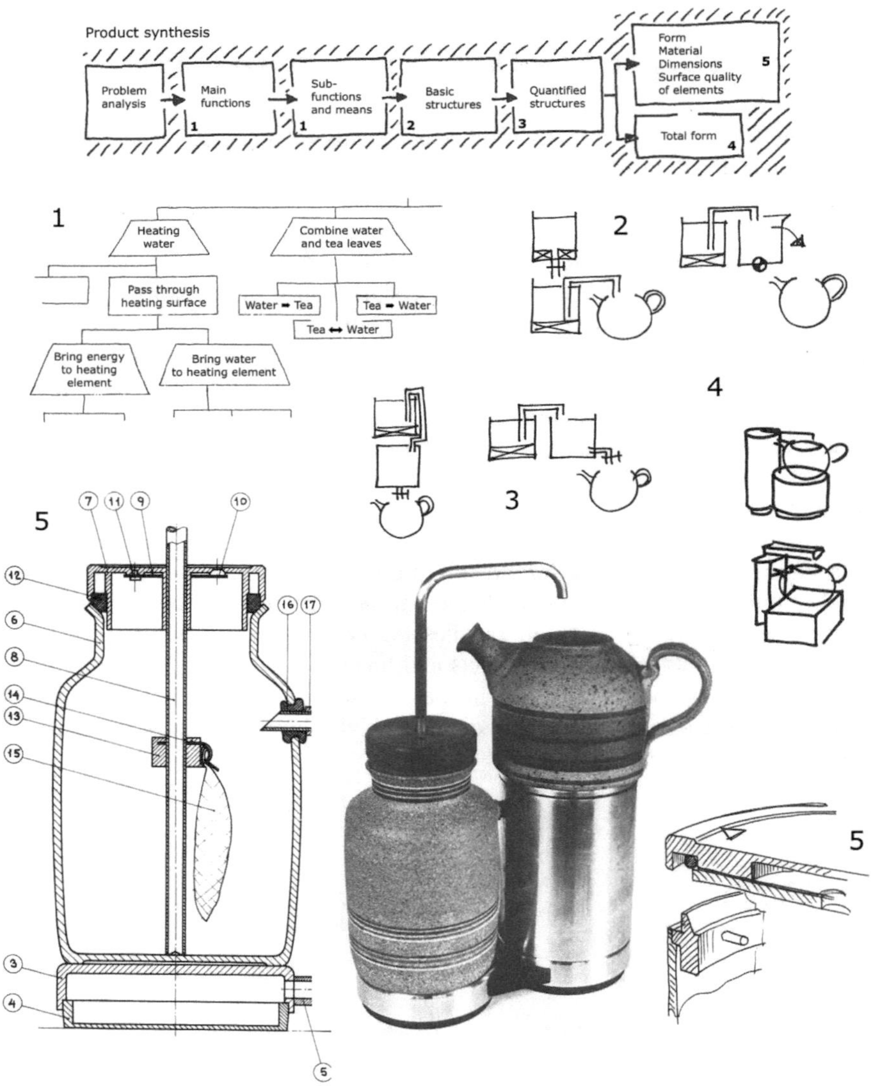

Fig. 8.17 Product synthesis as proposed by Tjalve (1979), illustrated using a copy of Tjalve's figures of the design of a tea machine. The numbers of the sketches refers to Tjalve's model of product synthesis at the top

their functional relations giving the principle of the product. Here, alternative solutions or combinations lead to different basic structures. **Quantitative structure** describes the transition from organ to part structure. The organs physical volume and physical interactions are determined, and alternative spatial structures are considered. These determine the operation of the product, production, assembly, and installation.

8.8.1 Top Down or Bottom up?

Product synthesis models mirror the artefact's nature. For example Tjalve's model generally points to many of the design degrees of freedom that also exist in the design of non-mechanical products, while most models from literature concern mechanical products. All these models describe a *top down* sequence of activities from a functional perspective, where the authors note that feedback loops may occur. These models tend to be visually sequential with the feedback loops or derivations from the sequence poorly described. However, as we saw in Chap. 5 these loops are almost unavoidable because of co-evolution and intrinsic features of design. For example co-evolution shows how increasing insight into possible solutions gives better insight into the problem and thus forces the designer to refine their approach. We explore some of these issues in the following example.

Example:
Egg sausage machine. In the late 1960s a Danish egg producer came up with the idea of producing hardboiled egg in the form of a sausage. The aim was partly to deal with the needs of its catering business and partly to better utilize its raw materials in producing a refined product. Hubka and Andreasen worked together on the development of the machine. After a number of unsuccessful attempts to find a principle for a continuous process it was decided to use discontinuous forming and coagulation. Here, a set amount of yolk and white were combined and heated, causing coagulation from the outside in. Once this process was complete the forming tube was removed. The end result was an egg sausage that gave well-proportioned identical slices. This is illustrated in Fig. 8.18.
Experiments showed that the originally expected machine layout, using a hinged form, was impossible because the white stuck to the form surface. This problem was solved using a spring-form, however, this changed the design of the form parts, demanding a new layout. A second change based on experimentation was that the original concept, with a forming tube surrounded by the heating chamber, was not as simple as first imagined. Here it was found that it was necessary to interrupt coagulation by cooling. As such, cooling chambers were defined and the form with the coagulated sausage moved from

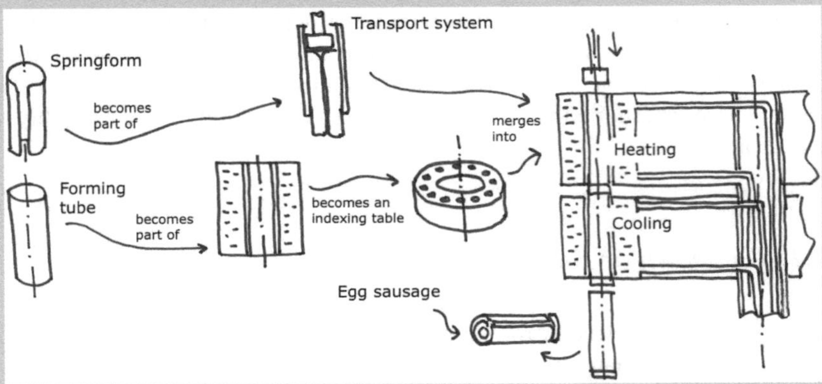

Fig. 8.18 Opportunistic considerations in the design of an egg sausage machine

heating to cooling zone. Finally, in order to achieve the desired level of productivity it was necessary to arrange 48 tube subassemblies in a carrousel.

This leads to two realizations. First, something that was originally thought to be a minor detail, the separation of the sausage from the form, ended up being decisive in the machines layout. Second, the physical realization of the organs via the part structure solved multiple functions. For example, the spring-form not only solved the separation issue but also was also integral to the transport system moving the sausage between heating and cooling zones. Further, the heating and cooling chambers became an integral part of the larger production carrousel.

Necessary organ materializations were utilized for the through flow functions: the spring-form becomes a transport system, the form tube becomes part of the heating and cooling chambers and part of the transport system, the heating and cooling chambers becomes a carrousel.

This example shows how detailed considerations can have significant effects on the overall product synthesis and that this is realized by moving between the three domains. This is illustrated in the three parallel activity patterns in Fig. 8.19. In all three domains there is progression from abstract to concrete and from undetailed to detailed. Further, both the satisfaction of the need via the final use activity and the complete specification of the part structure can be seen as end points. This is idealized in the cause chain shown in Fig. 8.16 and the example progression shown in Fig. 8.19.

Many factors influence the design of the product. Broadly speaking any single factor or issues can from the basis for new product conceptualization. A typical example of such an influential issue is sustainability. A company may see redesign for sustainability as an opportunity to develop a more sustainable brand image and increase its market share. However, sustainability can be linked to almost 'everything' in the company, e.g. product range, suppliers, manufacture, product life

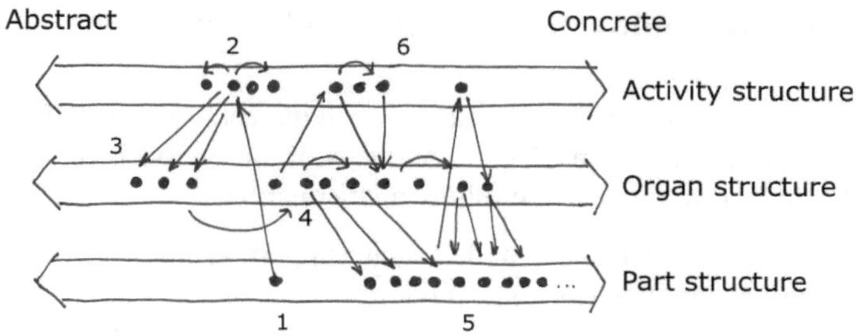

Fig. 8.19 Interrelation between the three domains in a product synthesis case. The *arrows* illustrate the progression from the egg sausage example: *1* spring-form > *2* activity considerations > *3* sub-system functions and *4* organs > gradual detailing of new sub-activities *5* and new organs *6*

aspects, and material consumption. Thus, it is up to the company's ecological conscience and its creativity to select a focus area and perform innovation.

We postulated a "three domain progression" as being a fruitful understanding of product synthesis in this section. Although this can seem complex it is a recognition that all three aspects need to be considered if freedom and potential solutions are not to be lost. As such, we propose some heuristics to help the designer with this progression.

> There is no reason to follow or expect a strict sequence in the product synthesis activity. It can be much more powerful, in terms of innovation and value creation, to be able to fluidly focus on ideas in each domain as they arise.
> Even though *Exploration* and *Concept Synthesis* create the 'conceptual core' it is never too late in product synthesis to add new dimensions if they enhance the product, value or business. These can also reveal new perspectives and thus change the conceptual core.

Product synthesis takes different forms depending on the type of design being undertaken. For example consider the three types introduced in Chap. 1: new, incremental, and platform. In **new product design** synthesis is dependent on Exploration and Concept Synthesis for confirming the need is valid and for ensuring the tractability of the innovation project. In **incremental design** past designs are used as a starting point and thus substantial contextual and design information can be brought forward into the new product. However, this should be balanced against a strong description of the new product concept, ensuring it has a raison d'être in and of itself. Finally, in **platform-based design** each relevant product variant belongs to a family of products. The definition of variation and commonality between the family members is designed to ensure broad market coverage, as well as good utilization of company assets. Thus, the individual product is bound to certain game rules, dictated by the characteristic activities, organs, and parts, associated with the product family.

8.9 Product Modelling: The Product's Chromosome

Many models are created during the design activity, each with different purposes and techniques as discussed in Chap. 3. As many of these models relate to the synthesis of the design we propose the following distinction:

- **Design models**: intermediate models that support synthesis, communication, and verification activities via, e.g. simulation.
- **Product models**: models that support the progression and final specification of the design via, e.g. technical drawings for manufacturing purpose.

Roth (1988) noted that many partial models, not necessarily linked to each other, characterize design work. Figure 8.20 illustrates how design activity can be seen as levels of functional decomposition leading to a growing number of related models, and how Roth's imagination of scattered models might look. In order to bring these views together we developed what could be seen as a genetic design model system (Mortensen 1999). As proposed by Ferreirinha et al. (1990) we call this core product-defining model a Chromosome Model, shown in Fig. 8.21. The Chromosome Model is intended to give a generic descriptive model of the product, mirroring Domain Theory. The composition of activities, organs, and parts is modelled in hierarchical patterns and interrelated as shown in Fig. 8.21. The model has gradually been changed (Andreasen 1990; Jensen 1999; Mortensen 1999). As

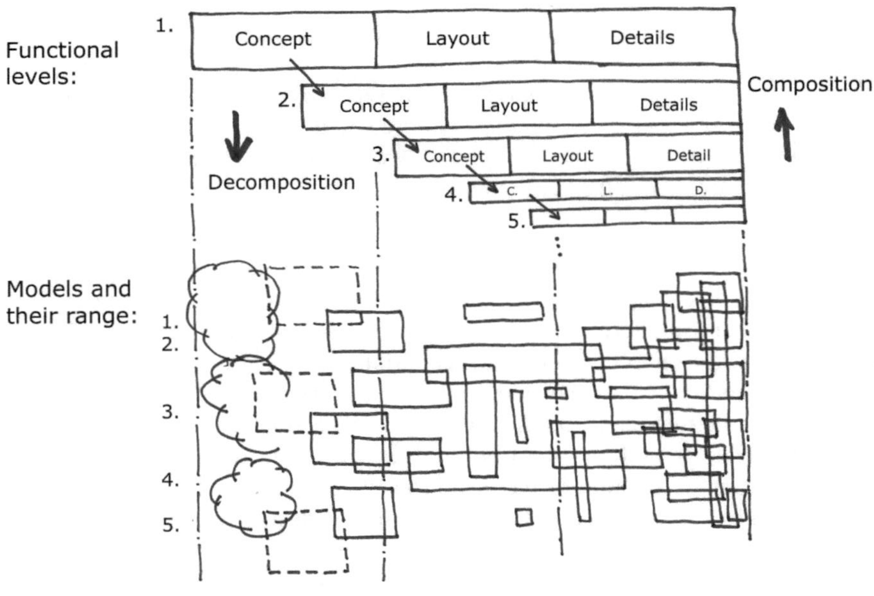

Fig. 8.20 A scattered pattern of design models in product synthesis. The *top picture* shows the top-down determination of functions and synthesis, while the bottom picture shows Roth's description of scattered models in time and coverage

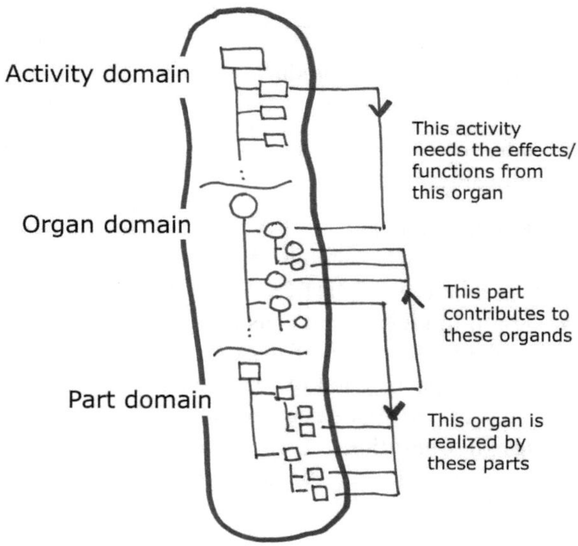

Fig. 8.21 The chromosome model

such, the model captures the gradually synthesized design entities. When we want to clarify a certain sub solution we may sketch and detail it in a design model until its characteristics can be added to the overall product model.

In contrast to clarifying a certain sub solution we can explore and justify certain properties of the design via a view model. This is composed of the design characteristics, parameters related to the surroundings and the influences (stimuli). Figure 8.22 shows these relations with respect to the product model and other intermediate design models.

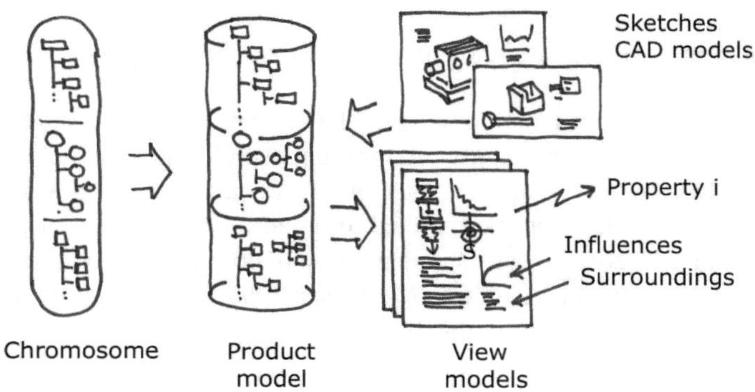

Fig. 8.22 Ideal application of the chromosome model for creating a structural product model in the three domains. The product model gradually builds up from design considerations and delivers characteristics to view models of certain properties

Definition: A **view model** is a model derived from the product model able to articulate a certain product property.

In order to work effectively there are three major preconditions for these models functionality:

- **Design language,** i.e. a vocabulary for thinking, reasoning, conceptualizing, and specifying solutions in all three domains based on semantics and syntax. This should be applicable for human reasoning and computer operations.
- **Design models,** i.e. models for structures of activities, organs, and parts, carrying the specification of these structures. This also allows for more or less formalized specification of the relationships inside and between domains, as well as the property statements of entities.
- **Design operations,** i.e. methodologies to support gradual synthesis in all domains, e.g. via methods for synthesizing, composing, evaluating, and simulating.

The application of the Chromosome Model's structure in the definition of the product means that the following basic views are present in the design of the product: Its use activity, its functionality (described by the organs), and its embodiment (described by the parts).

8.9.1 Applying Product Synthesis Models

Domain Theory is a model-based theory, which means that it is based on words and concepts taken from the description of the domains and the phenomena related to the domains. However, other concepts are also plausible and therefore we find several other proposals for models of product synthesis (Pahl and Beitz 2007; Hubka 1976; French 1985; Pugh 1991 and many others). What we think is most important for a designer though, is how productive a theory is when used in practice.

The concepts outlined in this chapter have been chosen because they have shown time and again to be highly relevant and useful in practice in, e.g. conceptualization, Design for X, dispositions, modularization, and platform thinking. Further, because modules are both linked to organs and composed of part domain entities Domain Theory is very suitable for supporting design reasoning (Chap. 11) and modelling modular systems (Mortensen 1999; Harlou 2006). We return to the some of these topics later in this book, including the view model and property reasoning in Chap. 12.

8.10 Conclusion

The module 'Product Synthesis' is the design activity we find described in the textbooks as engineering design. Our approach here is to use Domain Theory as the basis for explaining, modelling, and articulating the design process. This leads to a pattern quite different from traditional 'sequential' engineering design models, rather Domain Theory provides the basis for 'spelling' the product. This theory has shown its worth in a large number of industrial modularization and platform projects, as well as in supporting new design and management procedures in many Nordic companies.

In Chap. 3 we introduced Cross' 'designerly ways of knowing' and the 'constructive mode of thinking' as core characteristics of design. In complement to this we see Domain Theory as articulating basic constructive traits of designing, i.e. the need to consider the three views, each with their own properties but all necessarily integrated in the final design. Thus, product synthesis is based on theories about the nature of products and the relationships between activities and products. As such, Product Synthesis encapsulates Exploration and Concept Synthesis in a development project, practicalities and the nature of the project determine what is most appropriate. In the following chapters we encapsulate Product Synthesis in the larger modules, Product Development and Product Life Synthesis and thus add the organizational conditions for utilization and realization of Product Synthesis' results.

References

Andreasen MM (1980) Syntesemetoder på systemgrundlag (synthesis methods based upon system theory). Ph.D. Dissertation, Lund University, Sweden

Andreasen MM (1990) Designing on a designer's workbench. Unpublished notes from WDK Workshop Rigi

Andreasen MM, Hein L (1987) Integrated product development, IFS (Publications)/Springer, Berlin Facsimile edition (2000). Institute of Product Development, Technical University of Denmark Copenhagen

Arthur WB (2009) The nature of technology—what it is and how it evolves. Penguin Press, London

Ashby WR (1968) An introduction to cybernetics. Methuen Press, London

Blanchard BS, Fabrycky WJ (1998) Systems engineering and analysis. Prentice Hall, New Jersey

Cross N (2008) Engineering design methods. Wiley, Chichester, GB

Ferreirinha P, Grothe-Møller T, Hansen CT (1990) TEKLA, a language for developing knowledge based design systems. In: Proceedings of ICED 1990 Dubrovnik, Heurista, Zürich

French MJ (1985) Conceptual design for engineers, 2nd edn. Design Council and Berlin/Heidelberg, Springer, London

Gero JS (1990) Design prototypes: a knowledge representation schema for design. AI Magazin 11(4):26–36

Harlou U (2006) Development of product family based upon architectures. Contribution to a theory of product families. Ph.D. Dissertation, Technical University of Denmark, Copenhagen

Hede Markussen T (1990) Betragtninger over metoder i menneske/maskin-samspil (Considerations on Man/Machine interaction). Instituttet for Konstruktionsteknik, DTU Copenhagen

Hubka V (1973) Theorie der Maschinensysteme (Theory of machine systems). Springer, Berlin

Hubka V (1976) Theorie der Konstruktionsprozesse (Theory of the design process). Springer, Berlin

Hubka V, Eder WE (1988) Theory of technical systems. Springer, Berlin

Jensen T (1999) Functional modelling in a design support system—contribution to a designer's workbench. Ph.D. Dissertation, Technical University of Denmark, Copenhagen

Mortensen NH (1999) Design modeling in a designer's workbench. Contribution to a Design Language. Ph.D. Dissertation, Technical University of Denmark, Copenhagen

Pahl G, Beitz W (2007) Engineering design: a systematic approach, 3rd edn. Springer, London (First edition 1977)

Pugh S (1991) Total design—integrated methods for successful product engineering. Addison Wesley, Wokingham

Roth K-H (1988) Elementarmodelle der Maschinneelemente—Möglichkeit zum Konstruieren durch den Rechner (Elementary models of machine elements—feasibility of computer based design). Konstruktion 40(1988):309–316

Suh NP (1990) The principles of design. Oxford University Press, USA

Tjalve E (1979) A short course in industrial design. Newnes-Butterworths London. Facsimile edition (2003) Systematic design of industrial products. Institute of Product Development Technical University of Denmark, Copenhagen

Chapter 9
Product Development

Abstract The encapsulation element Product Development is normally organized as a procedure in companies. It is seen as an 'all inclusive' perception of design but often misses the emphasis on exploration, conceptualization, product synthesis, and lifecycle.

The encapsulation element Product Development is normally organized as a procedure in companies. It is seen as an 'all inclusive' perception of design but often misses the emphasis on exploration, conceptualization, product synthesis, and lifecycle.

Product Development leads to the establishment of sales and production, i.e. new business ready to be exploited. Because the synthesis of sales system and production are closely related to product synthesis, the question of integration becomes central to this activity. For conceptualization this means added complexity of concerns and influences, and more generally that the complex organization of the product development should be established. We bring an understanding of conceptualization's relations to this complex organization.

9.1 Expansion to a Complete Company

Product development is the fourth module in our **Encapsulation Design Model** introduced in Fig. 5.8. Product development describes the linking activity that draws together market research, product synthesis, manufacture, and sales. In this

M.M. Andreasen et al., *Conceptual Design*, DOI 10.1007/978-3-319-19839-2_9

module user needs are actually satisfied. Seen from the conceptualization perspective, the aim of this chapter is to explore the organizational implications and procedures underpinning successful product development, and especially how they influence conceptualization. In particular we explore two major elements:

- Where is conceptualization positioned in relation to the product development process?
- How does product development influence conceptualization and vice versa?

Dissolving these points, we will emphasize the role of conceptualization in supporting integration. This is not only organizational but also intrinsic to the concepts and wider design process. This multifaceted role demands what we call dispositional thinking. We deal with this at length in Chap. 13 but here it is sufficient to understand this type of thinking as the ability to arrange the product, its realization, and its use activities to best satisfy the user and lifecycle actors.

Product development is normally used to describe everything from project initiation to product launch. This means that conceptualization is implicitly found in companies' procedures or scholars' models—often described as integrated product development. In this chapter we use this integrated model in order to better understand the link between product development and conceptualization (Fig. 9.1). This is achieved in three steps. First, Sect. 9.2 explores the **nature of product development**. Second, Sect. 9.3 identifies the **game rules for conceptualization** by explicitly splitting out conceptualization. Finally, Sect. 9.4 brings these together by explaining the organizational dimension that we call the **product development machinery**.

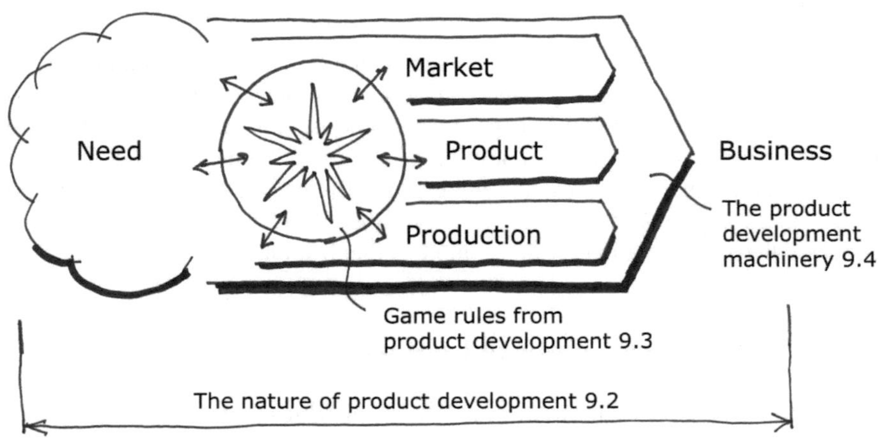

Fig. 9.1 The model of integrated product development used to structure the chapter

9.2 The Nature of Product Development

Broadly, new product development is concerned with the creation of new products. Here we refine this scope as follows:

Definition: **Product development** is a company's activity associated with creating new business based on developing and launching new products. The activity is initiated by need and market research, as well as ideation, and ends with production, distribution, and sales.

In addition to new product development there is a range of alternative approaches to developing new business including copying products, buying patents, licences, designs or consultant support, and buying other companies. Although these can be lucrative, they are less concerned with conceptualization, hence our focus on product development. Here, product development is composed of elements incorporating both innovation and operational activities. Ultimately, in order to successfully produce and sell products the development activity needs to utilize knowledge from across a company. The complexity of this organizational perspective is illustrated by Hales and Gooch (2004) model of a development project as part of a much wider context. Hales' layered model of project context is given as an example of this in Fig. 9.2. Here, the core design activities are depicted as the vertical sequence starting with 'competition'. These activities are nested within six layers: design, project, management, company, market, and environment. This graphically depicts the many conflicting influences on the design activities and product development. In summary, the design process and the organization are fundamentally interconnected and need to be managed holistically if a successful outcome is to be reached. As such, this section explores how that can be achieved from a design perspective.

9.2.1 Integrated Product Development

In order to understand the advantage of thinking about product development as integrated with the wider organization let us consider disintegration. We have already discussed the many incremental steps towards specialization and departmentalization in industry. Here, manufacture can be achieved with almost no contact with development, or recycling with no contact to sales. When this type of disconnect occurs, we do not need to look far to find product and business failures. In order to combat this organizational disintegration, product development actively integrates methods and procedures such that relevant issues from all stages are taken into account during the design activity. This manifests in product development through the explicit integration of two other development activities,

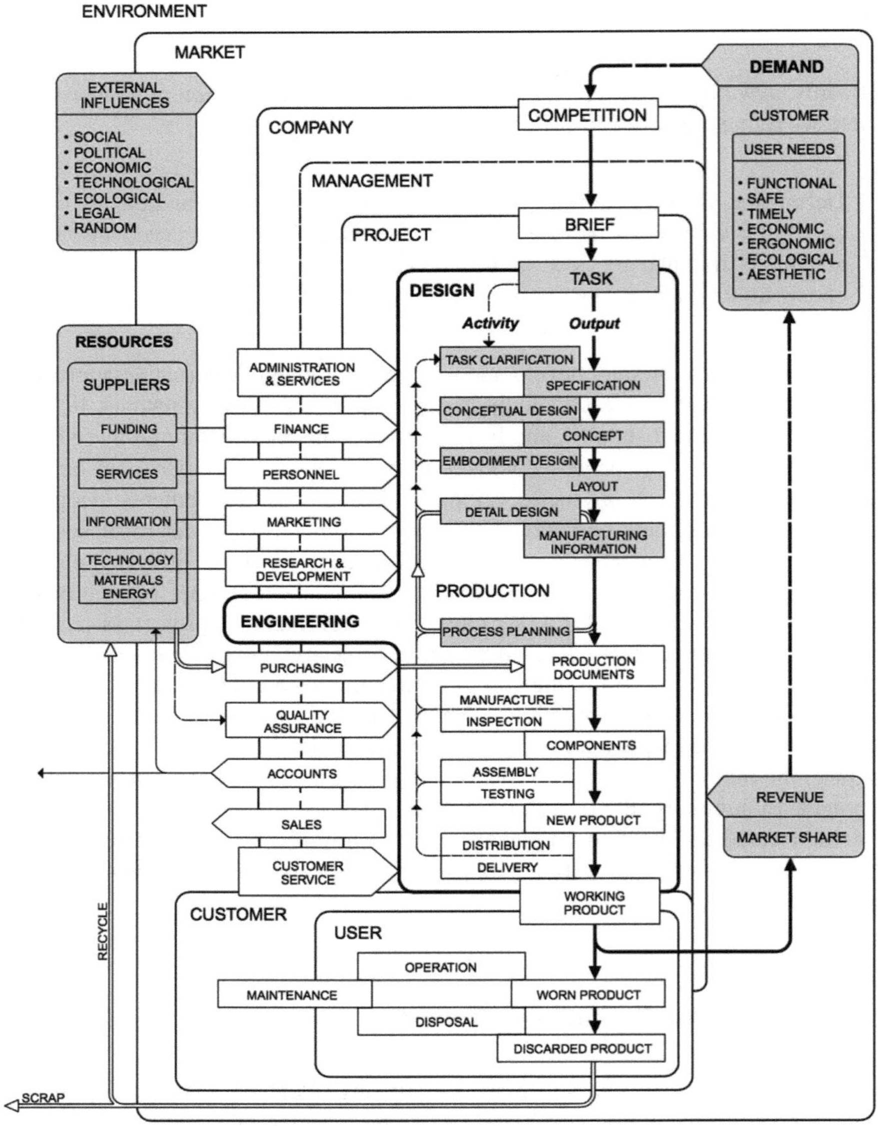

Fig. 9.2 Model of product development activity and organization in context (Hales 1993; Hales and Gooch 2004)

establishing the production and sales requirements or needs. Where these activities are fully integrated, the 'best' business result is possible. This is the reasoning underpinning the Integrated Product Development model advanced by Andreasen and Hein (1987) (Fig. 9.3). The model's terminology differs slightly from this book's terminology.

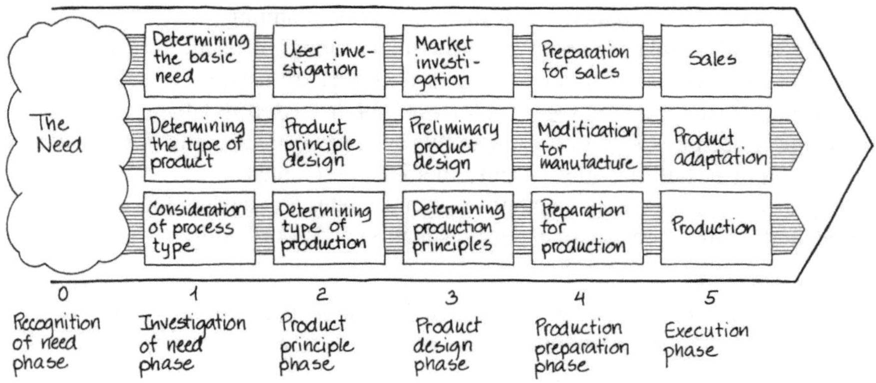

Fig. 9.3 An ideal model of integrated product development combing the three development activities: sales, product, and production, Andreasen and Hein (1987)

We see this model as exemplary in its clear progression and therefore use it as the basis for our explanation of product development. The model explicitly brings integration to the fore as a core part of product development where no one element can succeed alone if the best result is to be achieved. The model spans from need to execution and is widely represented in industry. The main virtue of this approach at a practical level is that it defines the roles of marketing and production in the early design phases, helps in aligning the milestones of each activity, and shifts the focus to the process as a totality where all aspects must perform concurrently.

9.2.2 Use of Procedures

There are many proposals for models of product development, comprehensively reviewed by Clarkson and Eckert (2005). These include descriptive models, e.g. Hales and Gooch (2004), prescriptive models, e.g. Cooper (1984), and combined models, e.g. Ulrich and Eppinger (2004). The models are widely used in industry, although for different purposes and in different forms. Here, model use can range from setting a common mindset to specific procedures. Such procedures typically form the basis for developing a project plan, as well as detailing activity and time plans. We have already introduced procedures in Chap. 5 but revisit them here in order to explore their use in product development.

Development project procedures serve several roles: they become carriers of best practice, they help transfer experiences from past projects, and they support more cohesive management across projects. In the organizational context they help to highlight input from marketing, sales, and production, as well as other specialist areas, such as finance, quality or environmental experts. This can facilitate the distribution of electronic and mechanical tasks, the planning of special milestones associated with regulatory approval or the management of relationships with other companies in a network. In a company, procedures are usually thought of as universal and exceptions are avoided. However, as discussed in Chap. 5, to be most effective, procedures should be adapted to each project's specific context. Part of this adaption is tailoring the specific methods associated with a procedure. For example, one project may demand a greater focus on design for manufacture, while another may need more extensive ideation and coordination.

A company's design procedures mirror its practice and should be tailored to the issues and context affecting the specific company.

An example procedure is shown in Fig. 9.4 from the company Bang and Olufsen. This procedure reflects changes made after a new development strategy, focusing on lead-time reduction, was introduced. The main adaptions from more generic procedures are the increased focus on prototyping and the reduced number of phases. In particular, the start of each phase is carefully managed with a critical review of requirements. Overall these changes reduced the lead-time from 127 to 72 weeks.

Of particular note in the Bang and Olufsen case is their recognition of the importance of concept definition and subsequently product design. This enabled them to more effectively judge project progress and plan the related organization processes accordingly. This key relationship between conceptualization and development is expanded on in the next section.

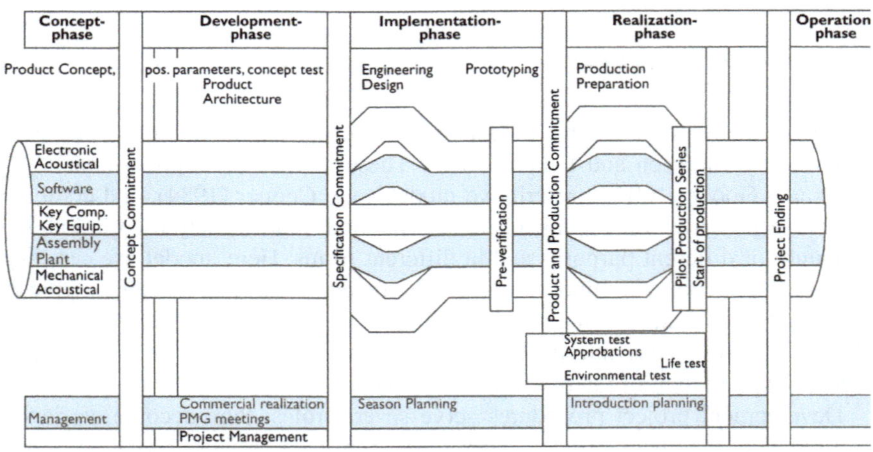

Fig. 9.4 Bang and Olufsen's product development procedure. *Horizontal lines* depict activities while *vertical bars* denote milestones (Kirkegård et al. 1996)

Fig. 9.5 Two product development cultures: ideation and execution (Andreasen et al. 1989)

9.2.3 Conceptualization in Product Development

Although it is typically advised that conceptualization be integrated into procedures, many companies resist this. Here, the decision to initiate a new project is considered so important and ill-defined that companies often prefer to isolate these 'front end' activities in order to reduce risk and attempt to ensure quality in the scoping work before initiation. This tendency leads to our description of two distinct activities Exploration and Concept Synthesis—each addressing one aspect of concept integration in product development.

Depending on how these activities are included in a company's procedures, two cultures appear (Andreasen et al. 1989). The first is an innovation culture where conceptualization thinking is fully integrated. This type of culture is characterized by its ability to create new business potential, address user needs at a low cost, create tractable concepts, and best utilize a company's strengths and weaknesses. The second culture is more execution- and sales-focused. Here, realization of the product is primarily achieved through production and marketing. As such, this type of culture relies on its ability to identify the basic idea underpinning a product and leverage this through marketing. Overall the focus is on cost reduction and optimization of overall work processes including quality and efficiency.

Both of these cultures provide advantages at different stages of the project and thus should be integrated as suggested in Fig. 9.5. For example, the first culture is poorly suited to logistical optimization while the second can stifle technical innovation. As such, design teams sometimes attempt to shift culture during a project, often through staff exchange and tightly controlled milestone reviews. Alternative structures include the use of specialist conceptualization teams who 'consult' on a number of projects. This conditional dependency between conceptualization and product development is one of the key rules when developing your product development game plan.

9.3 Game Rules for Conceptualization

There is a growing recognition of the huge influence new products have on the composition and operation of a company. As such, it is key that we understand these influences and the basic rules by which they affect a company. In particular, we seek to answer: how can management ensure a positive, successful direction? In this section we discuss the main 'rules' to be considered in the conceptualization and how they impact company success.

First, it is critical that a product's **identity** aligns with that of the company's wider corporate and design identity. This includes aspects, such as quality, service, and warranty support. Without this alignment new products can damage not just their own sales but the wider brand and perceived integrity of all the company's products. In particular this requires close collaboration between the designer, project manager, and top management. The major exception to this is where a company is specifically trying to change its identity through, for example, rebranding a new product. An excellent example of this type of alignment is outlined below in order to demonstrate the real world impact of these rules—when successfully employed they impact every aspect of a companies' operations.

Example:
The Philips corporate identity. A case in point is the way Philips manages its brand identity throughout their product portfolio. Philips is a large manufacturing company of products in the area of healthcare, lighting, and consumer lifestyle. Their corporate identity is focused on three core values (Philips 2013):

- Philips is a caring brand that puts people and their needs first.
- For Philips, innovation is the lifeblood of the company.
- At Philips, innovation is about making meaningful impact on people's lives.

The Philips brand identity is intended to be recognizable throughout their portfolio by Philips consumers and users and includes graphical elements, the products, and services that form the brand line, as well as communication in terms of photography and tone of voice. Philips uses the brand identity to "celebrate the company's longstanding heritage as a leading international technology company and reconfirm its passion for delivering meaningful innovations that matter to people", says Thomas Marzano, Head of Brand Design, Philips. For product developers who work in or for a corporate environment it is imperative to not only design to serve human needs, but to do so in a way that fits the corporate identity. Modern organizations want their products to be perceived as part of the brand and its corresponding values. This also works the other way around; through good design, products serve to express and communicate an organization's brand identity and increase its perceived value by the customer. Figure 9.6 shows Philips' humanized environments for a hospital's scanning equipment.

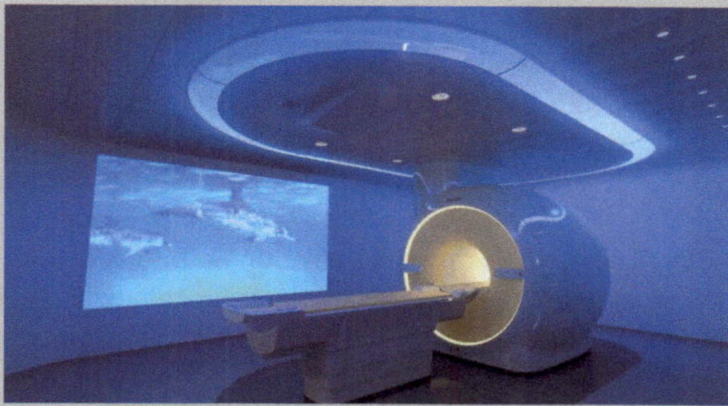

Fig. 9.6 Establishing a friendly environment for a hospital's scanning activities, Courtesy Philips Healthcare

Building on this, new product launches heavily influence the **strategy** of a company. As such, product development must account for the overall strategy it is contribution. This is typically characterized by the development of strong links between top management and the design teams. In practical terms, design teams bear a responsibility for understanding and addressing the strategic areas discussed in Chap. 2. This is particularly important with regard to market and production in an increasingly global product development domain. Closely related to this is **policy** alignment. Here, company integrity, goals, and performance are realized through direct action. These can include employee conduct, equal rights, and human relations considerations. In the context of conceptualization policies related to, e.g. branding, product testing, quality assurance, and supplier relations need to be considered as a core part of product development.

Finally, effective product development exploits a company's **resources** to their limits. Essentially, the ambition of new development should be to leverage the knowledge available in the company in order to outperform competitors. Bringing in external resources can also play a key role where networking or open source strategies are favoured. An important consideration here is knowledge management and the ability to monitor and adapt to changes in the technological state-of-the-art or developments by competitors. If this is effective, new innovation opportunities can be identified early.

Innovation not only concerns new products but also company identity, business, production, marketing, and sales. Product innovation can be an important driver for wider company level innovation and, as such, should be aligned with corporate strategy. However, this is a two-way relationship: strategy should inform development but at the same time designers and their managers have the responsibility to articulate new possibilities or potential innovations. This give and take is illustrated in the following example.

Example:
Organizational innovation. As part of a wider consultancy project Andreasen et al. (1989) developed a new approach for fostering innovation, illustrated in Fig. 9.7. This dealt with four main areas: the frames and surroundings, i.e. the company's situation in the market, legislation, etc.; the goals that the company wanted to achieve; the tasks that are currently executed or planned; the development system, where product development took place. The aim of this approach was to explicitly identify the interdependencies between these four elements in order to better align them with respect to the company's overall innovation strategy.
In order to realize the aim of improved innovation, four steps were proposed: *believe*, *is*, *should be*, and *became*. The first step describes the current 'official' picture of the company. This is labelled *believe* because the official picture is often far from reality. Here, this picture was developed from organizational diagrams and interviews with management. The next step focuses on establishing what *is* or the true picture of the company. This 'true' picture was built up by empirically mapping the surroundings, goals, tasks, and development system, based on analysis of current projects and interviews. Next, the *should be* step was used to define the ideal outcome desired by the company based on the previous analysis. This included the identification of key performance and product portfolio gaps. Finally, the *became* step closes the loop and acts as a measure of what actually changed in the company after the consultancy process was complete. Ultimately this process was widely used and resulted in three key conclusions:

- Diagnosis of company issues is possible through empirical analysis and offers a robust basis for proposing improvements to both the company and development system.
- The main management tools rely on alignment between vision and goals, which are then supported by specific, actionable tasks.
- Any changes to the development system should be associated with explicit measures so that the feedback and improvement loop is integrated in the development process.

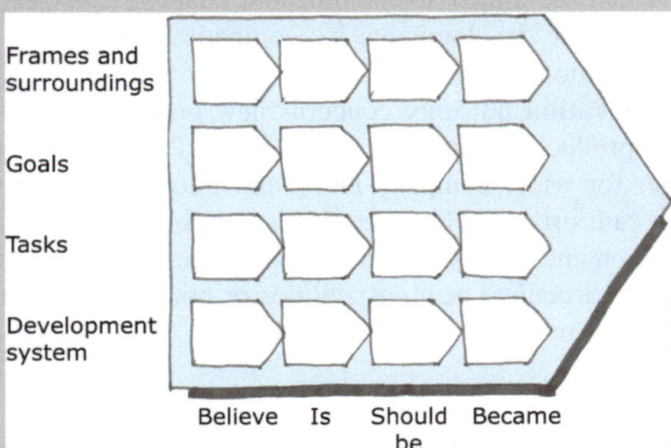

Fig. 9.7 Basic pattern for an organizational innovation process

Product life thinking focuses on alignment between new product development and product life elements including after-sales service, maintenance, and disposal. Although this sounds simple on paper, in reality the stakeholders later in a product's life are often unknown at the product development stage and thus significant care should be taken in considering these elements. We discuss this further in Chap. 13, but suffice to say here, product lifecycle considerations cannot be ignored in a successful development process.

Finally, the last factor we will highlight here is **integration**. This is both crucial and multifaceted—linking to all the other points in this section. Ultimately, effective integration and alignment of these factors is what makes or breaks a successful product development process. As such, this brings us back to the concept of integrated product development. In this integrated paradigm the designer plays a central role summarized in the following:

> One of a designer's key roles is as an integrator and aligner of design effort.

The 'rules' outlined in this section serve to guide designer's thinking when they are planning how conceptualization should be best integrated with the wider process and company organization. Integration is a challenge for the staging, not only on a team level, treated in Sect. 4.4, but also on the level of the whole product development machinery, which describes the tangible structure of the product development process.

9.4 The Product Development Machinery

As we have discussed throughout this book many parts of a company contribute to the development of new products, not just the development department. Instead the development function can be seen as an amalgamation of inputs as illustrated in Fig. 9.9a (Andreasen et al. 1989). These inputs can be further decomposed into seven distinct sub-systems as shown in Fig. 9.8b (Sant 1988) and summarized below:

- **Organization structure** defines the arrangement of tasks, responsibilities, and staffing.
- **Decision structure** links strategy, tactics, and operational decisions to the tasks to be carried out and the associated organizational units and results.
- **Social system** defines the formal and informal goals, norms, and values, underpinning staff's activities and cooperation.
- **Methods and tools** define the approaches used to complete the product development tasks.
- The **knowledge structure** collects and develops knowledge by connecting the internal and external knowledge sources used during development.

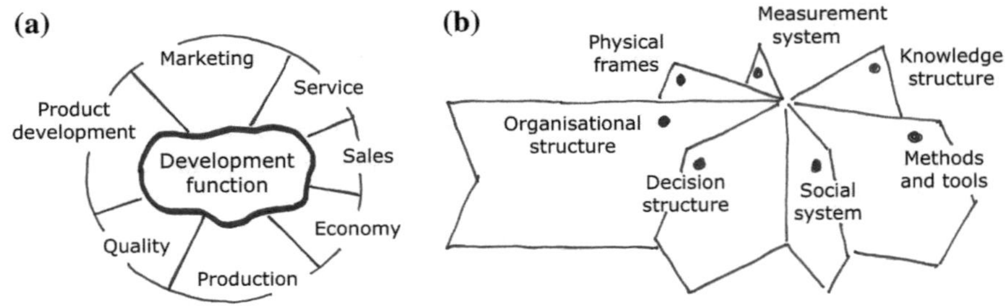

Fig. 9.8 A company's development function (**a**) and the seven sub systems in the development system or machinery (**b**)

- The **measurement system** is the means by which strategies, goals, and subgoals are monitored. This includes and integrates operations level key performance indicators.
- **Physical frames** denote the environment where the development activity takes place.

These seven subsystems form the 'machinery' through which product development is realized. As such, we now explore the implications of each one in greater depth.

The first subsystem to consider is the **organizational structure** as this is the core around which tasks and staff are arranged. In the context of a project, organizational structure is dynamic, changing as the company matures. For example, companies often start with an entrepreneurial approach before becoming more specialized as functions are split into decentralized divisions. Typical steps in business development are: introduction, growth, maturing, and liquidation. At each step there are certain high-level goals related to company output and competitive advantage that are reflected in its organization.

Traditionally, companies execute product development in the form of a project. This means that tasks are defined with respect to time and output, often in a cross-functional organization. A well-known example of this approach is the matrix organization; where the company's various functions deliver staff to teams that each has a project leader. A quirk of this structure is that staff often experience conflicting management between the function leader and the project leader. A number of other approaches are also found at different steps in company development. For example, experimental, opportunistic activities are more common in the entrepreneurial stages, while splitting research and development into specialist groups is usually adopted when the product is mature and optimization is the main driver.

Closely tied to the organizational structure is the **decision structure**. Decisions typically follow the formal hierarchy in a company and, as such, are closely related to organizational approach. The decision structure is what transforms goals and strategies into concrete plans to be realized as specific product development tasks. A strategy group or a product committee usually manages product development, while new products are dealt with by thorough product planning activities

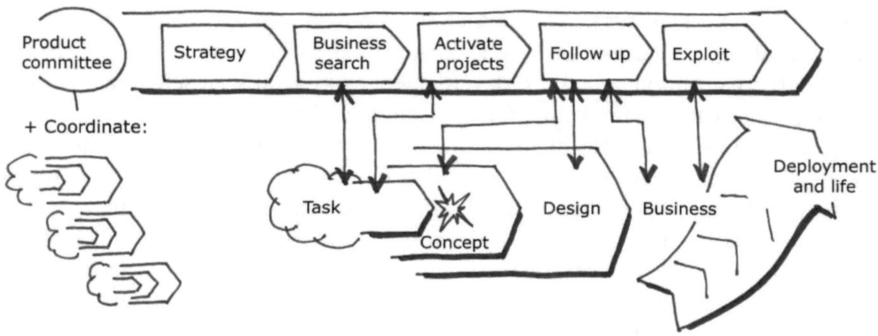

Fig. 9.9 The product committee plans and coordinates the development activities

(Andreasen et al. 1989). In order to support new product planning there is typically a following group (sometimes made up of top management) that is responsible for ultimate approval of concepts and launch decisions. A key concern here is in the effective integration of exploration and concept synthesis activities between product planning and product development. The relationship between these various bodies is illustrated in Fig. 9.9.

Although the formal structures outlined above play a core role in shaping product development, one of the most important sub-systems is the **social system** linking people. The social system describes relationships, competences, political power, and collegial networks in the company. These relationships are often invisible to outsiders and transcend formal role descriptions. The success of the social system is critical to effective performance and cannot be underestimated. For example, consider the shear volume of books written on company culture and 'winning' teams.

While the social system may dominate staff interaction, **methods and tools** dominate the technical aspect of product development. This is also true of models to a lesser degree. As we discussed in Chap. 4 these elements are inseparable from a company's problem solving approach and knowledge. Tools provide supporting procedures for engineering, integration, and management tasks. Methods and tools affect every aspect of product development from planning to environmental impact analysis. As such, they need to be carefully tailored to each project in order to be combined and executed effectively in the wider community of practice. In particular conceptualization is dependant on creative mindset, and communication tools.

The **knowledge structure** is a mental construct describing the knowledge elements of product development. This includes how knowledge is collected, structured, communicated, and utilized in development. It is not enough simply to store knowledge, if it is to be used it must be easily available, readily applicable, and concrete. In particular it is important to consider how knowledge should be articulated in procedures and methods. For example, in the conceptual part of new product development, application knowledge is closely related to a designer's awareness of creativity in a tacit form. This is then transformed into ideas and concepts that can be challenged and assessed. In a company, knowledge structures are interrelated with organization structures, development approach (from scientific to craftsmanship), and with the marketing focus (from broad branch to customer insight).

Finally, all of these subsystems are in some way reflected by changes in a company's performance. As such, the last subsystem we will deal with is the **measurement system**. This is often treated as a simple measure of economic balances, frequently made with ridiculous precision in comparison the large number of elements not measured or controlled, e.g. development cost in relation to turn over, number of new products, and innovation. In this sense the vitality of a new product can be seen as a balance between the projects' business results in the first three years after launch, the actual person-hours used, the number of corrections to components or production, and the estimated production performance verses reality. Successful measurement takes into account both individuals' and teams' performance without losing sight of the overall strategic goals. In particular measurement should be used as a feedback mechanism for directing changes and ensuring that things are in fact improving. However, a word of caution is that measurement must always be considered holistically. For example, design influences all aspects of the product lifecycle. As such, using design-focused measurement might cut costs at the design stage only to incur serious problems in, e.g. product quality during production, resulting in extra costs exceeding those savings made during development. In the conceptualization context measurement is about alignment with strategy and the overall plan for innovation. This concerns the amount of effort to be invested in conceptualization activities and how the outputs of these can be assessed. This dimension is normally related to the goal formulation for the product but can be expanded to reflect the team's performance in order to account for more social dimensions.

Ultimately, the model shown in Fig. 9.8b should be used to develop a deeper understanding of the many factors influencing the successful progression of product development. However, the nature of the culture in each company will determine the magnitude of influence each element exerts on the overall process, e.g. a focus on strong staff performance measurement or the promotion of certain design support tools. As such, the designer must weigh these sub-systems against both the company and the type of product development to be undertaken.

9.4.1 Types of Product Development

In the context of conceptualization, the main aspects of product development we must consider are the types of concepts, products, and development projects we find in industry. In this regard we group projects into the three types described here for simplicity.

In the development of **new products**, often called innovative design, conceptualization takes the form of an explorative, experimental activity. This applies in both entrepreneurial and established contexts. Here exploration and concept synthesis are in focus and, depending on the situation, are augmented by product design and product development in the realization of the concept.

More usually a project will build on past solutions or technologies via **incremental design**. Even in novel new product development elements of this type of design are almost unavoidable, as all technologies build on some established elements in their realization (Arthur 2009). Here, the main challenge facing companies is in establishing a sufficient competitive advantage. This could be through product branding, reduction in resource use or in reduction of risk through the use of past partial solutions. In this context conceptualization requires insights from the existing product catalogue, company assets, and precise market information. Although these elements are needed to allow the conceptualization activity to remain targeted, care should be taken that creativity is not stifled. The key risk here is that product development becomes a non-reflective upgrade process of mindlessly customer-driven design, leaving no room for innovation or significant change.

The final type of project we deal with is the **platform-based design**. In the broadest sense platforms describe a common core from which multiple variants can be created. This core can be anything from a specific technology, key design principal or specific visual design. The main challenge in this context is developing a sufficiently innovative platform such that its lifespan is adequate to develop a range of products without being overtaken by competitors. Further, it is necessary to constrain the compatibility of new products to the common platform in order to reap the benefits of platform rationalization. This constraint must be balanced against the demand for innovation in the company. Here it is easy to lose sight of the platform's competitive power, when its dominant influence on the designer is constraining their work, especially where there is a conflicting demand for innovation. Thus communication and alignment of expectations in project execution are key.

Example:
Handpresso's development. Following up on the examples related to Figs. 3.8 and 7.1, we want to explore the established business. Nielsen Innovation is a consulting company, which decided to establish production and sales of their new product. In order to do this they established a network of producers and market organizations. The product was launched at a show in Milano 2008 and 300,000 have now been sold in more than 50 countries. The brand is supported by the basic innovative idea and by winning seven international design prizes. Today the company launches new products like their device for making coffee in a car (Fig. 9.10a). The inventor's approach to design is inspired by Leonardo da Vinci's statement: "Simplicity is the ultimate sophistication". Early in conceptualization the innovation company had the dream: A Handpresso integrated in a Swiss army knife (Fig. 9.10b).

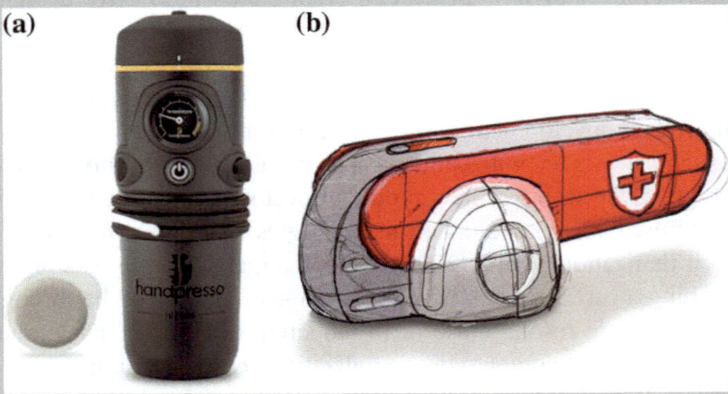

Fig. 9.10 **a** The new car espresso maker **b** The inventor's ultimate dream: a Swiss army knife with integrated Handpresso, *courtesy* Nielsen Innovation, France

Although it is beyond the scope of this book to further explore product development by its self, we do dissolve the question of how functions, properties, and dispositional reasoning can be aligned with these different types of development (Chaps. 11–13).

9.5 Conclusion

Product development forms the fourth module in our Encapsulation Design Model and has been extensively discussed in textbooks, such as Ulrich and Eppinger (2004). However, this discussion has had a tendency to focus more exclusively on product development's engineering aspect. As such, our view of product development, as part of a wider process and underpinned by conceptualization, takes a broader perspective, including those elements which 'cannot be engineered' yet are still inarguably part of product development, e.g. market, customers or sales. In particular, our view of product development coupled explicitly with the exploration and concept synthesis modules allows us to more fully explore the product life synthesis and the creation of products that are fit for life. In doing this our discussion of product development has focused on its wider relationship with conceptualization and the other aspects needed to tailor development activity for a product's whole life. In the next chapter we bring these elements together in the final module of the Encapsulation Design Model, Product Life Synthesis. This brings a product's lifecycle to the fore and explicitly integrates this with the design process.

References

Andreasen MM, Hein L (1987) Integrated product development, IFS (Publications)/Springer, Berlin. Facsimile edn. (2000) Institute of product development, Technical University of Denmark, Copenhagen

Andreasen MM, Hein L, Kirkegård L, Sant K (1989) Udviklingsfunktionen – basis for fornyelse (English: The development function—the foundation of innovation). Jernets Arbejdsgiverforening, København

Brian AW (2009) The nature of technology—what it is and how it evolves. Penguin Press, London

Clarkson J, Eckert C (eds) (2005) Design process improvement, a review of current practice. Springer, London

Cooper RG (1984) Third generation new development processes. J Prod Innovation Manage 11:3–14

Hales C (1993) Managing engineering design, Longman Scientific and Technial UK

Hales C, Gooch S (2004) Managing engineering design, 2nd edn. Springer, London

Kirkegård L, Ryding OJ, Aagaard NP (1996) Produktudvikling – med Bang & Olufsen som eksempel (Product development—with Bang & Olussfsen as an example) Børsens Bøger, Copenhagen

Philips (2013) Refreshing philips brand identity. http://www.design.philips.com/about/design/designnews/Refreshing_Philips_brand_identity.page (Accessed on 20 Nov 2014)

Sant K (1988) Fastlæggelse af Udviklingssystemet (Determination of the product development system). Institute of Product Development, Technical University of Denmark, Copenhagen

Ulrich KT, Eppinger SD (2004) Product design and development, 3rd edn. McGraw-Hill/Irwin, New York

Chapter 10
Product Life Synthesis

Product Life Synthesis is not only determined by the design activity but also by the context in which the product is deployed. This type of synthesis also takes place during the product's lifecycle activities where life systems are established and utilized. In this last design phase we identify three key influences from the finalized product development activity: **need satisfaction** and **new business** are established, and the product's manufacture and utilization leads to **environmental impacts**. When these are better than earlier effects we have innovation. Understanding innovation has its roots in understanding lifecycle conditions. In this chapter we focus on these effects and how designers are able to influence them.

10.1 The Lifecycle: Where Innovation Happens

The Product Life Synthesis describes the lifecycle of a product created during the design activity. The lifecycle is the ultimate result anticipated during exploration and goal formulation. Product Life Synthesis encapsulates the entire design activity, at first as ideas, goals, and design entities, and after launch as sales and reality. There are two main foci for the designer's efforts: creating the best possible fit between the product and the life conditions; and establishing new, specific life conditions, e.g. a disassembly system. The core of Product Life Synthesis is of course the users' deployment of the actual product, where the need satisfaction is now observable and where users gradually develop their perception of the product's value.

M.M. Andreasen et al., *Conceptual Design*, DOI 10.1007/978-3-319-19839-2_10

Lifecycle insight provides the data necessary for creating the product's 'fit for life'.

During the lifecycle, users add contextual conditions that fit the product to their aims and ideas for use. This can range from small practicalities to complex life conditions developed over time. For example, a new car is confronted with society's road system, companies' supply of fuel, public parking possibilities, and government taxation. Other products are launched with fewer concerns, only being modified as negative effects are observed as a result of their use. For example, mobile phones have created unforeseen demands for rare materials, as well as social problems where parents and children struggle to manage access to phones. The impact of a new product typically becomes evident relatively early in the lifecycle, giving rise to substantial remedial efforts to mitigate failures or support success—consider the 'patching' typical of computer games. The lifecycle supports synthesis through opportunities related to utilization, sales, and need satisfaction, e.g. service, reuse, and leasing.

Eekels (1994) proposed a model that highlights three important consequences of a new product's launch, sale, and deployment, here in our illustration using Eekels' terminology (Fig. 10.1) (compare with Fig. 3.21): **business results**, **need satisfaction**, and negative **side effects** and waste.

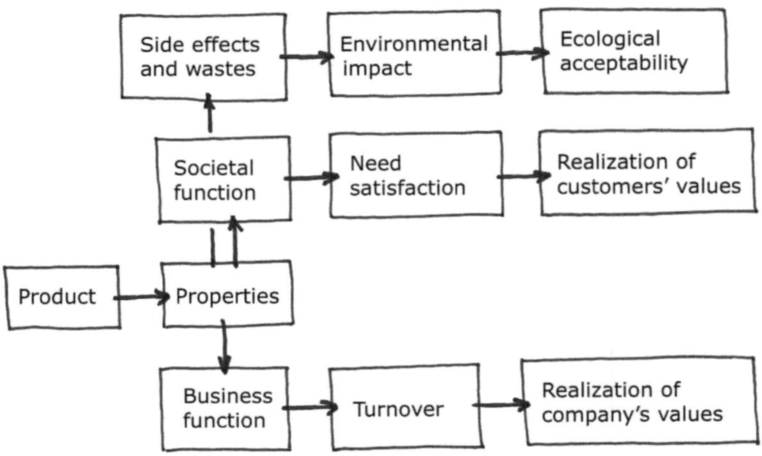

Fig. 10.1 A product's effects: business, need satisfaction, and negative side-effects

10.1.1 The Design Process Paradox

When we consider the complexity of the lifecycle with its use activity, activity chain, life phase systems, and stakeholders, we find a design process paradox: the logical direction for reasoning would be following the arrows in Fig. 10.2 (lifecycle and need insight > deployment process > business, balancing user need and market size > design feasibility > concept consideration > ability to cope with the task), however, current models advise that we start with the task and progress from there (i.e. we try and predict the ideal solution and design activity). In this way the goal formulation can easily become a substitute for finding real insight.

We dissolve this paradox using the Encapsulation Design Model where lifecycle is the result, rather than the product. The Link Model allows us to balance the designer's goal perception and the user's value perception (see Chaps. 11–13). This is reflected in Jeppesen's (Andreasen et al. 1973) encouragement for designers to personally confront the need situation as a condition for the design activity, Fig. 6.6. Understanding and focusing on the really important conditions helps the designer deal with the lifecycle's complexity. Unfortunately, we have seen too many projects start with a concept that is 'believed to be good' but ends in disaster due to poor understanding of users' needs and values. As such, this chapter will explore the nature of the lifecycle and its relationship with conceptualization and product design.

- Section 10.2 seeks to answer: what is **the nature of a product's lifecycle**?
- Section 10.3 then looks at the relationship **between a product and its lifecycle**. We use 'fit for life' and the theory of dispositions to discus how we can create the ideal product lifecycle.
- Section 10.4 finally deals with **design of, and for, the lifecycle**. This activity helps us establish a balance between established lifecycle patterns and new activities.

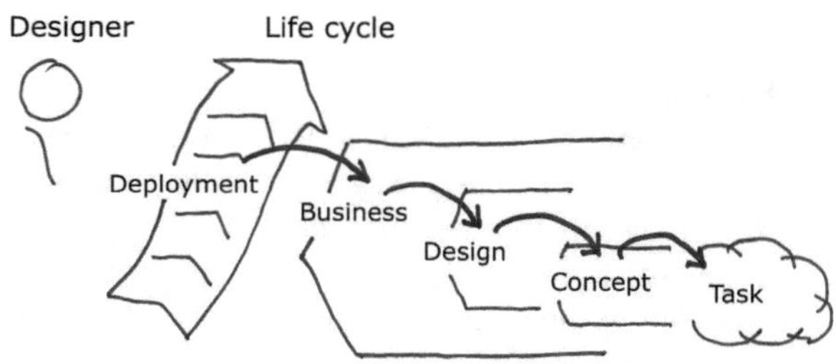

Fig. 10.2 The flow of ideal insight from lifecycle to task

10.2 The Nature of a Product's Lifecycle

The activity in which a product performs its role is at the heart of the relationship between product and activity perspectives, e.g. cycling to and from work or the cook's use of an oven. From this specific operation activity we can expand our scope to include use activities, and ultimately, the totality of the product lifecycle.

Definition: A product lifecycle is the totality of activities related to an individual product's life, from its establishment through its deployment to its disposal.

During most lifecycle activities the product is not passive, instead **lifecycle systems** are acting: production, assembly, distribution, sales, maintenance, reuse, and disposal. This gives us a key characteristic of a lifecycle: it is populated by **actors** with overlapping areas of interest. For example, users' focus on the quality of the use activities while manufacturers' focus on the buyers' need satisfaction and value perception. Further, many actors are focused on specific activities with respect to the product and its use: distribution, installation, maintenance, accessories, and so on. Finally, in the closing phases of the product's lifecycle, actors focus on, e.g. upgrade, remanufacture, partial reuse, recovery and recycling of raw materials, safe disposal, etc.

Traditionally we only consider a single product's destiny in the hands of a single user. However, from the designer's perspective this is an idealization. Designers must also consider variations in products' destiny and satisfaction of customers' needs. Although our discussion here may make these activities seem well known and easily understood the reality is far murkier. This is driven by interfaces with many other products and systems' lifecycles. For example, a television's lifecycle interfaces with various broadcast services using a range of distribution channels (e.g. satellite, cable, internet), with repair shops, and with public disposal systems, all seen as lifecycles.

Example:
Food supply network. Figure 10.3 shows a service network where meals are delivered either ready-to-eat or with hot components. This system interacts with a system for providing appliances, e.g. vending machines, microwave ovens, software for meal configuration, packaging facilities, and distribution. The identity and variety of the food is related to the organic food manager and delivery requests are partially linked to a system for dietary management based on individual patients' prescribed diets.

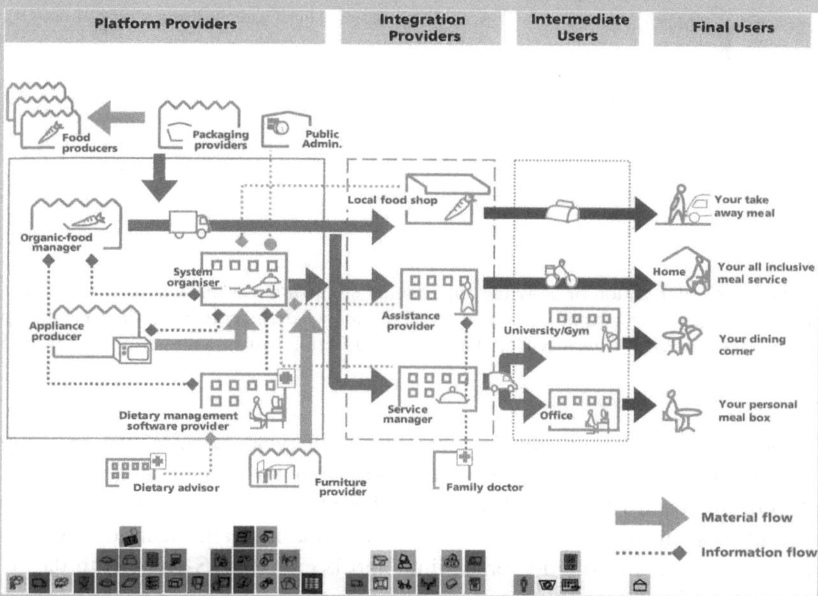

Fig. 10.3 Interacting lifecycles: food delivery, appliances, and medicine courtesy Luisa Collina (Manzini et al. 2004)

The lifecycle is distinct from a **value chain**, which is the chain of activities leading to value from suppliers, manufacturers, and service companies (Porter 1985). Value is related to company activities and functions linked to clarification, product synthesis, production, sales, distribution, and after sales service. The value chain view can be expanded to the 'composed value chain' leading to customer satisfaction. Key questions for a designer are how their company contributes to this chain and how they can best profit from it.

Example:
Food value chain. One value chain in the example above (Fig. 10.3) is the food. Here, key links are production, packaging, delivery, and the combination of prepared meals and customer information. However, the illustration has several value chains, for example, where one institution is composed of furniture supply, system design, and supply coupled to the food delivery.

It is interesting to compare the value chain perspective with the 'normal concern' fraction of the total lifecycle cost, see Fig. 10.4. The additional cost elements are all linked to service dimensions, where there is potential for new business for the product supplier.

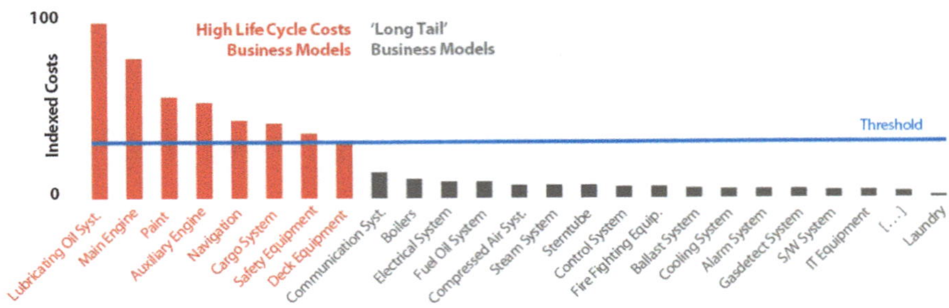

Fig. 10.4 The composition of activities on a tanker ship, showing the activities where the ship owners themselves take care of costs, and the activities where a service supplier has the opportunity to offer a service. *Courtesy* Christian Palm and Lars Balstrup with permission from Torm A/S

Another chain phenomenon is a product's lifecycle, seen as the total aggregated volume of a specific type of product, from the first launched to the last produced, mirroring supply and demand. This cycle typically follows the characteristic phases: introduction (infancy), growth, maturity, and decline. This can be described in terms of volume versus time and is called an **S-curve**, similar to the development of new technologies.

10.2.1 A Typical Lifecycle Activity

Each element in a lifecycle is an activity where there is interaction between the product and the activity. Only in the use activity is the product the operator (Chap. 8); in all other activities the product is the operand and specific lifecycle systems are the operators. Olesen et al. (1996) highlight this interaction between product, life phase system, and actor (responsible for, and interested in its proper, efficient execution). Activity results are primarily outputs from the activity, e.g. distributed product, efficiency, environmental effects, and cost. This is illustrated in Fig. 10.5 where 'service' is also included. It is important to note that these four interacting entities can be related to more or less composed systems. For example, a distribution **activity** might be made using a lorry with a local driver (**actor**) that crosses international borders and pays tax (**system**), while carrying compressors (**product**). However, the delivered service might be the shipping, operated by a company responsible for export, insurance, and transport taxes.

A new product's effects, i.e. activity results, can be difficult to assess but serve as important pointers for both product and business innovation. For example, a business partnership could be established with the lorry company above to improve efficiency or the distribution system could allow the development of improved packing technologies.

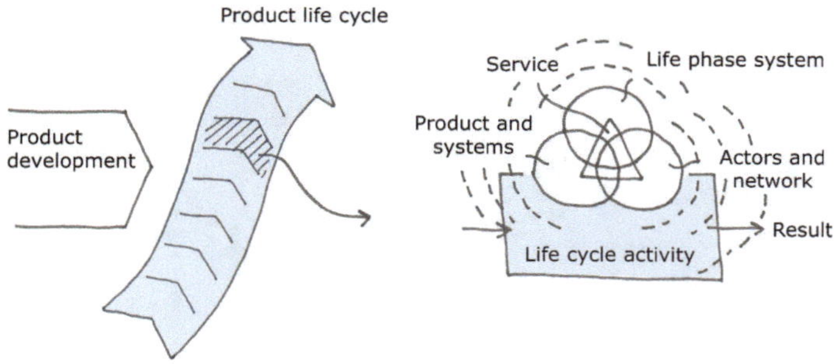

Fig. 10.5 Lifecycle activities as interactions between product, life phase system, service, and actor

> Understanding a product's life activity networks can be a key driver for innovation and can empower business opportunities through new service offers.

10.2.2 Actors and Users

In a basic sense, products are closely related to a user who operates it during an activity. Further, when we follow a product from production to disposal many actors influence, or are influenced, by the product's destiny. These actors should be considered 'customers', i.e. people who react to the product's appropriateness and help form product perception. Understanding the activities and values of these actors has a major impact on the success of the design work and the product.

When we use 'actor' it not only refers to people related to a product's existence but also represents a personification of multiple issues and concerns related to this, e.g. ethics of suppliers, use of natural resources, strength of network alliances. In product life synthesis and therefore also exploration it is important to identify these actors and take their situation and views into account during product development (see Chap. 4). As noted in Chap. 4 socio-technical and actor network approaches are key supports here. These aim to analyse the actors related to a certain task or system by describing a network of interactions. Actors can also be non-human, e.g., companies, legislation or artefacts that define relationships (e.g. dependency on petrol).

> **Example:**
> An actor network for incontinence care. Incontinence (bladder and bowel weakness) affects 5–7 % of the population worldwide and up to 80 % of residents in nursing homes. Therefore organizing care in a community involves a complex web of suppliers, distribution, nurses, doctors, supervisors, and

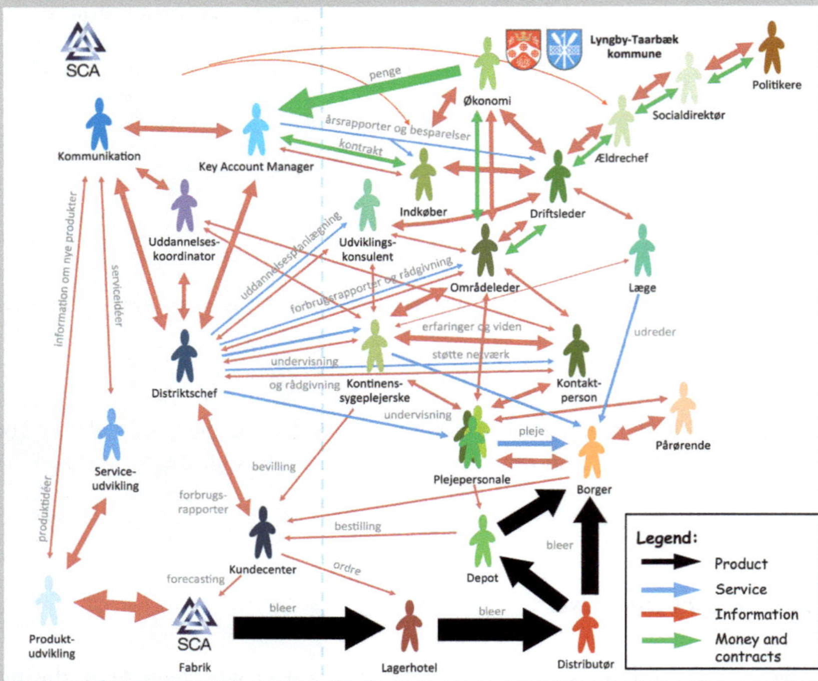

Fig. 10.6 The actor network for incontinence care in a Danish community (Tan 2010)

municipal staff, as illustrated in Fig. 10.6. The company *SCA, Hygiene Products A/S* aims to become a core supplier in this area. Here, incontinence products account for 1 % of costs in a nursing home but 13 % of all operational costs. The actor network shows the company's intended interaction with the area. It is interesting that the model differentiates products, service, information, and money, as well as the interactions' nature (text on the arrows) (Tan 2010).

10.2.3 The Expanded Business View: Service

Over the last half of the twentieth century the backbone of development (task/concept/design/business/deployment) has expanded the scope of product development to include marketing and sales viewpoints. Key to this are the lifecycle activities and the life systems that define the arena for new business. Many of these activities are services, i.e. actions supporting the product's use, McAloone (2011).

Service has received less interest in engineering design and product development companies who have traditionally seen themselves as product suppliers. This is despite services' dominant role in society, employing 75 % of the workforce (ILO 2007). Service industries comprise "*wholesale and retail trade, restaurants and hotels, transport, storage and communications, finance, insurance, real estate and business services, and community, social and personal services*" (Tan 2010; ILO 2007). Such services have research and development processes comparable to new product development. Further, we can see service as the core entity where product delivery is only one part. These two features set the stage for combining product and service development into Product Service Systems (PSS), defined as follows and inspired by Goodkoop et al. (1999).

Definition: A **product service system** is a marketable set of products and services able to jointly fulfil a user's need. One company or a network can provide PSS.

Many industries recognize the business potential of this service-oriented perspective or *PSS strategy*. For example, the aircraft industry has been service-focused for many years. More recently, SKF (a ball-bearing producer) has shifted to include engineering consultancy services, e.g. condition monitoring, industrial sealing, lubrication, and vibration analysis. Similarly, Electrolux delivers professional washing machines and supplies '*laundry systems*', helping to start new launderettes or upgrade old ones, installing, training, financing, etc.

Service primarily describes the non-material part of a customer offering. Tomiyama (2005) identified the key elements: "*A service receiver receives service contents from a service provider through a service channel. Service sent by the service provider changes the state of the service receiver, which is the most important feature of service as* [*an*] *activity*". A service's mode of operation is affected by the activity in which the service acts, leading to the model in Fig. 10.7 (Matzen et al. 2005). Service is executed in parallel to the user's activity (a) and is supplied via service channels related to an operator's products (Ts), manpower (Hu), information (I), and management (M). These could be, e.g. helping materials delivery, data base information for analysis, and quality management. Modelling service as an activity chain (b) can build on this book's activity modelling but alternatives exist, e.g. service blueprints.

Many products are characterized by a repeating deployment pattern in the form of production or use cycles, frequent similar utilizations (e.g. car use) or repeated preparation, upgrading, adjustment, and maintenance. Thus activity, from a service perspective, should be seen as cyclical, i.e. a **Customer Activity Cycle** (Vandermerve 2000; Matzen 2009), see Fig. 10.8. A distinction is made between activities related to **Pre**: the customer is deciding what to do, **During**: the customer is doing it, and **Post**: the customer is sustaining it.

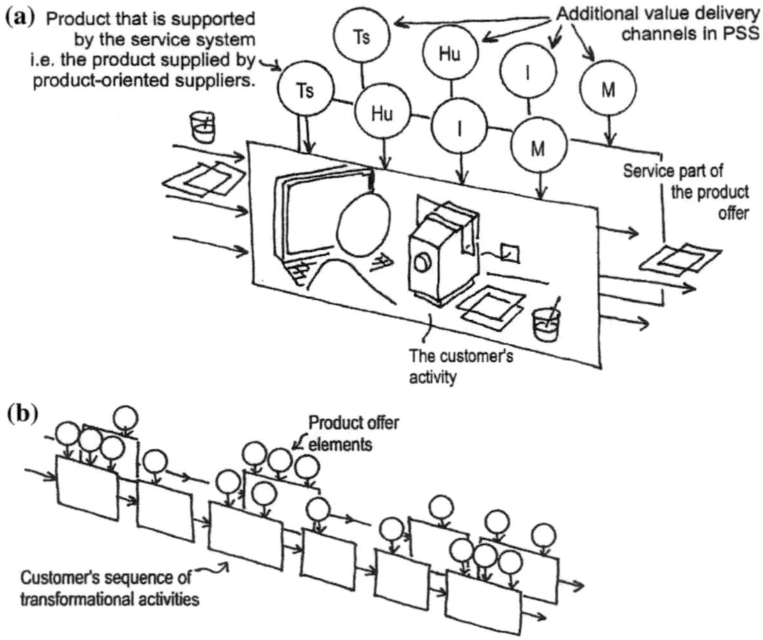

Fig. 10.7 Adapted transformation system model showing service parallel to the user's activity

Example:
The customer activity cycle

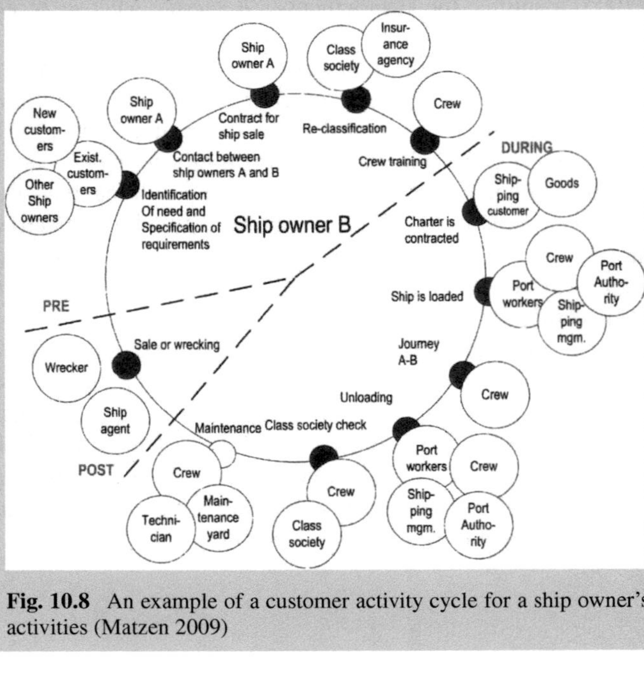

Fig. 10.8 An example of a customer activity cycle for a ship owner's activities (Matzen 2009)

Figure 10.8 shows a model of a merchant ship's activity cycle, where the ship owner is the core actor. The sequence of activities follows the lifecycle from contracting, commissioning, and operation to re-approval for sales or wrecking. The network of actors related to the activities is identified and plotted, here with the aim to identify a service-oriented actor network (Matzen 2009).

In Chap. 5 we identified several design entities that interact during need satisfaction. Lifecycle and service activities are also related to these in an interwoven pattern of activities by the customer and service provider; containing the design entities:

- The **life phase activity** including actors' and users' tasks, as realized by the customer.
- The **life phase system** in the customer's staging.
- The **service executing activity** where the service is delivered in the customer's staging.
- The **service channel** via which the service provider delivers the service.
- The **service offer** seen both as the delivered service and the financial agreement.

A service-oriented approach to design means the designer must consider *two* lifecycles: the **product lifecycle** and the **customer's activity cycle**, Fig. 10.9. Based on these patterns, services are identified to support the product business and customers' actions. These two lifecycle chains form the landscape for identifying services in a PSS approach.

Adding service to the product-oriented approach described in this book means adding new design degrees of freedom. The designer now has the opportunity to enhance the product's utility and value through service. Further, ideas for service can point to products that satisfy needs in new ways. Key questions here are: how can the design predict these lifecycle elements and how can they be influenced?

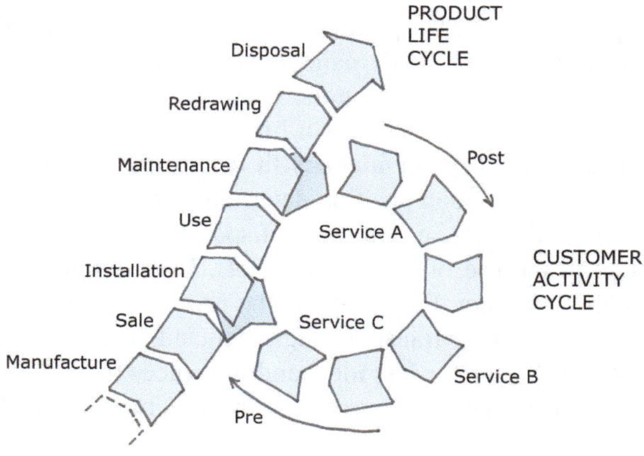

Fig. 10.9 The two lifecycle chains

Service-thinking supports innovative thinking and allows for a bigger influence on the design's proper use and its effects.

10.3 Between Product and Lifecycle

Lifecycle synthesis, from the designer's perspective, means to identify and judge how to make a product best **'fit for life'**. This is tricky as many actors have opinions on the degree of fit. The **user** pays for the product and expects a long, problem free service period as well as ethical, problem free disposal. The **manufacturer** is focused on sales, determined by how well a product fits the user's need and financial situation. Adding service can enhance attractiveness, show ethical and sustainable responsibility, and influence the user's buying decision, but also comes with its own cost and complexity issues. The **actors** include competing companies ready to replace the product, service suppliers, material suppliers, repair companies, re-manufacturers, etc. All these must be respected during the product's development.

It is the designer's task to understand this pattern of interests and issues to identify the nature of lifecycle activities and their actors, and translate their needs into a balanced goal formulation.

Many aspects are only gradually revealed through the design activity and can thus lead to new directions and design results.

The key question is how 'fit for life' can be identified as a product-lifecycle relationship. This has two fundamental relationships: operator and operand. In the first the product is the **operator**. Here, the product serves the use activity and relationships are defined by the product's functions in line with Domain Theory. The main stakeholder is the user and their focus on utility, usefulness, and value. This relationship is dealt with during the product design activity. In the second the product is the **operand**. Here, the product is the subject of numerous activities, e.g. being assembled, distributed, sold, maintained, and disposed. The product is linked to multiple life phase systems and actors. These relationships are core to product life synthesis.

When a product enters a lifecycle activity there are three aspects of fit to be considered. **Type or identity fit** determines the activity's feasibility and meaning. The product's composition determines its fit with a range of supply and manufacturing activities, e.g. product disposal might not fit established disposal systems due to the use of new components. **Functional fit** links the product and the life activity system. For example, the product might demand a certain diagnostic tool during repair or transport might demand an eyebolt for hoisting. Finally, **property fit** can be more or less ideal and influences the efficiency of the life activities. This fit is articulated as, e.g. ease of manufacture, assembly, transport, installation, and minimal environmental effects.

Fit is not an absolute state; it depends on 'soft' human systems that either compensate for or hinder fit. In response designed artefacts carry affordances, i.e. a range of functions that can be activated or utilized by the users (see Chap. 11). Similarly, the systems against which fit is judged dynamically change over time. The Theory of Dispositions (Olesen 1992) gives an essential understanding of these fits between product and lifecycle. Olesen defines a disposition as:

> **Definition:** "*A* ***disposition*** *is the part of a decision taken in one functional area that influences the type, content, efficiency or progress of activities within other functional areas*".

Decisions regarding product characteristics and their influence on the activities are primarily based on considerations of fit. For example, if a design is tailored for robotic assembly, it will have few assembly axes and have prepared surfaces for grippers; these can then be assessed by the efficiency of the assembly activity. If manual assembly is used instead the misfit will reduce efficiency. Thus, the designer disposes the choice of assembly method. In Chap. 13 we explore dispositional reasoning further, as such, we merely introduce it here.

Olesen et al. (1996) propose a model where product elements are related to life activities and effects or consequences are identified. The aim of the model is to identify ecological effects but it is also useful for other types of relationship influencing the life activity. Figure 10.10 illustrates the model, in which a product's composition is confronted with the lifecycle activities. This raises the question: what components cause the effects in this life phase? Effects can both add value or be harmful negative effects, and relate to the activity's elements: life phase system, product, actor network, and service. The example composition illustrated here comes from Fig. 10.6.

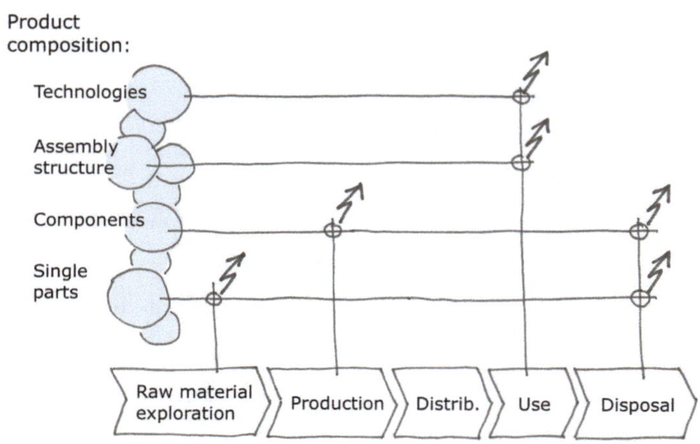

Fig. 10.10 Product composition confronted with lifecycle and potential effects

10.3.1 Actor Roles

Lifecycle synthesis introduces many different actors each influencing the life activities. As such, fit to actors is just as challenging as the fit to life phase systems. At the heart of the product lifecycle is the use activity, with the user as actor. In this activity we see a composed articulation of the product's goodness with contributions from the product (its functions and properties), the activity (its progression and properties), and the user (their experience, wellbeing, and perception of the result). A key problem area is recovery, e.g. a family's daily consumption, leading to problematic amounts of waste. Here, waste represents consumption of natural resources that should be recovered and reused. However, recycling depends on sorting, collection, and industrial support, where consumer sorting is the critical link. Recovery is key to reducing material consumption and the negative effects of waste disposal. Common recovery loops are shown in Fig. 10.11:

- **Revitalization** of products based on repair and exchange of parts.
- **Remanufacture** based on partial reuse and exchange of worn out or out-dated components. This aims to re-establish the product's initial quality state.
- **Cannibalization** of used components to re-compose products, leading to a second service period. These products typically have a lower quality than before.
- **Recycling** based on separation, recovery, and reuse of materials.

It is important to realize that users are relatively independent, meaning the product has a limited influence on their use and life activities. One response to this can be to design products 'with a meaning', e.g. by adding specific signs, narratives, style elements or user societies highlighting ecological concern. This type of thinking can lead to both improved user activity, as well as inspiring new product perspectives and innovation.

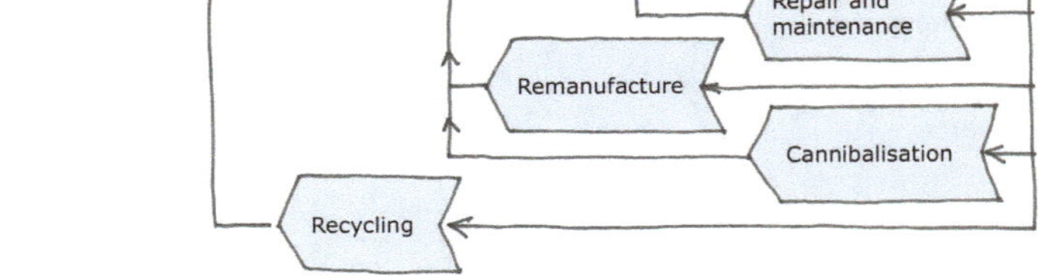

Fig. 10.11 Recovery loops related to lifecycle

10.4 Design of, and for, the Lifecycle

Fundamental to the nature of a product lifecycle is the relationship between product and lifecycle. A key question: is the lifecycle designed, i.e. as a product entity, or simply existing, i.e. the product life synthesis just fits the product to this lifecycle? In either case we must understand the lifecycle, particularly with respect to sustainable behaviour. This is highlighted by Kimura and Suzuki (1996): "*For sustainable product development, it is essential, to first design total lifecycle in order to make reuse/recycling activities, more visible and controllable, and then to design products appropriate, to be embedded in the lifecycle*". This seemingly crazy idea points to the moral that designers who do not know how the product will solve its tasks, what is necessary for creation and disposal, etc., should not be designing new products. Thus we need product life insight; ideally early in the design activity but in practice this comes gradually.

The activities change their nature across the chain because the product is an active operator during its operation activities and a passive operand during the rest of the lifecycle. Of particular interest for the user is the product's **service period,** i.e. the period over which it maintains its productivity, usability, and value for the user before disposal. Together with the influences discussed in Sect. 10.3 we find the following heuristics useful.

- The most probable ideal lifecycle should be identified together with the subsequent changes required in the life systems.
- Based on this ideal lifecycle the product should be designed for best possible fit and effects.

It is important to remember that there are many possible lifecycles and thus the designer has the chance to direct the product into a preferable cycle. We call this lifecycle composition *mode of life*.

10.4.1 Mode of Life

When we discuss mode of life we are dealing with the high level characteristics of the lifecycle activity as a design entity. This is composed of a number of key issues:

- **Customers and markets,** e.g. product users, customer segments, offers, prising, and dynamics.
- **Sales and distribution channels,** e.g. outsourced, geographically determined, physical shop- or web-based shop.
- **Offering types,** e.g. sales, leasing, return agreements, upgrade, and combinations of offers, such as, availability guaranties or renewal arrangements.
- **Position in the value chain,** e.g. component sales, systems sales, OEM agreements, and contracting 'total solutions'.

- **Product portfolio policy,** e.g. offering broad combined programs, customer specific products ('one-of-a-kind'), and systems based on standard components.
- **Competitive advantage**: key sales arguments, e.g. price, quality, environmental properties, delivery, service, and customer support.
- **Recycling strategy,** e.g. arranging internal disassembly, upgrade, and recovery of materials.

These elements are generic and mode of life must be adapted for each company, project, and situation. In particular, the adoption of a PSS strategy can lead to new opportunities (Matzen 2009) because of the added design degrees of freedom. Specifically, McAloone and Andreasen (2002) identify five opportunity areas in PSS: re-invention of core business, increased competitive advantage, greater control over the product lifecycle, increased insight into the product and its use, and alternative approaches to sustainability.

Identification and development of new services is a key part of 'designing the design' (see Chap. 5). Here, the early development stages mirror those of product design, i.e. they build on exploration and concept synthesis. However, the nature of a service is different from that of a product and therefore concretization diverges from the typical product design stages. Tomiyama (2005) describes the 'Service Engineering' process as led by service goal requirements and with the aim to create new services. For both products and service the designer must imagine the lifecycle and identify specific aspects that can be formed into assets.

10.4.2 Lifecycle Identification

Key to this imagining and understanding of the lifecycle is creating the fit for life. This has two roles in this context. First, to identify **similarities** with possible, existing lifecycles to help inform the fit between product and lifecycle. Second, to identify the product's **effects** in each life phase and compare these to the company and actor's goals and values. For example, consider the effect of exceeding weight or size requirements on distribution.

The first role is key to reducing risk and uncertainty in new projects! It is hard to imagine any new product being so unique that it can not be related to existing lifecycle patterns, either in terms of functions and use, existing market offers and customer segments or sales patterns. This gives the designer a foundation for considering a wide range of alternative concepts or life phase systems. For example, an external distributor, responsible for packaging and distribution, might replace a traditional distribution system.

> Identifying and modelling products' lifecycle is a key driver for innovation and finding new possibilities.

Olesen et al. (1996) highlight how both the product and the lifecycle contain multiple design degrees of freedom and thus a high degree of uncertainty. It takes a disciplined approach to map the many potential fit problems, identify dispositions, and judge their pros and cons. Here, Olesen warns against a component-oriented approach because this hides higher-level design degrees of freedom and their effects. Modelling is an important support for lifecycle identification (see Chap. 8), providing an overview and helping clarification, comparison, stakeholder identification, etc. When the lifecycle is unclear scenario techniques can support understanding an imagination.

Example:
Product life gallery. Besides the use activity there is a long line of activities related to the product's life phases. In order to create innovative products we must understand how the product behaves in and influences each life phase. We can use insights from the ideal activity to create products and services with competitive edge. Based on this philosophy we teach an exercise where students create a product life gallery for a piece of luggage. Using this they then propose new products, e.g. disposable case, luggage-detecting system, a case financed by advertisements, etc.

A comprehensive modelling technique is shown in Fig. 10.12 (Tan 2010) in the form of a lifecycle poster. Besides the lifecycle model this brings together actor network, activity cycles, key meetings, interpretation of customer values, trade-offs, and analysis of environmental effects.

Product life galleries and lifecycle posters are useful tools for supporting a lifecycle approach, helping identify service opportunities and establish product service systems. These models can be used at any level of concretization from speculative scenarios to detailed operational models.

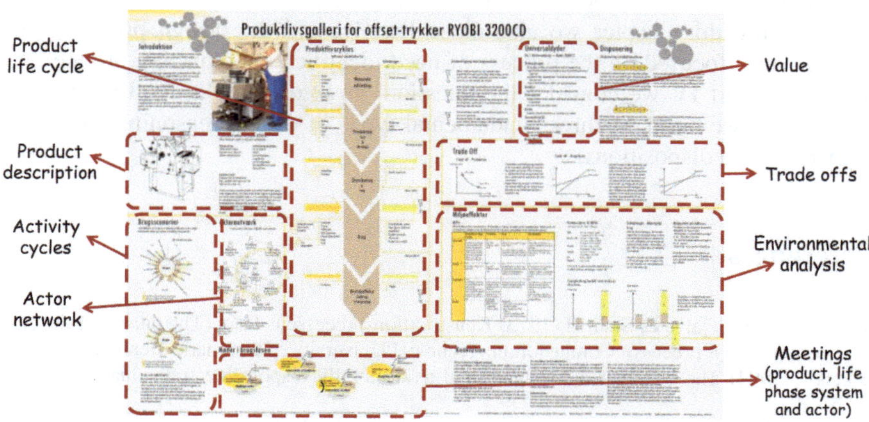

Fig. 10.12 An example of a lifecycle poster (real size A0) for an offset printing machine, (Tan 2010)

10.4.3 Product Life Synthesis

A key outcome of product life synthesis is the comparison between the design result and the ideal description of the lifecycle. This closes the total design activity as described in the Encapsulation Design Model. However, activities after launch are also the focus of the company's sales, marketing, and service departments. These are key to interpreting user activities and supporting the search for new product development opportunities.

A good outcome for product life synthesis is a balanced understanding and utilization of the possibilities surrounding the fit between product and lifecycle. However, it should be noted that this utilization can only be partial because of the complexity and scope of variation between the lifecycles of individual products. The problems companies face are the identification, utilization, and documentation of this search.

10.5 Conclusion

Our discussion of Product Life Synthesis gives a new perspective on product creation and its relationship with the user's deployment and appreciation of the product. We return to these topics in Chap. 13 in our discussion of 'Design for X', which links to lifecycle issues such as production costs, usability, and disposal. Successful lifecycles depend on the product's fit for life, linked to the designer's dispositions. Further, successful product deployment brings together all elements in the Encapsulation Design Model. This chapter has explored how a product service approach can help link users and companies and support a holistic approach to user orientation and satisfaction, including sustainability. This is substantially beyond what we find in traditional product launches.

In closing this chapter we also close Part III and the design process. Until now we have focused on the process of need identification, concept, product, and service. In Part IV we consider the question: how to create a good product? In solving this we build on three fundamental types of design reasoning: functions, properties, and dispositions.

References

Andreasen MM, Tjalve E, Stahl H (1973) Konstruktionsprocessens faser (The phases of the engineering design process). Compendium. Technical University of Denmark, Denmark

Eekels J (1994) The engineer as designer and as a morally responsible individual. J Eng Des 5:7–23

Goodkoop MJ, Halen CJG van, Riele HRM, Rommens PJM (1999) Product service systems, ecology and economic basics. VROM/EZ, Dutch Ministries of Environment (VROM) and Economic Affairs (EZ), No 132

ILO (2007) Key indicators of the labor market. International Labor Organization (ILO)
Kimura F, Suzuki H (1996) Design of right quality products for total life cycle support. In: 3rd international seminar on life cycle engineering, CIRP 'Eco-performance 96', Zürich, Switzerland, pp 127–133
Manzini E, Collina L, Evans L (eds) (2004) Solution oriented partnerships. University Cranfield, Cranfield
Matzen D (2009) A systematic approach to service oriented product development. Ph.D. thesis, Technical University of Denmark
Matzen D, Tan AR, Andreasen MM (2005) Product/service-systems: proposal for models and terminology. In: Meerkamp H (ed) Proceedings of 16th symposium 'Design for X', Neukirchen, Lehrstuhl für Konstruktionstechnik, TU Erlangen-Nürnberg
McAloone TC (2011) Boundary conditions for a new type of design task: understanding product/ service-systems. In: Birkhofer H (ed) The future of design methodology, vol 10. Springer, Berlin, pp 113–124
McAloone TC, Andreasen MM (2002) Defining product service systems In: Beiträge zum 13th Symposium 'Design for X', Lehrstuhl für Konstruktionstechnik, TU Erlangen, pp 51–60.
Olesen J (1992) Concurrent development in manufacturing—based upon dispositional mechanisms. Ph.D. thesis, Technical University of Denmark
Olesen J, Wenzel H, Hein L, Andreasen MM (1996) Miljørigtig Konstruktion (design for environment). Institute of product development. Technical University of Denmark, Copenhagen
Porter ME (1985) Competitive advantage: creating and sustaining superior performance. Simon and Schuster, New York
Tan AR (2010) Service-oriented product development strategies. Ph.D. dissertation, Technical University of Denmark
Tomiyama T (2005) A design methodology of services. In: Samuel A, Lewis W (eds) Proceedings of ICED 2005, engineering design and global economy, Melbourne
Vandermerve S (2000) How increasing value to customers improve business results. Sloan Manag Rev 42(1):27–37

Part IV
Reasoning About the Good Product

Chapter 11
Function Reasoning

This chapter opens Part IV: reasoning about the good product. In this part we deal with function, property, and dispositional reasoning. Goodness is first and foremost related to the user and need satisfaction, but also to actors in the lifecycle and to the manufacturer's perception of good business.

Both activities and products carry functions and in the composed product we find functions related to organs as introduced in Chap. 8. Ultimately, these all build on the user's perception in order to assess functions' appropriateness and value. This is illustrated in the Link Model, which is the reference model for all three chapters in Part IV. Creating proper functionality in a product has a huge influence on users' perceived value.

11.1 Functions

A product's functions determine its identity and are key to satisfying users' needs. The starting point for function reasoning is thus in the need, intention, and goal for what the product should do and what the user can do with the product. This allows us to identify solutions that we can use to realize these functions and compose the product. Thus, functions determine a product's utility, usability, excitement, productivity, and value. It is critically important to be able to identify and build in functions; we call this competence **function reasoning**.

M.M. Andreasen et al., *Conceptual Design*, DOI 10.1007/978-3-319-19839-2_11

We have already met the term function: as the core of product ideas, as part of creating concepts, and as the label for organs during product synthesis. In practice identifying a product's functions is a major task. A balance must be struck between the user's/client's wishes, future actors related to the life cycle, the product's use and life cycle activities. Faulty composition of functions can negatively impact goal satisfaction, and where additional functions must be added late in the design process can have a major negative impact on overall project progression.

Example:
Rescue lift. When fires strike large buildings it is essential that most occupants are able to quickly rescue themselves by evacuating the building. However, wheelchair users or people with walking difficulties are often not able to manage the many stairs involved. Thus, a device is needed to help rescue these people. We can imagine devices such as a sledge, operated by volunteers, which can slide safely downstairs. Early ideas might focus on analogies, e.g. sledges, tracked vehicles, multi-wheel vehicles, multi-legged walkers, people with straps at both end of the device, or people linking hands to collaboratively carry someone, Fig. 11.1. Based on these we can start to identify sub-functions: able to navigate corners, support a range of possible occupants, etc. Further, we might ask who should reset the device ready for the next rescue? How? Identifying these use functions is thus related to both the users' and the helpers' roles. Gradually we can refine these elements to describe detailed use activities and product characteristics. Thus, our reasoning on functions is closely linked to our reasoning on properties.

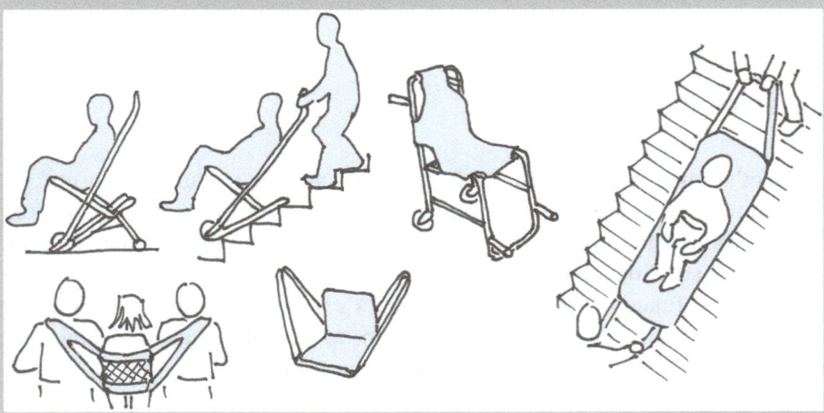

Fig. 11.1 Design sketches for rescue devices

This example highlights the difference between our imagined use of a device and the functionality built into the real product. This dualism is central to this chapter and links to our previously stated design philosophy: *both the* ***product*** *and its use* ***activity*** *are carriers of functionality and thus both are something to be synthesized (being design entities)*. Vermaas (2014) explains this as a confrontation between the traditional engineering-based **rational view** and the user-oriented **intentional view**. In the terminology of this book we distinguish between *the idea with* the product and *the idea in* the product:

We must distinguish what we want to do with the product, its **use function**, and what we want the product to do, its **action function**.

The main entities that carry a product's functions are the activity and the organs. These functions are achieved using natural phenomena to create effects. Understanding these building blocks from a functional viewpoint is essential. As such, we base our understanding in the concepts '**mode of action**' and '**mode of use**' introduced in Chap. 8, which build on natural phenomenon creating effects via state changes.

At the heart of understanding functionality is understanding mode of action and mode of use, i.e. the **behaviour** and its **effects**.

Creating a composed product is based on cause/effect reasoning about the interaction of its elements, i.e. the activities and organs. Understanding these interactions is the prerequisite for creating a network of functions, which forms the foundation for composing activities and organs.

The network of functions based on cause/effect patterns leads to the identification of the product's composition, i.e. its **structure** of activities and organs, and their **interaction**.

In daily language the term function is vague. When we say that something **is** a function, we focus on behaviour not the organ or activity realizing the function. Thus, what we call a 'function' in daily language is in fact the apparent effect of an artefact's properties.

- *Functions* are **not** artefacts; they are the artefact's **properties**.
- *Effects* are **not** functions; they are how functions **influence** the world.

Functions are ubiquitous in design methodology because, as designers, we must be able to describe a concept all the way from vague idea to concrete product. Functions are one of the few elements that are able to be described at any level of precision and thus can form a foundation for the whole process; from a single idea to a network of functions, and from simple 'on/off statements' to questions of good functional properties. Despite this flexibility many approaches use highly constrained functional models, meaning that many possibilities are lost (Eisenbart 2014). We adopt a fundamentally different approach in this chapter.

- Section 11.2 first explores how **functions can be used to link task and product**.
- Section 11.3 outlines functions' role in cause/effect reasoning based on **understanding the core elements**: **activities and organs**.
- Section 11.4 shows how functional interaction can be determined and leads to the composition of the organ structure: **the function network and product composition**.
- Section 11.5 focuses on production by looking at **creating the part structure**: **from function to embodiment**.
- Section 11.6 looks at how **cross-disciplinary function reasoning** is supported by the discipline independence of functional language.
- Section 11.7 examines the consequences of **function reasoning during synthesis**: **design-type dependency**.

11.2 Functions for Linking Task and Product

From one perspective 'function' is simply a label we use to discuss an artefact's abilities, action, deployment, etc. Alternatively, 'functions' are objective things carried by an artefact.

> **Definition**: **Functions** are a **product or activity's** ability to do something actively or be used for something, i.e. deliver an **effect**.

Based on this definition there is no such thing as a 'passive function'. Thus identifying the active element is key to function reasoning. This means that we can differentiate two function types: use and action. A **use function** describes when effects are obtained through a product's use, while **action functions** describe when the product or its organs deliver effects. There are only four types of effect: material, energy, information, and/or biological. It is through these effects that an artefact interacts with its environment, as illustrated in Fig. 11.2.

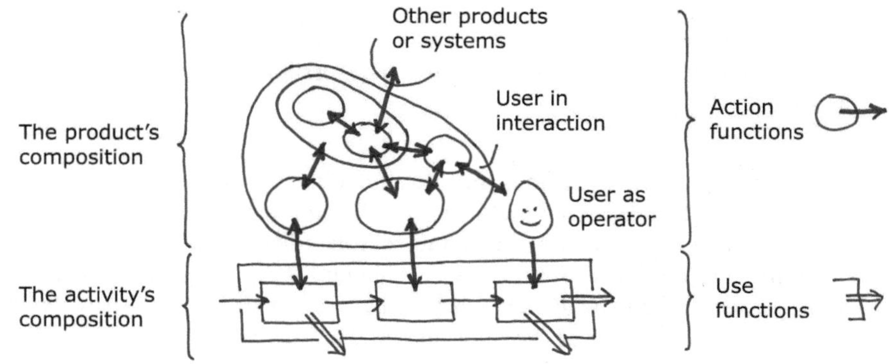

Fig. 11.2 Use functions and action functions, all based on effects

Example:
Electric drill. An electric drill is typically used to drill holes using a spiral bore. In daily language we might say '*the drill's function is to drill holes*', but as designers we must be more specific. First, we can define the technical elements: the spiral bore coupled with feed force and rotation is able to make holes. Here, the drill's motor/gearing delivers the rotation, but an operator is required to position the device, activate it, and deliver the feed force, which is transmitted through the machine. This means we have two functions: the **use function** 'bore holes', and the **action functions** 'create rotation' and 'transmit feed force'. We can also expand these if we consider alternative uses for the electric drill, e.g. mixing paint or polishing surfaces, where other tools replace the spiral bore.

11.2.1 Other Types of Functions

Vermaas (2014) describes two perspectives on functions: structural/behavioural and intentional. The structural/behavioural interpretation focuses on product composition, while the intentional interpretation focuses on aim, use, and value. In addition to these, three other major interpretations of function exist. First, **potential** or **latent functions** describe when alternative product interpretations create effects linked to action functions (using a coffee brewer to make tea) or use functions (using a toy torch to charge a mobile phone). Second, **emergent functions** describe when effects emerge for users exceeding the designers intended product functionality, e.g. repurposing an old phone as an interactive child's toy. Finally, **affordances** describe the complete set of a product's use functions, including both unforeseen and planned functions (Brown and Blessing 2005).

The product should be designed such that it is used as planned for its intended purpose, where additional uses and possible harmful effects are prevented.

We see functions as active and thus they determine the utility and deployment of the product. As such, we do not consider the following features to truly be functions: **aesthetic**, **semantic**, **social**, **economic**, **status**, **ideological** or related to life activities like **maintenance** (Monö 1997). Although these terms relate to functions, properties, features, or patterns they actually describe the value perception viewpoint (Miles 1972) where they link the product to certain conformities. For example, cars are taxable objects in the eyes of the tax authorities, but this is not a function of a car. Similarly, a car motor's function is to 'create force'. Due to its mode of action it gets hot, creates exhaust gases, consumes fuel, etc. but none of these can be described as functions unless their state changes are utilized, e.g. using the heat to warm the passenger cabin. Instead the motor has properties, e.g. efficiency or rotational speed that define its suitability for a given task. In our definition products manifest themselves through their properties, effects, and functions. As noted above (Fig. 11.2) these effects can be material, energy, information, and/or biological.

Example:

Danger signals. In most Danish railway stations there are marker lines denoting a safe distance from the platform edge. This is created using plastic yellow discs that are glued to the platform (Fig. 11.3). What is their function? Is it a passive effect? Is there an activity and an output? From an

Fig. 11.3 Platform with warning sign (saying: The *yellow dots* may not be passed if there is no train) and marked risk area

information perspective, the signs 'programme' the passengers to connect the discs and their colour with danger. This changes the behaviour of the passengers, who are less likely to move into the danger area. Thus, the active effect is of an information nature created by the discs, but the generated effect is found in the passenger's mind and subsequent behaviour.

Functions are at the core of people's perception of products and activities and link to their usability, utilization, operation, and purpose. This is the foundation for the Link Model, which helps us answer the question: What functions should we build in?

11.2.2 Link Model

The Link Model is illustrated in Fig. 11.4. This describes the situation when the user actually uses the final product and forms a perception of its value. On the left we show the designer imagining a user's need and value perception to create a goal formulation. The key assumption described in this model is that the user's perception of value must be closely linked to the designer's reasoning from goal to synthesis. In this way the model presents an idealized interpretation of practice; the designer can add valuable new dimensions but can also make naïve or wrong goal formulations.

A user's perception of need satisfaction and **value** is based on the **recognition of functions** in the product and use activity.

The model shows how the designer sees the product and the use activity as systems with structure and behaviour. Although the actual workings of a product are

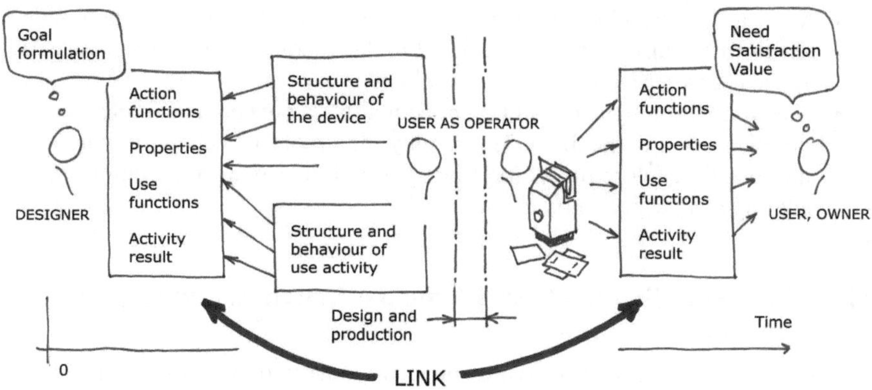

Fig. 11.4 The link model showing the user's value perception and the designer's reasoning

often a 'black box' the user still experiences the functions as tacit insights. The key here is the **activity result,** i.e. how the output from the use activity satisfies the need. At the core of a product's deployment is how the user interprets its use. With respect to the previous danger signals example, the authors have seen children jumping from one plastic disc to another as a game—quite a different interpretation of the product from the intended safety warning. This leaves us with two perspectives when designing the product.

- The **'idea with'** is use and user dependant leading to multiple, ambiguousness functions linked to the actors, situations, and life phases. For example, in the egg sausage machine (Chap. 8) the 'idea with' is the sausage delivered to the consumer. The user is the sausage producer and the key qualities should be in accordance with the customers' taste.
- The **'idea in'** is based on the designer's intent. For example, the egg sausage producer might instruct the design team in how the manufacturing machine should function, e.g. automatic operations or how the sausage should be packed and delivered. This leads to a certain composition of the machine's functions.

These different perspectives allow us to consider both the known degrees of design freedom and to propose new ideas where the product 'creates a need' unknown to the user. For example, text messaging on mobile phones was originally designed so that operating companies could notify users of important information, e.g. payment options. Accordingly, text messaging was initially a free service. However, users found that they could also use the service to send messages to other users, something that has become extremely popular—even superseding tradition voice calling. Thus, a new need was discovered together with many new market opportunities. The demarcation between what the product is able to do and what the product is actually used for is essential.

- A product cannot be understood without use and context. Remember that both the bicycle and bicycling are inventions. Both the product and its use activity carry functions that are asking for original solutions.
- A key to successful synthesis is being able to articulate the ideal product's functions but also being able to adapt to the uncontrollable effects created by the user after launch.
- Function reasoning can lead to the feeling that a concept is complete once the functions are identified. This is only the tip of the iceberg, it is important to fully explore the action functions and their realization before considering a concept complete.

11.3 Understanding the Core Elements: Activities and Organs

As highlighted by the final heuristic above we must understand how things work and how their functions are realized. This comes back to **state changes** as the fundamental means for understanding design, i.e. creating a product's functions and properties, and understanding the processes of need satisfaction and product realization. More practically, we can describe this in terms of function, behaviour, and structure.

11.3.1 Products and Activities

Artefacts describe objects created by humans, while products include additional ideas such as sales and fulfilment of a need.

Definition: **Product** is a general denominator for materialized, executable artefacts, i.e. artefacts able to carry behaviours and realize functions and properties through a **use activity**.

Activities (with the synonyms: processes, operations, and transformations) describe a sequence of related state changes over time, resulting from a product's use and mode of action.

11.3.2 The 'Core of Everything': Mode of Action and State Change

Design is focused on creating products that help us achieve desired results, e.g. making a cup of coffee. State changes are the mechanisms that underpin these results, e.g. brewing coffee.

Designers' core activities are understanding **natural phenomena**, reasoning about **cause and effect**, and designing things that utilize these.

A precondition for effective reasoning is to understand what creates state changes. In design this is realized through **mode of use** and **mode of action**. Figure 11.5 gives an overview of these phenomena: mode of use and mode of action are arranged about a *natural phenomenon*, where external *effects* and *interactions* between *action conditions* (in the form of bodies) occur. These interactions create *state changes* that define the activity or organ's *behaviour*, observable from the outside as *functions*. Certain state changes create *effects* that influence the surroundings.

The term 'body' should not be interpreted too 'mechanically'. Natural phenomena occur in certain media, fields, and materializations that are linked to the mode of action and the action conditions. Bodies and their interactions become the link to the part structure, which mirrors and realizes the action. Further, the media themselves (e.g. liquids, gases, or fields) also belong to the part structure. In this model actions link activities, organs, and parts. An example of this is illustrated

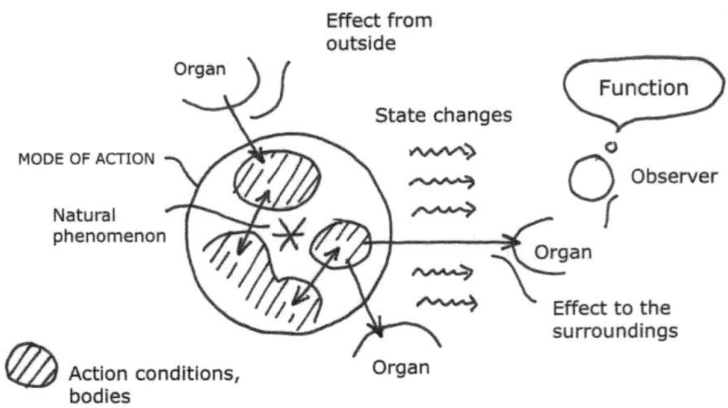

Fig. 11.5 Mode of use and mode of action in an organ (contrast with Fig. 11.7)

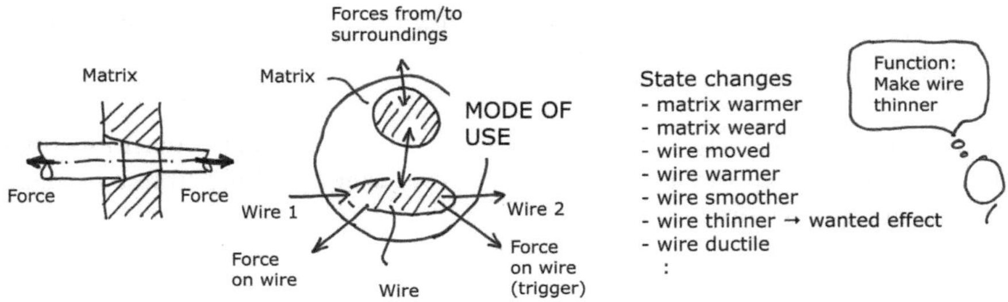

Fig. 11.6 Mode of use in a wire drawing machine

in Fig. 11.6 where a wire is plastically drawn. Here the action conditions are the *matrix*, the *wire,* and the *effects* (the forces), and the wire as operand.

In order to understand 'action' we need to understand *effect*. Here there are three elements. First, state changes do not happen spontaneously, they need triggering effects. Second, once the state change occurs it delivers the desired effect. This can then go on to trigger subsequent state changes. In this way we make up a cause/effect chain, where *mode of use* and *mode of action* interactions correspond to cause/effect chains.

Definition: An **effect** is a state change in a mode of action or mode of use, which leads to interaction with other entities.

A condition for interaction is that there is something for the effect to act on. In Sect. 11.2 we defined function as 'the ability to deliver an effect' and now we can complete that definition in terms of mode of action and mode of use.

Definition: **Mode of action** and **mode of use** are phenomena where **effects** from the surroundings and interactions between the **action conditions** realize **natural phenomena** resulting in a desired **effect**. One or more effects trigger the activity or organ.

Hans Christian Ørsted placed a magnetic needle in an electrical current and observed the effect: the needle turned. Through this **action condition** he discovered the **natural phenomenon** of electromagnetism. From here, the development of a modern electric motor is a matter of creating better action conditions, as determined by the designer. One natural phenomenon can lead to many applications as shown in Fig. 11.7: (a) producing plastic foil, (b) melting, (c) pump for evacuating, (d) bearing, and (e) centralization and lacquering (Rodenacker 1970). Natural phenomena are bound to activities and organ's bodies, e.g. the wire and the matrix in Fig. 11.6, and thus link to the part structure. Simply put, all patents are really about arranging actions.

Definition: **Action conditions** are the arrangement of external effects and interactions between bodies, which create the physical conditions for utilizing a natural phenomenon to create state changes, and subsequently effects.

Although Domain Theory claims that materialization only belongs to the part domain we must also consider materialization in the activity and organ domains. Natural phenomena like physical effects do not exist without a materialization, here in form of bodies.

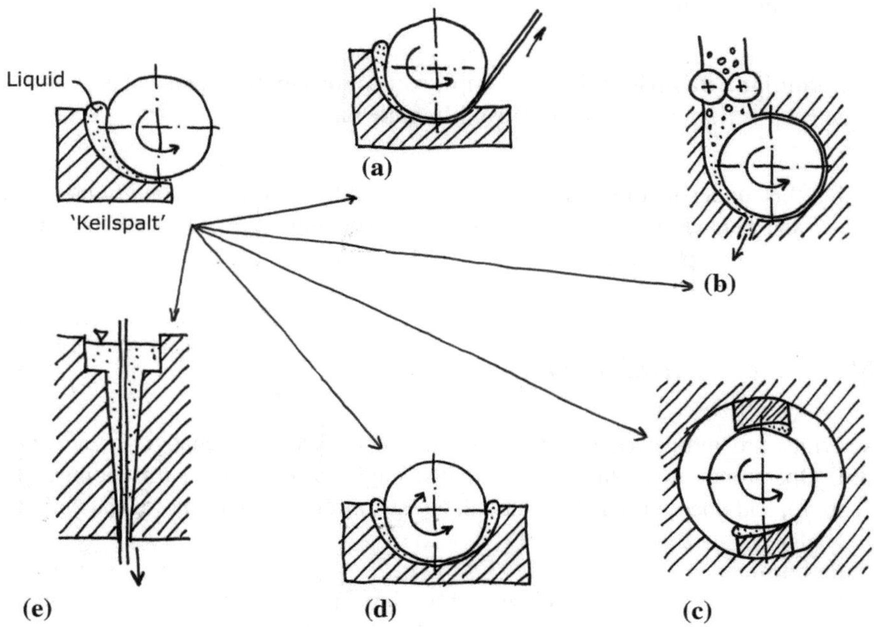

Fig. 11.7 Applying Bernoulli's flow phenomena to various rotating and fixed bodies. **a** Producing plastic foil. **b** Melting. **c** Pump for evacuating. **d** Bearing. **e** Centralization and lacquering

In nature state changes happen **autonomously** in complex processes that are influenced by their environment, e.g. weathering and erosion of mountains. In a societal context we find that state changes are driven by **human intervention** and are related to technological rather than natural systems, e.g. agriculture. Further, these are linked to products and tools. These differences are found in both how a product is used and in its entities, organs, and parts. For example, petrol can burn in a natural process but it takes human intervention to burn petrol in a car. Air and fuel injection is controlled together with the ignition mechanism to produce the maximum amount of mechanical rotational energy.

Definition: **State** is a description of an entity in terms of **parameters** (physical quantities), e.g. temperature, pressure, composition, phase, momentum, etc.

All state changes are related to natural phenomena, while human involvement is based on insight into natural phenomena. In a car engine the state change is obvious: petrol and air are combusted to create pressure on the pistons. The movement of the pistons ultimately drives the rotation of the wheels (the action function) through mechanical transmission. This is a precondition for the car's use function, i.e. moving people from A to B. While effects concern the **behaviour** of physical objects, functions concern human perception. Both the designer and user are reasoning about functionality.

Definition: **Behaviour** is the complex of **state changes** that occur in an activity or device based on **natural phenomena**.

Using this understanding of action and its elements (action conditions, state, effect, and behaviour) we find that they are not simply LEGO pieces but active entities with complex interactions.

11.3.3 How Activities Work

Activities are influenced by **operators** that drive the state change in the **operand**. Operators can be humans, products, methods, or active surroundings (see Chap. 8), and operands can be material, energy, information, or biological entities.

Example:
Microwave oven. The use activity of a microwave oven is based on a mode of use, in which food entities are under the influence of electromagnetic radiation creating thermal energy by dielectric heating. The heating requires

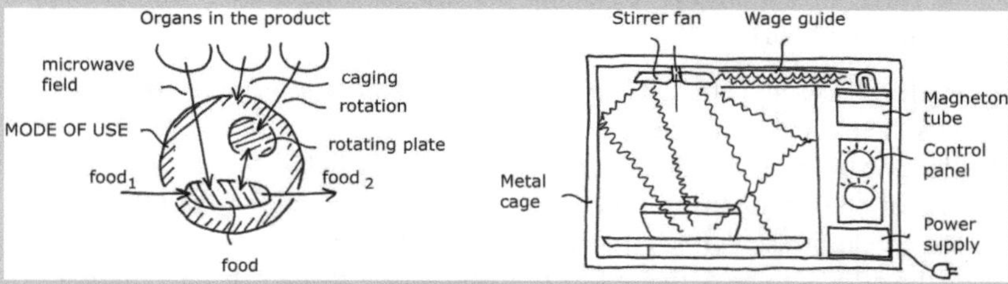

Fig. 11.8 The mode of use, use activity, and effects in a microwave oven

the entities to move in order to achieve an even heat distribution. The product creates the wave effect via a magnetron, with the waves being guided to the food entities. The whole system is enclosed by a Faraday cage to avoid irradiating the surroundings. Figure 11.8 shows the use activity's mode of use, its interactions, and the product's affects, i.e. the product's action functions.

At the core of an activity's action is the interaction between the operation and the product. This interaction is facilitated by an **interface organ**, which is typically a combination of action conditions from the product and the operand, e.g. the rotating plate in a microwave is an organ together with the food. The result of the activity is the operand in its output state, e.g. the microwave has the output 'hot food' and the use activity has the function 'deliver food'. Thus cause/effect and function reasoning are used to synthesize activities. First, the cause/effect chain starts from the activity result, e.g. brewed coffee, drawn wire, or hot food. Here, it is necessary to identify the relevant natural phenomena linking input and output via state changes, and subsequently create the action. Based on our interpretation of action we are now able to define the use activity more precisely.

Definition: The **use activity** is an arrangement of mode of use. This brings together **natural phenomena** and state changes, **effects** from the product, humans, and active surroundings, and **action conditions**.

Let us explore this definition using the example of shoe making. A key activity is sewing the upper shoe to the sole, where the upper and the sole are the operands. The natural phenomena are a combination of penetration and positioning, respectively, materialized in the interaction organs of the needle and the 'tooth rack' (to move and position the sole). One effect of the product (the sewing machine) is the reciprocal movement of these interacting organs. The effect of the user (the operator) is the supporting movement of the shoe assembly by the operator's hands. The effect of the active surroundings is simply that gravity helps position and drives the shoe parts. Based on this reasoning the design of an activity can be supported by the following heuristics:

- Mode of use and the activity result (the activity's primary use function) must be explicitly articulated.
- Identification of the state change in the activity helps define the role of the natural phenomenon, and therefore directs the search for solutions.
- The mode of use defines the effects to be delivered by the product, as well as the effects to be delivered by humans and/or the active surroundings.
- When the use of the product is determined the goal formulation can be expanded to include additional desired action functions.

In Chap. 8 we saw that the operation activity has the *product as operator* while all other product life activities have the *product or its components as operands*. In effect, during product life activities the wider product life systems take the role of the primary operator. This illustrates the close relationship between activities and products, which is fundamentally created by the product's organs.

11.3.4 How Organs Work

Most products are composed of several organs, each realizing a certain function, and all interacting in a cause/effect structure. Further, some organs interact with the operator, the surroundings or other products and systems. Thus, although the final product is composed of parts we must understand its functional composition in order to allocate the parts. Organs explain this functionality and the reasoning behind organs' mode of action. Only very simple products carry a single function, e.g. hand tools, typically several organs are interacting. The final product is therefore seen as a structure of interacting organs. Taking the microwave example (Fig. 11.8) the interface organs are shared by the product and the activity. From a product perspective these are 'half organs', where we only see the other half in the deployment of the product. Going deeper into understanding organs' composition consider the pencil sharpener in Fig. 11.9.

In Fig. 11.9 interface organs 'cutting' and 'guiding' are interacting with the pencil, while a third 'gripping' organ is interacting with the user. These 'half organs' depend on external effects and therefore the functions are termed 'allow cutting', 'allow guiding', and 'allow gripping'. The body of the sharpener interacts with all three interface organs and positions the pencil relative to the knife. This embodiment requires an additional connecting organ 'hold knife', which could be realized by a screw.

In order to understand an organ's mode of action we can look at organs' interaction and the action conditions using the Moka Pot example, Fig. 11.10. The brewing organ's mode of action is composed by its action conditions: the two filters and chamber walls that enclose the ground coffee, which together with the top filter raise the pressure. The organ requires effects from the boiler chamber: hot water and pressure, and delivers effects to the serving chamber: coffee flow.

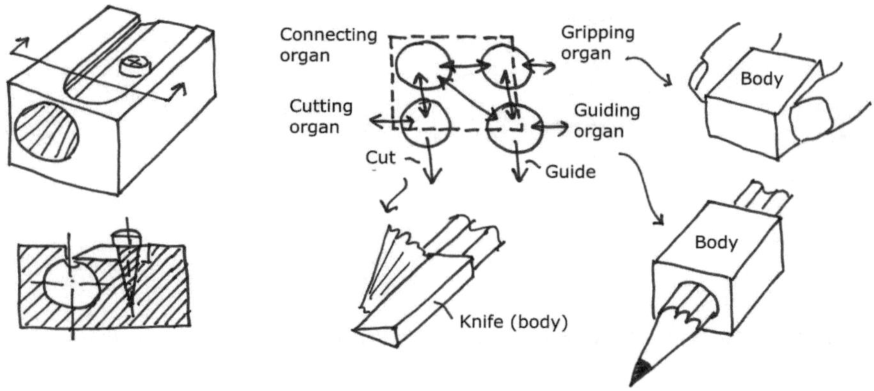

Fig. 11.9 A pencil sharpener delivering cutting and guiding effects to the use activity

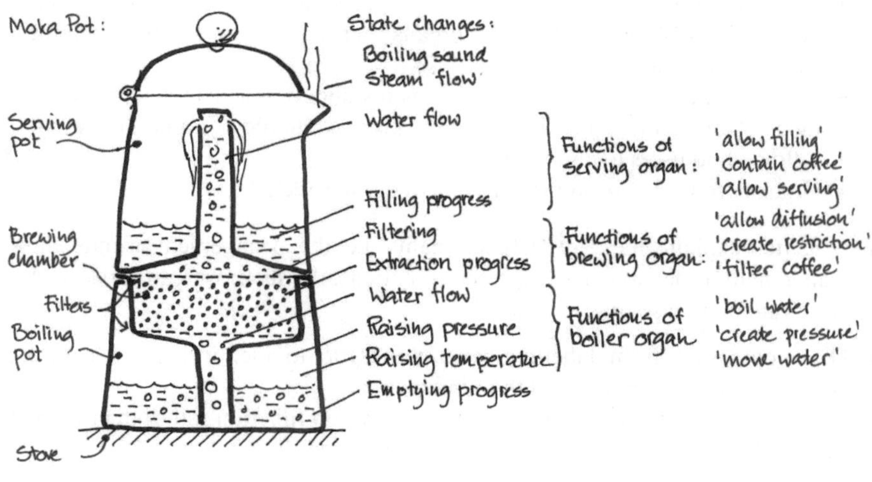

Fig. 11.10 A Moka Pot's boiling and brewing chambers with the brewing organ highlighted

In the pot's deployment the organs undergo state changes as shown in Fig. 11.10. It is notable that the brewing organ creates the effect 'counter pressure' allowing the pressure to build up, as well as the pressure for transporting the coffee to the serving pot. This cause/effect chain of interactions is illustrated in Fig. 11.11. Parallel to our definition of activity we can define an organ as follows:

> **Definition**: An **organ** is a functional unit of a product where the arrangement of **mode of action** is based on **effects** from other organs, **action conditions**, and **interactions**, by which it creates **functions**.

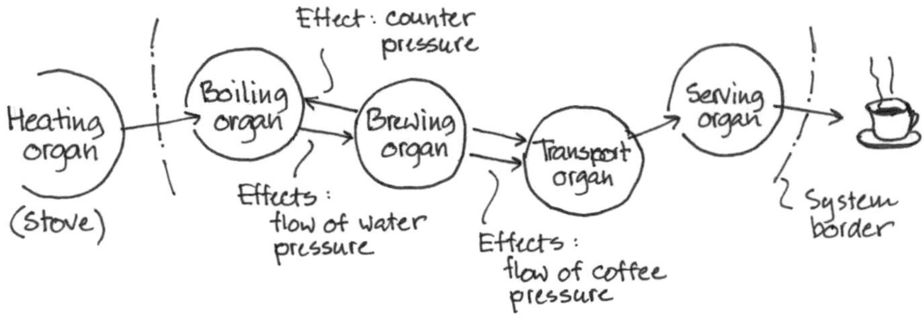

Fig. 11.11 A Moka Pot's brewing organ and its cause/effect interaction with other organs

Based on these elements we propose a number of heuristics to guide the determination of organs from function reasoning:

- The reasoning should follow the progression: label the organs purpose > identify the nature of the organs > arrange the interactions via the cause/effect chain.
- We can articulate functions via statements about effects: 'transmit movement', behaviour: 'when I…, then the product should…', and conditions: 'allow blind users to choose a setting'.
- Organs often require supporting functions, see Sect. 11.4.

This section explains the core of functionality, i.e. the arrangement of interactions to realize natural phenomena and create functions. As such, we can broadly characterize function reasoning:

- The progression from labelling what things do to identifying the nature of things, specifically organs and their interaction.
- A solution neutral form of reasoning, focusing on function rather than embodiment and allowing for alternatives to be considered.

11.4 The Function Network and Product Composition

A function describes what a thing should do, e.g. 'create force' and an organ is a concrete solution able to deliver this effect. Thus, reasoning from function to organ keeps the solution space open, i.e. it allows us to articulate what we want before we jump to a solution. This makes it extremely worthwhile to alternate between finding solutions and reasoning about functions. In this way product composition (the organ and part structure) is supported by the following patterns:

- Domain Theory explores **horizontal and vertical causality**.
- The organ structure follows the rules **function/means tree** and **supporting function types**.

- The logic of cause/effect relations follows rules including **structural state transitions**, **flow reasoning**, and **interaction**.

Although this section focuses on determining functional synthesis the designer must also consider inputs from property and dispositional reasoning, which we deal with in the following chapters.

11.4.1 Horizontal and Vertical Causality

Hubka and Eder (1988) identify two types of causal chains that determine the scope of a product's abilities, illustrated in Fig. 11.12.

1. **Horizontal causal chain**: this describes activities related to the product, including the activity result, the operation activity, the use activity, and product life activities; see also Fig. 8.8.
2. **Vertical causal chain**: this describes the interactions between a product's organs that create the active effects necessary for a product's use activities.

Between these two chains and the task allocation (to the product, the human, or the active surroundings) necessary for the activity chain we can fully define a product. To illustrate this task allocation and product delimitation we translate the general pattern in Fig. 11.12 (a) to the specific case of a printer. (b) Here, the line of activities shows the users' tasks: adding paper and removing completed printing. The task of controlling the printing is allocated to a computer. Between the interface organs (concerning feed, data transfer, etc.) and the product's interaction with the activity chain there is a causal pattern of functions to be determined in the design activities.

The focus on a product's primary purpose raises the question of how many additional functions can be added, e.g. the many different roles served by a smartphone. The simple answer is that as long as these functions are perceived as value adding by the customer there is no limit. Note, this must be considered holistically, as at some point a device becomes hopelessly complex with the individual functions ceasing to add value. These functions are called **auxiliary functions**.

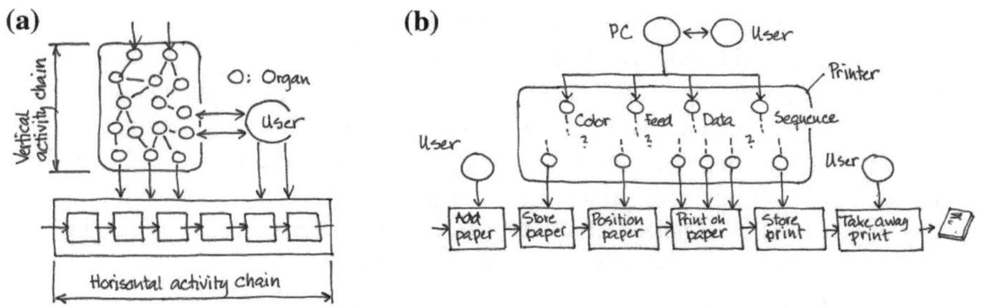

Fig. 11.12 Horizontal and vertical causality (**a**) a generic model (**b**) printer example

An extreme example is found in various software-based smart devices, where the average user only sees a fraction of the device's true functionality.

- Derive the product's task and action functions from the use function and activity. Additionally, check the proposed action functions are able to perform the desired use activities.
- The user's perception of a product's value is related to the activity result, interaction with the user, the user operations in the use activity, and the function properties.

11.4.2 The Function/Means Tree: A Hierarchy of Functions

Functions point to organs that must be arranged in an interacting pattern, i.e. organ structure. Typically, this is based on a **main function** as the foundation. For example, a can opener's main *use function* is 'open can' but what is the main *action function*? If we imagine successive cutting operation along the rim of the can we need the product to realize the functions 'position cutter', 'progress cutter', and 'move cutter'. In this situation we cannot identify one main function that determines all additional functions, rather a set of functions to be realized. We can also see that main functions are seldom sufficient, additional functions must be added. It is important to distinguish between the causal reasons for additional functions and the nature of the functions themselves, which also affects the organs' interaction.

Hubka and Eder (1988) observe that when the existing sub-solutions are insufficient there is a need for more functions. Thus, composition becomes a pattern of alternating functions and solutions in a causal hierarchy. Andreasen (1980) described this as Hubka's first Law and identified two types of solutions: activity chains (related to, e.g. flow of materials) and part detailing. Hence, why we use 'means' as a general denominator for solutions to a function. Hubka's first Law can be visualized as a hierarchical tree of functions and means, the **function/means tree,** e.g. Fig. 11.13 (Tjalve 1979; Andreasen 1980).

Example:

Toy torch. Consider the various solutions in a toy torch as illustrated in Fig. 11.13. At the highest level the principle is selected—a hand-driven dynamo supported by a battery. Other top-level principles can thus be considered, e.g. a solar-powered battery. At the next level LED lamps and a dynamo are selected. However, we can again look for alternatives, especially because the selected mechanical systems are costly. Going further down we decompose the rack and gear system. Here , an alternative is a crank input that simplifies the transmission. New concepts are created by combining means and adding functions as necessary. We can explore alternative products at any level of the tree, depending on the level of change desired.

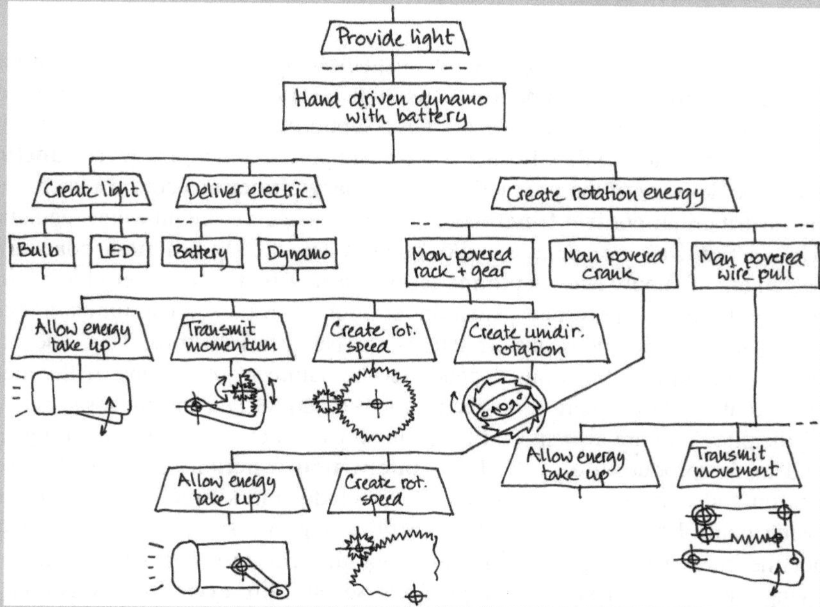

Fig. 11.13 The functions/means tree for a toy torch with illustrative alternatives

The primary role of the functions/means tree is analytical, however, it is also useful in synthesizing a new product. Here the tree is gradually built up in conjunction with synthesis. This helps structure the design process and gives an overview of the product's progression but does not fundamentally explain or directly support synthesis.

The strength of the function/means tree is in helping order solutions and point to possible alternatives.

When we are focused on adjusting or improving existing solutions the functions/means tree is a good starting point for deciding which branches should be kept and which should be renewed. The tree structure shows causality (the reason for a means) but ideation methods are needed to actually create the means. In this way we can use the tree structure to identify the impact of alternative decisions on lower levels. However, causality does not give a complete picture of the organ structure, as such, we need to consider supporting special function types.

11.4.3 Supporting Function Types

Due to the nature of organs (i.e. built on natural phenomena) there is normally a need for a structure of supportive functions as described by Hubka's second Law (Andreasen 1980). The role of these supporting functions is illustrated in Fig. 11.14. This leads to five distinct types of supportive functions: **work functions** concern the delivery of the product's effects; **energy functions** concern energy and signal distribution; **control functions** concern sequencing, monitoring, regulation, and interaction with the operator and/or other systems; **Help functions** concern the management of supply and supporting tasks; **frame functions** concern the creation of a body, frame, support, encapsulation, etc. for the product. The function/means tree does not describe non-causal relationships between functions. Thus, the existence of these support functions depends on the nature of the means. If the means are self-contained then there is no need for sub-functions and sub-means.

Typically, there are also special functions for interaction with surroundings and linked to the product life cycle. First, **interaction functions** concern the user's operation and control of the product. This includes input functions, e.g. the user pressing control buttons, and output functions, e.g. a display informing the user about the current setting. In some situations this can also include functions for carrying, sitting on, or storing the product. Second, **life cycle functions** concern specific life activities, e.g. transport (an eyebolt for a crane lift), storing (partial disassembly for stacking), installation (changeable connectors, feet), maintenance (diagnostic support, exchange of components), and disposal (material identification labels). Identifying these life cycle functions is a borderline problem involving different competences. This brings us back to the importance of visualization as a means for bringing together different inputs (Chap. 7). Functions identified by life cycle thinking cannot normally be fitted into a hierarchy or flow pattern of product functions but may be linked to specific organs or sub-systems where they act, e.g. an eyebolt is fixed to a product's frame.

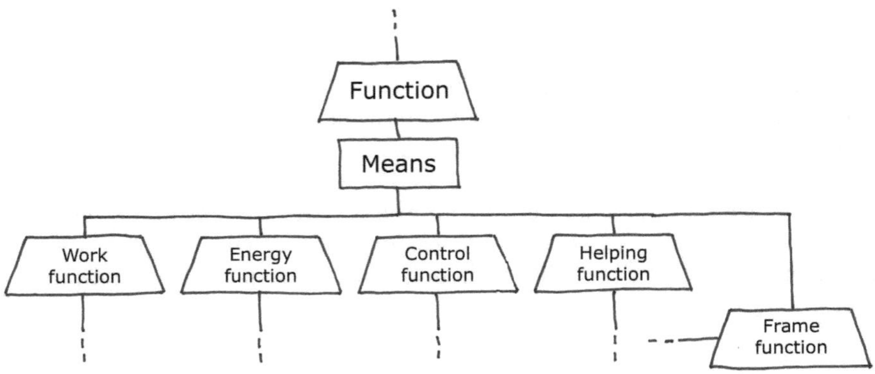

Fig. 11.14 A function in a product supported by a 'family' of secondary functions

11.4.4 Structural State Transition

We saw in the Moka Pot example (Fig. 11.11) that a precondition for brewing coffee is the loading and assembly of the pot. Further, after use the pot needs to be disassembled for cleaning. Thus, the pot goes through two **structural state transitions** during use. These transitions require features such as connecting threads, sealing gasket, and detachable brewing chamber. Structural state transitions can be modelled as illustrated in Fig. 11.15. Here, a traditional car gear set is shown in four states. This state transition diagram shows a typical gearshift sequence, where gears 4 and 6 can slide but are fixed rotationally. In each state an arrow is used to denote the part of the gear that is active. Generally, state transitions play an important role in menu-based products (i.e. products with multiple selectable structures). In the software domain much of a product's functionality can be achieved via state transitions.

When structural state transitions are used to create multifunctionality or as necessary conditions for preparing the product for its use activity, the question is: how should the user be informed about the transition and what should trigger it? This again highlights the importance of respecting cause/effect relationships and task allocation.

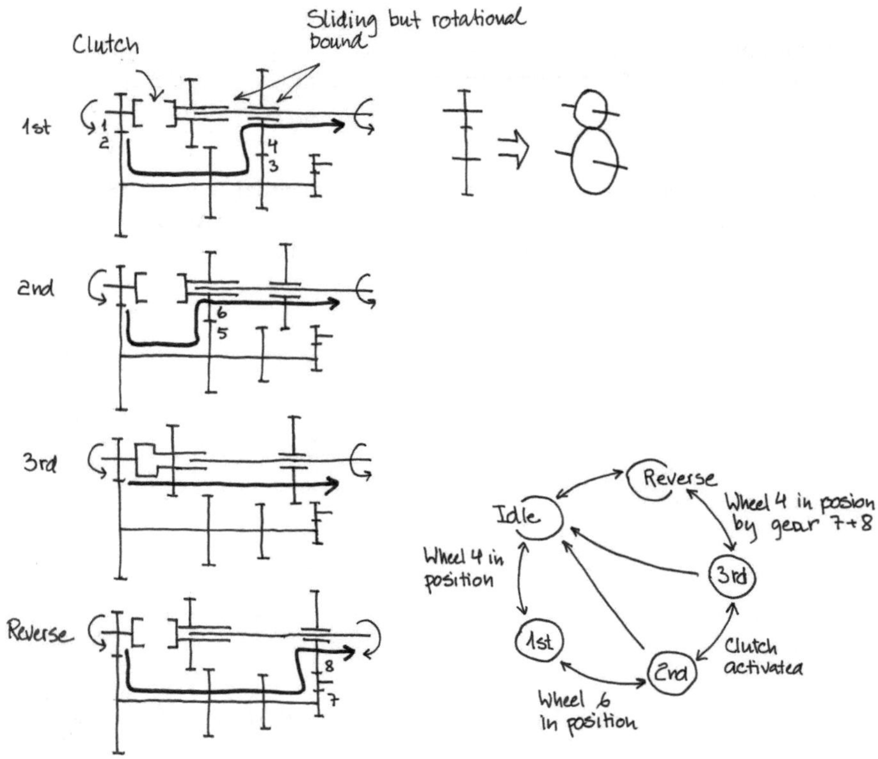

Fig. 11.15 Example of structural state transitions in a gearbox

11.4.5 Flow Pattern Reasoning

Some operands (material, energy, information, or biological) create a **flow pattern** in the organ structure, i.e. a chain of organs bound together by one operand. Here the similarity between activities and organs is clearly illustrated. Dym and Little (2000) explore this with respect to internal signal and energy transformations in a radio, Fig. 11.16. Here, the flow pattern is a consequence of the necessary signal transformations linking the functions.

11.4.6 Interactions

One consequence of using Domain Theory and its three system types is that we are confronted with two types of interactions: **between** activities, organs, and parts; and **inside** activities, organs, and parts, described by actions and leading to mode of use and mode of action. This dual nature means their definition is quite general.

Definition: **Interactions** are the propagation of effects between and inside organs, explaining the behaviour of the organ structure and each organ's mode of action.

All of these interactions are defined by effects (material, energy, information, and/or biological), the product composition, and its activity. At the core of these interactions are the natural phenomena deployed in the product via the chain of

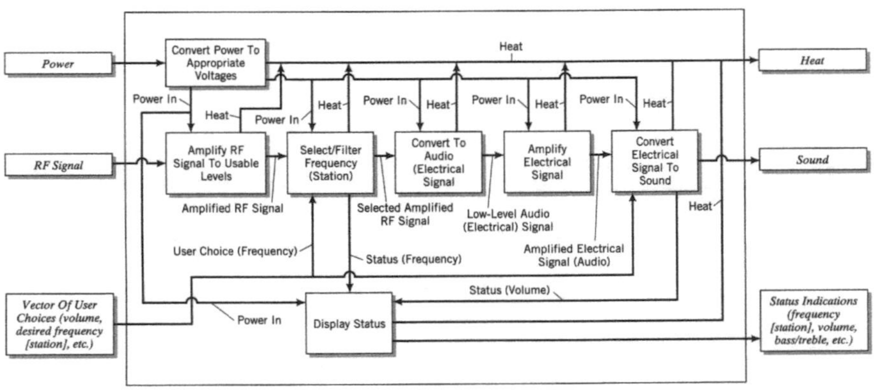

Fig. 11.16 Modelling signal and energy transformations in a radio (Dym and Little 2000)

state changes and tasks. Unfortunately, distinguishing what is inside the product (organs) and what is outside (activities) is seldom straightforward. However, operands and operators are of a fundamentally different nature: operands are passive and are changed, operators drive the changes in the operands. Finally, organs are realized in the part domain where the interactions are materialized. Still this does not give a simple mapping, i.e. an organ can be based on many parts and a part can contribute to several organs.

In this chapter there are several examples of interactions between and inside activities, organs, and parts. In Fig. 11.9 we examined a simple concretization of an organ structure in a pencil sharpener. The interactions, in the form of forces, were managed by merging all the bodies except the knife, which required an additional function and an assembly organ. This example also demonstrated interactions with the surroundings: the operator's finger and the pencil. Finally, this also illustrated how a function can be necessitated by the embodiment: the requirement for connecting the metal knife to the main body.

Example:
Combustion engine. At the heart of a car engine is the cylinder where petrol is burned. The piston head and its assembly to the crankshaft make it possible to transform combustion energy into rotational momentum in the shaft. The action conditions in the combustion chamber organ are linked to the design of the cylinder and its interaction with the organs: air supply, ignition, exhaust, etc., see Fig. 11.17. It is also important to realize that of these interactions the link with the transmission organ is bidirectional, i.e. the piston can drive the crankshaft but the crankshaft can also drive the piston.

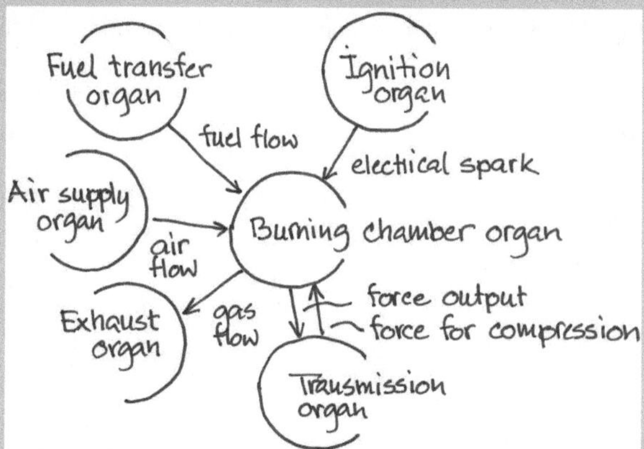

Fig. 11.17 A car engine's combustion chamber organ and its interactions

In addition to the complexity highlighted above the effect can also be composed, e.g. the interaction between the brewing organ and the serving organ in the Moka Pot is a mix of water and steam, where the water contains coffee oil. Here the user observes that the transfer has started via the boiling sound, i.e. they are informed about the brewing. As such, this effect is composed of material, energy, and information interactions. An interaction is 'what is happening' not 'how it is happening', thus we need something to carry the effects that compose the interactions. This 'something' is realized in the creation of the part structure.

11.5 Creating the Part Structure: From Function to Embodiment

Organs and parts were introduced in Chap. 8 where we defined a part as the basic material element of a product. Now we consider how parts are synthesized. Parts have their origin in organs' bodies, i.e. the action conditions are brought into one or more parts. Again consider the pencil sharpener (Fig. 11.9), the guiding and gripping bodies are merged into one part to which the knife body is attached to create a three-part structure.

Domain Theory gives a three-dimensional understanding of 'how products are used, work, and built'. Thus a part is characterized by form, material, dimensions, and surface qualities (and sometimes also state when this influences how a part is assembled).

Example:
Toy torch finger grip. In a toy torch we find 'force take up' and rack organs (introduced in Fig. 8.15). The part structure related to these organs is shown in Fig. 11.18, which can be used to identify the bodies of the two organs. (a) Shows

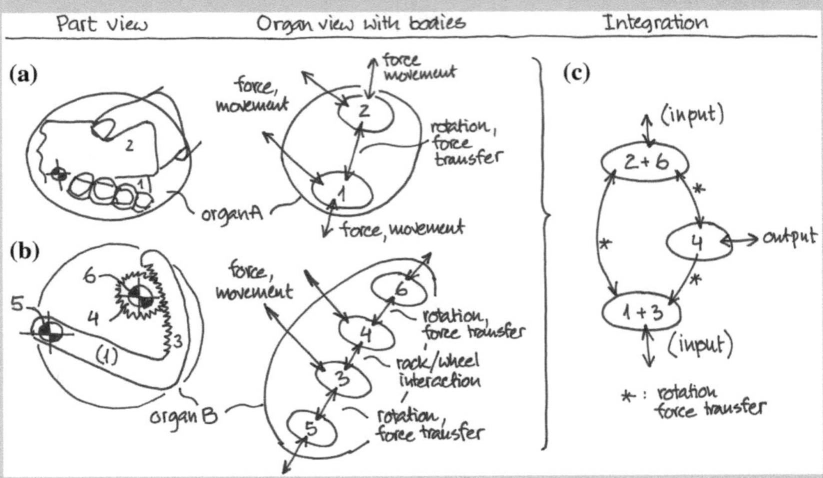

Fig. 11.18 Two organs in a toy torch and their merging into one part

the force take-up organ, which has two bodies that interact with the user and each other via a rotational relation. (b) Shows the rack organ, which has four bodies that create the rotation force on Part 4 and the required fix-points 5 and 6. We can explain the actual part structure by merging these bodies (c). It is possible for the two organs to share the rotational relationship by using Part 2 as the frame for the device and integrating bodies 1 and 3 into one part. Here, we have decided to transmit movement further down the effect chain, from Part 1 via Part 3 to the rack mechanism. Part 1 + 3 also carries bodies related to a number of other functions (return spring, end stop, and locking (Fig. 8.15)).

The example shows how a part can materialize the bodies from one or more organs. This is one of the reasons why parts are not fundamental elements of organs or organs themselves. Rather, the organs' natural phenomena and action conditions define **what** each part does, e.g. transmit force, conduct current, or keep position. Then the interaction pattern of the action conditions defines **how** the parts relate and interact, e.g. creating a hinge or sliding against each other. The result of this synthesis is parts defined by their characteristics.

Definition: A **part** is a material element of a product. The part materializes the bodies and their interactions and is characterized by its form, material, dimensions, and surface qualities.

The form and material characteristics are key. For completeness in the transformation from organs to parts we must include the operands as parts, e.g. the water in a pump or the air in a fan. Without these entities we are not able to explain the product and its embodiment.

11.5.1 The Embodiment Activity

The Encapsulation Design Model shows the sequence: Concept Synthesis, Product Synthesis, and Product Development. In order for transition from organs to parts we must consider both conceptualization and product development perspectives. Pahl and Beitz (2007) give a comprehensive treatment of the embodiment activity, starting with a concept and ending with what they call 'overall layout' and definitions of the component's shapes and materials, i.e. the part structure.

The transition from organ to part view depends on the degree of concretization required to verify the concepts/solutions used in the product's composition. During product synthesis the detailing of the organs can reach such a level that it is appropriate to start the part detailing, i.e. the activity changes from function reasoning to reasoning about manufacture. This typically happens heterogeneously following the set-based description of the sub-systems (see Chap. 7).

Definition: **Embodiment** is the design activity that determines the complete part structure of a product. This is based on the organ structure and the satisfaction of manufacturing requirements. This activity results in a full justification of the product's functionality and properties in its final manufactured state.

It is outside the scope of this book to go deeper into the many principles and issues of the embodiment activity but we will cover the main synthesis aspects: the creation of the part structure and defining the parts and their interfaces.

11.5.2 Part Structure Considerations

The following help us understand how function reasoning can be used to determine parts:

- Understanding the organs' mode of action is a precondition for determining parts.
- Several design and realization conditions can emerge from the transition from organs to parts, e.g. additional supporting functions and part features. These originate in organs' mode of action and manufacture.
- Reasoning should alternatively focus on understanding the organs' functionality and the parts' materialization.

The use of bodies to structure reasoning is an idealization. Hubka advised that we should work out 'manufacturing neutral, geometric solutions', i.e. creating the necessary bodies but giving them an unfinished form that can be gradually concretized. Tjalve believes the opposite, that one should identify preferred manufacturing means, e.g. thin plate forming, to give direction to the neutral solution, which he argued could otherwise be misleading. This is then gradually resolved through the creation, selection, and synthesis of variants to produce the final embodiment.

Complexity is a key cost driver because it has a wide-ranging impact on the number of operations, controls, suppliers, etc. Complexity reflects both the number of entities and their relationships, and the designer's perception of composition and difficulty. An important means of reducing complexity or more specifically weight, space, cost, etc., is to create parts that contribute to multiple organs. This can either be planned or happen opportunistically during the process. We showed a simple example of this in the toy torch (Fig. 11.18) but we also offer these general recommendations.

- Seek to materialize functions such that they are independent, eliminating constraints from the organ structure.
- Seek to integrate opportunistically by letting parts contribute to more organs or functions but remember to check for negative consequences of removing independence.

Balancing the contradicting demands of functional independence and integration is a key design challenge and care should be taken to avoid contradictions or sacrificing the products properties. This contributes to the complexity of mapping organs and parts: one organ may be realized by many parts and one part may contribute to many organs. When parts (or sets of parts) contribute to multiple organs the design needs to be integrated, influencing how we determine the product properties. We discuss this further in Chap. 12 but highlight here that when we talk about function **integration or differentiation** it is critically related to the materialization of the organs. In particular independence is a virtue when we seek to optimize and operate each organ separately; however, many of the most elegant products are based on effective integration.

Example:
Injection-moulding machine. In an injection-moulding machine plastic granules follow a continuous path from mixing, through the screw feed, heating, pressure increase, to injection into the mould where the product is formed. The only moving part in this machine is the rotating screw that drives the granules through this process. This is illustrated in Fig. 11.19 where we see elegant function integration.

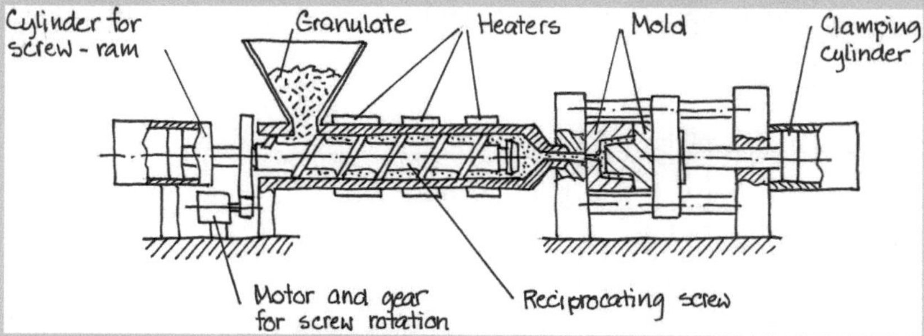

Fig. 11.19 Cross-section of an injection-moulding machine

The main problem facing the designer in materialization is its non-ideal nature, i.e. proper functionality is highly dependent on the parts characteristics. Unfortunately, these characteristics can easily diverge from their intended values due to tolerances on dimensions, material heterogeneity, surface quality deviations, etc. Further, manufacture often demands additional functions or part features, e.g. assembly interfaces or splitting a single part into sub-components, which further move the materialization away from the ideal case.

The composition of organs should be detailed until the interactions and interfaces can be defined with respect to the various solution alternatives.

The final result of product synthesis is the description of the part structure to be manufactured. Each part is specified with respect to form, material, dimensions, and surface qualities, as well as the assembly relationships. A traditional example is shown in Fig. 11.20, a fixture for milling. Although this model belongs to the part domain it also gives information on the activity and organ. Of note is that the features needed for manufacture (e) necessitate separating some parts. The component to be milled (shown with thin lines) is positioned such that the milling surface (denoted by a triangle) is in the correct position relative to the part's cylindrical bore. It is fixed in this position and transmits forces to the milling machine' body (not shown) during operation.

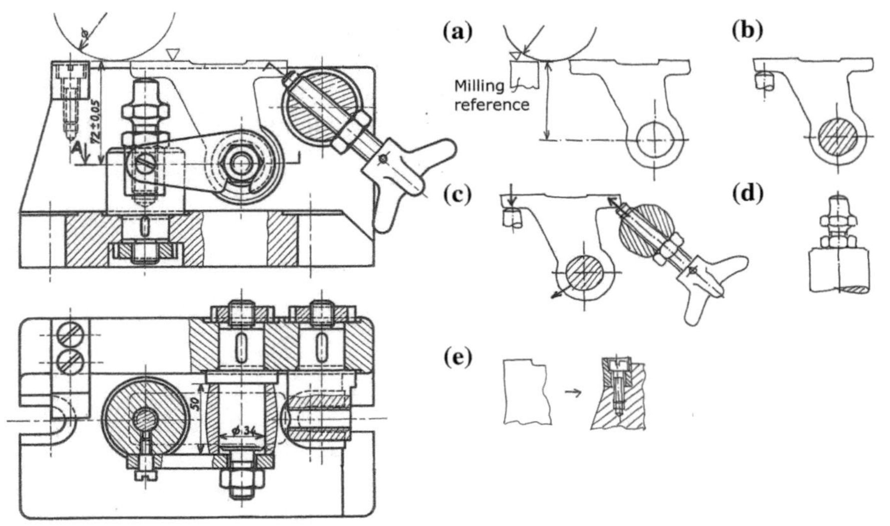

Fig. 11.20 Assembly drawing of a milling fixture, which can be read as activity, organ, or part structure. It shows the superimposed issues: **a** milling operation, **b** positioning, **c** fixation, **d** adjustment, **e** the sub-functions required for manufacture

Three overlapping issues complicate understanding of the device's functionality: positioning and fixation, fixture adjustment, and the fixture's composition (due to manufacturing practicalities). This follows the line of reasoning: what activities occur > what functions are required > how are these materialized. In this way the core functionality is retained while additional functions are included for adjustment, securing, and ease of manufacture.

Domain Theory helps us distinguish between the product's design in the three domains, each of which offer design degrees of freedom to be utilized in creating a good design.

11.5.3 Interfaces

There is no sharp distinction between interaction and interface in everyday language. However, we can differentiate them as follows: interactions are something that happens, while interfaces are something that 'is there'.

Interfaces describe features found in the part structure that have functions and therefore correspond to interface organs.

Interfaces are important in two situations. In the first, they are integral to creating a multi-part product that cannot be fully integrated, e.g. using standard components or creating independent modules. In the second, they are established after a product's realization as part of its deployment, e.g. connecting a trailer to a car.

Interface between activity and product. Typically certain organs interact with the activity. In the pencil sharpener example the knife and the guide interact with the operand (the pencil). These interface organs are shared between product and activity. A boat's propeller is its main interface organ. The mode of use in the boating activity thus relies on interaction between the rotating propeller and the water to create propulsive force. The effects from the boat are 'rotate propeller' and 'form propeller', i.e. deliver the form and effects to the water.

Interface with the user. Interaction between the user and product can take many forms but normally aims to transfer energy and/or information to control the product's mode of action. Information can be both input and output (see Chap. 13). In the toy torch example this interaction is in the transformation of forces from the users' fingers and hand into power for the torch (Fig. 11.18). Here, the housing and force uptake part create an interface with the user. These interface organs can also be seen as 'half organs', i.e. the hand is an action condition in these organs.

Interface with the surroundings. During the product lifecycle various functions interact with operators and life systems specific to each phase, e.g. maintenance, transport, or disassembly. For example, a phone has a specific power port (or wireless charging function) for charging during the use phase. More generally, typical interfaces include feet, hooks, and electrical connections.

All of these interfaces are designed as part of a product's functionality. However, when we move to the part domain we typically need to add interface organs to facilitate materialization, i.e. **interfaces in the product**. This connects the organ structure by adding the interface organs required for materialization, e.g. for transmitting liquid, current, or movement. For example, production processes mean that a car engine cannot be made in one piece but needs 'slicing' and therefore gaskets and bolts.

Industry has a number of preferred approaches for dividing products into subassemblies to allow parallel manufacture, easier handling, installation, etc. This is illustrated by the trend towards 'mass customization' using modularization and combinatory composition of product variants (see Chap. 8). A module is a product component containing one main function and all the necessary supporting functions, arranged such that it is independent of the rest of the product. Further, module's interfaces are designed so that a range of alternative modules can interact with the base module. However, these must all respect a common interaction and interface protocol, e.g. the USB connector type. When we interface different devices we often focus on connections that can be opened and closed to provide a more or less permanent interface.

The interaction between parts is the core focus of almost all machine element textbooks because of their general importance in realizing organs and creating standard product elements, e.g. Bluetooth. This is usually treated as classes of 'pairs' (German: *Elementepaar*): movable (e.g. bearings, sliding elements), connections (e.g. bolts, rivets, welding, gluing), transmissions (e.g. gears, chains, belts, couplings), and natural phenomena (e.g. friction).

11.6 Cross-Disciplinary Function Reasoning

Many products are created in multidisciplinary teams combining, e.g. mechanics, electronics, IT, chemistry, or biology. However, integrating specialists effectively is beset by communication issues.

- The natural phenomena may be such that the design team has problems understanding and translating the phenomena into a design.
- Integrating natural phenomena from different disciplines can cause difficulties in creating the interaction and interfaces between organs and between parts.
- The specialist who has insight into the product's application area often has little insight into design or the product's mode of action.

- A function can be realized by organs from different disciplines and thus solutions can be hard to compare and evaluate. It can also be difficult to predict the consequences of integration.

Companies typically have problems '*cutting the cake*', i.e. deciding how to compose a new product with the best balance between solutions from different disciplines. McAloone and Andreasen (2001) found that companies had an inability to articulate mechatronic concepts leading to a lack of coordination between concretization in the software, electronics, and mechanical teams. The problem was described as '*integrating the three heads*'. Salminen and Verho (1989) made an early diagnosis of this type of problem, which is still valid:

- Lack of common language between expert groups.
- Risk of clique formation among expert groups.
- Lack of overview understanding beyond disciplinary borders.
- Lack of responsibility definition and poor interfaces between areas.

Numerous authors have highlighted these issues as a problem of shared models and model integration. As such, function reasoning is a key means for supporting integration, and handling interaction and interface problems.

11.6.1 Function Reasoning as a Common Ground

Buur (1990) showed how the composition of mechanics, electronics, and information technology could be described via common use functions. Further, mechatronic products respect Domain Theory and can be modelled using Hubka's first Law in a function/means tree. Buur identifies structural state transitions as a core characteristic of mechatronic products. Creating a mechatronic concept is based on a negotiation between specialists, each able to offer potential organs for a product. Focusing on these entities as functional organs allows the team to compare highly diverse sets of proposals.

Example:
Interdisciplinary dialogue. The company Radiometer Medical ApS produces blood gas measurement instruments for a global market. Figure 11.21 shows the sequence of operations for one such instrument. During operation the instrument consumes sensor cassettes inserted by the user. It was found that a safety organ was needed to prevent the user from trapping their fingers in the closing lid. Different domain experts suggested solutions for this organ. The mechanical expert suggested a spring damping solution that would passively adapt the closing force if the lid were blocked. The electronics expert suggested an organ with active functions to sense blockages and then stop

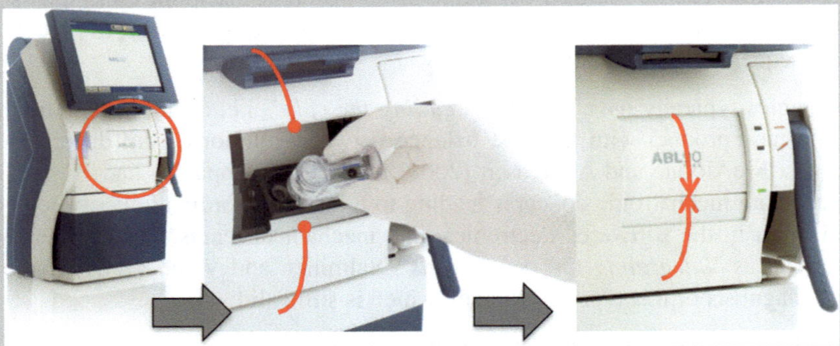

Fig. 11.21 Safe insertion of cassette into a blood gas analyser, courtesy Radiometer Medical ApS

the actuators. Finally, the software expert suggested a "tap when ready" button on the screen interface, to allow the user to decide when to close the lid. All suggested solutions had different problems including cost, weight, volume, complexity, and reliability. Ultimately, a decision was made to focus on the user experience and thus create a fully automated active sensing solution. This would (hopefully) minimize error rate and user risk at the expense of cost and complexity. Articulating the functions and alternatives allowed the different domain experts to create solutions that would meet the function of the safety organ.

The design activity becomes more difficult when boundaries between disciplines are core to the details of the design synthesis. Torry-Smith (2013) focuses on what he calls **dependencies** between the organs and parts: relationships that occur in the design activity as the consequence of design synthesis. Dependences are found between structural elements and the parameters related to these elements. Identifying and respecting these dependencies is key to effectively managing interactions and interfaces. In particular they allow the designer to identify failure modes and effects, identify the need for integrated design efforts, and judge trade-offs. Further, they can provide important insight into functional details. Inspired by Torry-Smith we propose the following checklist for avoiding dependency issues:

- The **organ structure** should be **complete**. In order to identify dependencies all relevant and necessary functions need to be defined: from the causal hierarchy, the product's delimitation and interactions, product life activities, and from properties' realization.
- The **causal chain** linking organ to organ should be complete in terms of type (material, energy, information, or biological), parameters and their values.

- **Interfaces** should be fully mapped and their effects understood in terms of interactions.
- The **logic of structural state transitions** should be complete from a functional perspective. If users are involved in the transitions these should also be explicitly articulated.
- Activity **sequencing** and modes of action should be articulated and complete.
- When organs are **integrated** at lower levels in the function/means tree this can cause problems especially when the supporting lower level organs are of a cross-disciplinary nature.
- Some organs can create **harmful effects** disturbing other organ's functionality, e.g. vibrations, humidity, or electrical fields.
- Some organs require **effects from the surroundings** that can disturb other organ's functionality, e.g. cooling, heating, or humidity.
- The **spatial arrangement** of organs can cause problems when multiple organs require specific positioning for, e.g. accessibility.

Dependencies highlight the need for a coevolutionary approach to synthesis and analysis, not only for identifying dependencies but also for refinement and correction. This closes function reasoning and brings us to how it can be used to support different types of synthesis.

11.7 Function Reasoning During Synthesis: Design Type Dependency

Function reasoning concerns a product's functionality and fit for life and is a decisive influence on product success. Finding solutions that realize the functions is a critical part; with innovative products often characterized by building in radically new solutions. We treat function reasoning's role in these considerations but start by getting to grips with function modelling.

11.7.1 Function Modelling

The current literature is very vague in its definition of function modelling (Eisenbart 2014). We counter this by focusing on the actual refinement of functions during the synthesis activity. At the start of the chapter we characterized functional descriptions as verb/noun statements, e.g. 'create heat' and 'show temperature'. When we come to composing a network of functions we face a reciprocal interaction between organ synthesis and function reasoning. The labels are still verb/noun but the effects are formulated as quantifiable parameters.

Some authors give a different interpretation of functions as 'how something is done', i.e. the same definition as the mode of action! We reject this usage and focus of modelling functions linked to a specific set of properties and function properties that describe behaviour. In contrast, modelling the mode of action deals with the natural phenomenon, action conditions, and influencing effects and uses the goal conditions to verify the resulting behaviours or effects.

One of the risks in design is that the product does not work as expected and does not satisfy its specifications, normally driving in intensive testing and simulation. This verification activity creates feedback loops in the design process, allowing for corrective redesign and refinement. Function reasoning does not magically make design a linear activity. Coevolution of goal and solution and the unknowns in problem and solution spaces cannot be avoided. In Chap. 3 we introduced three basic design types where we can examine the impact of function reasoning.

11.7.2 Designing a New Product

In this situation the designer metaphorically starts with a blank sheet of paper and a basic nucleolus of an idea, intention, purpose, or functionality. In particular Chap. 7 is relevant here. The additional concepts we have introduced in this chapter help the designer identify and manage the various degrees of design freedom. These support both the search for solutions and the concretisation of the product structure.

In this gradual concretisation the following functional questions must be answered: do the functions align with the goal? Can the action functions realize the use activity? and do the action functions create a logical, causal pattern that allows the product to work? During product design a pattern of functions is derived from the use activity, as well as from elements of property and dispositional reasoning (Chaps. 12 and 13). Function reasoning links the design to the desired properties. However, aiming to model every function in a product is almost impossible, instead we can draw on known solutions for lower level functions and progressively map the function/means tree as we concretise the design.

- It is important to verify the relevance and performance of the functions and their properties during the design process. Superfluous functions should be identified and removed.
- The value added by each function should be a key reflection point in learning from the design cycle: were they valuable and were they used?

11.7.3 Incremental Design

Incremental design builds on existing products, renewing sub-systems, or redesigning elements for higher performance and value. Starting with an existing design allows us to follow the strategy of Oja (2010), which we summarize as:

- Secure a product's value by identifying if it is possible to increase user value/ reduce costs.
- Clarify the existing product's behaviour and its relationship with the current concept.
- Build on this clarification and function/means tree to identify potential alternatives solutions.
- Identify and select sub-concepts but retain an overview of interfaces in the product.
- Synthesize the final solution by building on the structures described in this chapter.

The kernel in Oja's approach is to identify new value goals and map these to existing function chains and their properties. It is often necessary to clarify these properties via simulation or measurement for the new value goal. Oja's approach allows us to identify areas to be removed, substituted, or improved, perfectly suiting an incremental strategy. This also promotes the idea that better solutions are to be found through the integration of disciplines, e.g. mechatronic approaches. One of the risks of this type of strategy is that we impact many of the functional interactions between organs and change interfaces in the part structure. As such, additional care must be taken in tracking these changes. A structured way of achieving this is using "encapsulated" sub-systems (Andreasen 1998) where protocols are defined for interactions and interfaces. This prevents changes in one subsystem propagating across the whole product.

Example:
Product models for reuse. In Chap. 8 we introduced 'product models', i.e. the final specification of the total product including activities, organs, and parts. A functional overview of a product's composition is a good starting point when creating an incremental design. Harlou (2006) uses a modular view of products to create a product architecture. Modules are defined by their function and interaction with other modules. Harlou distinguishes between 'standard designs' to be reused, new elements 'to be designed', and potential 'future standard designs'. Figure 11.22 shows an example product architecture for 'BeoCentre 2' together with two new products to be added.

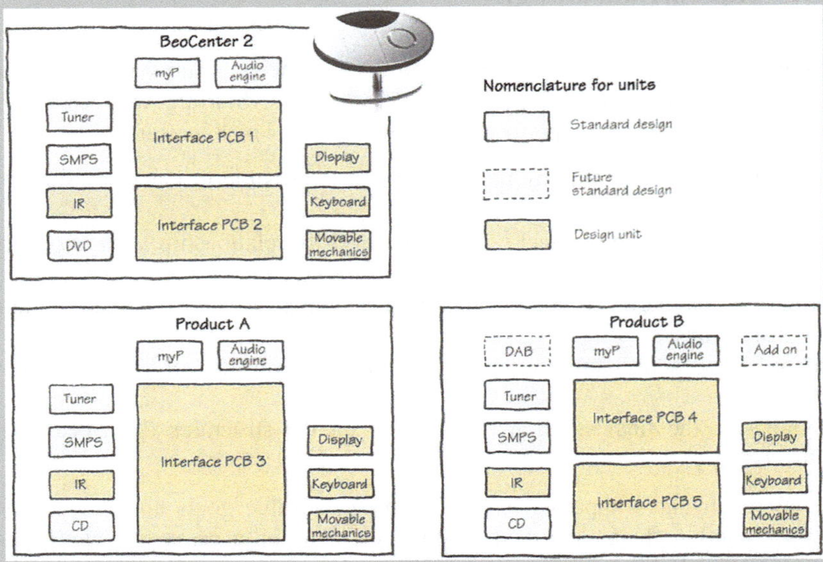

Fig. 11.22 A product family architecture used to define two new members (Harlou 2006)

The following mindset for incremental innovation emerges from these discussions.

- A key advantage of incremental design is identifying and realizing value enhancements based on careful evaluation and maximal reuse.
- It is crucial to make clear decisions about the focus of the new product, i.e. limiting the new sub-systems such that change propagation is controlled.
- Function identification and clarification of organ structure are key to limiting the propagation of change activities.

11.7.4 Platform-Based Design

The core of platform based design is a modular product structure, i.e. a product family definition where each member is a configuration of modules. A module is a functional sub-system, ideally with one distinct function but often containing additional support functions. Modules also comprise a composition of parts based on the function's scope and its defined protocol-based interactions and interfaces. Using this approach it is easy to configure product variants or new products. Each product is ‘spelled’ by its modules, with distinct functions, module variants, and performance values (Hvam et al. 2008).

A modular product family becomes a platform when we can reasonably claim that there is sufficient diversity in the modules that the family effectively covers the market. This is supported by customer-oriented individualization of products and a developmental focus on technological evolution and planned, incremental product innovation. Finally, modules can be aligned with company functions to rationalize internal resource use (Erixon 1998). This allows for module orientated: selection of suppliers, manufacture and quality control, logistics, repair and upgrade, and design development.

A generic product model controls the addition of members to a product platform. This generic model can be developed from the Chromosome Model introduced in Chap. 8, which can be used to create a Product Family Master Plan, see Fig. 11.23 (Harlou 2006). The illustration uses Harlou’s terminology, however, these views are synonymous with the activity view, organ view, and part view used in this book.

In platform-based design the design activity is on the one hand strongly constrained by the chosen modular composition and the strict interface rules, while on the other hand having a high degree of freedom in the design of individual modules.

- The core approach to platform- based design is function/organ identification and mapping, making function reasoning central.
- Platform-based design is a new type of design professionalism. Specific training and methods are needed if this strategy is to be executed effectively.

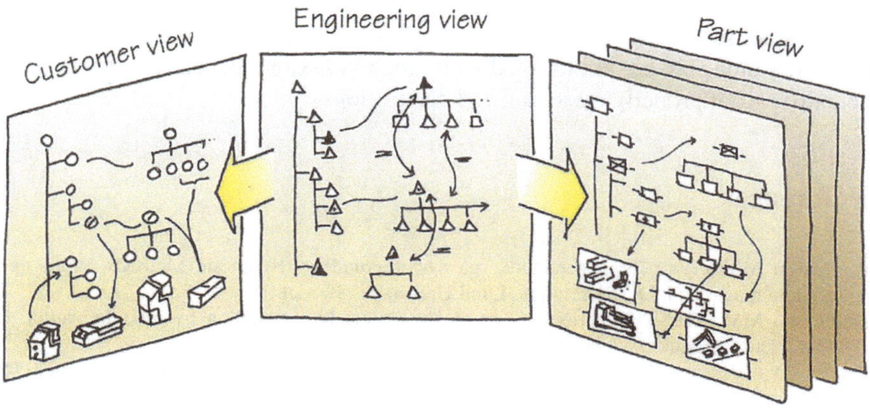

Fig. 11.23 Three views on modelling variety in a product family (Harlou 2006)

11.8 Conclusion

Function reasoning delves deep into the concerns of an engineer, answering the questions: what is this and how can it be used? The credo of past designers has been 'form follows function', but unfortunately there was no reason to believe this led to good and valuable products. Similarly, past engineers have focused on the 'idea in', i.e. developing good technical solutions. However, this does not ensure the product has a raison d'être.

Our approach balances the 'idea in' and the 'idea with', strengthening our ability to identify means that satisfy needs, give greater value and satisfaction, and ultimately improve the design. This is supported by function reasoning, which describes the totality of a product's synthesis:

- To determine what the product will be used for—its use functions.
- To determine what the product in itself is able to do—its action functions.
- To determine the organs and their functional interactions to realize the action functions, in parallel with the synthesis.
- To determine which functions should be added to achieve the products desired properties.
- To determine which functions should be added to achieve good properties in each life phase.
- To verify, during the synthesis, that the desired functions can be realized with respect to reliability and the desired functional properties.

> Functions are mental constructs that always require interpretation but are key to design reasoning.

As we shall see in the following two chapters on property reasoning and dispositional reasoning we are not finished with function reasoning, because functions are actually part of property reasoning and dispositions.

References

Andreasen MM (1980) Syntesemetoder på systemgrundlag (Synthesis Methods based upon System theory). Ph.D. Dissertation, Lund University, Sweden

Andreasen MM (1998) Reduction of complexity of product modeling by modularization. In: Conference Produktmodeller, Linköping, Sweden

Brown D, Blessing L (2005) The relationship between function and affordance. In: Proceedings of ASME design theory and methodology conference, longview CA. Paper No DETC2005-85017

Buur J (1990) A theoretical approach to mechatronics design. Ph.D. dissertation, Technical University of Denmark

Dym CL, Little P (2000) Engineering design—a project based Introduction. Wiley, New York

Eisenbart B (2014) Supporting interdisciplinary system development through integrated function modelling. Ph.D. thesis, University of Luxembourg

Erixon G (1998) Modular function deployment—a method for product modularization. Ph.D. thesis KTH, Royal Institute of Technology, Sweden

Harlou U (2006) Development of product family based upon architectures. Contribution to a theory of product families. Ph.D. Dissertation, Technical University of Denmark

Hubka V, Eder WE (1988) Theory of technical systems. Springer, Berlin

Hvam L, Mortensen NH, Riis J (2008) Product customization. Springer, Berlin

McAloone TC, Andreasen MM (2001) Joining three heads—experiences from mechatronic projects. In: Meerkamm H (ed) Proceedings of symposium design for X. Lehrstuhl für Konstruktionstechnik TU Erlangen Nürnberg

Miles LD (1972) Techniques of value analysis and engineering, 2nd edn. McGraw-Hill, New York

Monö R (1997) Design for product understanding. Liber Stockholm

Oja H (2010) Incremental innovation method for technical concept development with multi-disciplinary products. Ph.D. thesis, Tampere University of Technology Finland

Pahl G, Beitz W (2007) Engineering design. A systematic approach, 3rd edn. Springer, London (First edition 1977)

Rodenacker W (1970) Methodisches Konstruiren (design methodology). Springer, Berlin

Salminen V, Verho AJ (1989) Multi-disciplinary design problems in mechatronics and some suggestions to its methodical solution in conceptual design phase. In: Proceedings of international conference on engineering design ICED89 mechanical engineering publications, London, pp 533–544

Tjalve E (1979) A short course in industrial design. Newnes-Butterworths, London Facsimile edition (2003). Systematic design of industrial products. Institute of product development technical. University of Denmark, Copenhagen

Torry-Smith JM (2013) Designing mechatronic products—achieving integration by means of modeling dependencies. Ph.D. Dissertation, Technical University of Denmark

Vermaas PE (2014) Design theories, models and their testing. On the scientific status of design research. In: Chakrabarti A, Blessing LTM (eds) An anthology of theories and models of design. Springer, London

Chapter 12
Property Reasoning

If design is not to end up based on guesswork, we need to articulate goodness and know how it is related to the product's characteristics. The relationship between properties and characteristics is soft and difficult to reason about. Our approach is to confront these behavioural and structural descriptions. The Link Model points to the origin of property and value statements. Further, the Property Model, introduced here, articulates the property breakdown and the relationship with the product and use activity characteristics.

12.1 The Property Design Challenge

In general, the properties of an entity determine its value in the eyes of the user. In this way, users are able to articulate their opinion about a product's value via descriptions of its desired properties. However, these properties are the

M.M. Andreasen et al., *Conceptual Design*, DOI 10.1007/978-3-319-19839-2_12

consequence of the product's characteristics (introduced in Chap. 7) and thus the main challenge in design becomes: how to determine the characteristics such that the desired properties are achieved? To answer this question, we use **property reasoning**. In this chapter, we explore what elements make up property reasoning and how we can use this in the conceptualization of a product in practice.

Functions are the basic conditions for utility but properties use a broader definition, being influenced by many other factors, e.g. efficiency, durability, robustness, environmental effects, and cost. The user's perception of properties is subjective, reflecting a combined view of the whole product rather than individual properties. This is further complicated by the fact that the user is not the only stakeholder and property perceptions are also affected by the way in which the product interacts with other products, such as accessories, spare parts or other systems. For example, much of users' perception of a phone's properties come from how it interacts with other systems, e.g. synchronizing with a computer, fitting power chargers, achieving good Wi-Fi reception. The Link Model provides a useful foundation for property reasoning. Here, we use it to help align the user's experience and value perception, with the designer's interpretation of how the product should satisfy the need through the properties.

Our explanation of the design machinery (Chaps. 5–10) describe the design activity, which from a property perspective has three main elements:

- **Exploration** forms the foundation for the task and goal formulation, which define the desired properties and product requirements. This provides a target for the development activity.
- **Concept** and **Product Synthesis** support the designer in building the desired properties into the proposed design. These are continually refined during synthesis via user and market testing to confirm this is the *right product*; and analysis, modelling, simulation, and experimentation to confirm this is the *right materialization*.
- **Production**, **Sales** and **Refinement** form the feedback loop of product improvement based on real-world experience of the product's properties. Here, rapid response to identified problems and the gathering of feedback and insight for future product generations is key.

Example:

Injection pen. The company Novo Nordisk A/S develops injection pens for insulin and growth hormone. When using these pens the patient sets the required dose and makes the injection. The company has a long history of similar products where the *Norditropin Flexpro®* represents one of the latest generations. The Novo Flexpro has five main elements illustrated in Fig. 12.1: (1) thread for mounting the needle, (2) glass container for growth hormone, (3) driving spring—when the dose is set the spring is compressed, (4) activation switch—when the pen is activated the spring is released and forces the liquid out via an injection mechanism, and (5) push-button to reset the spring.

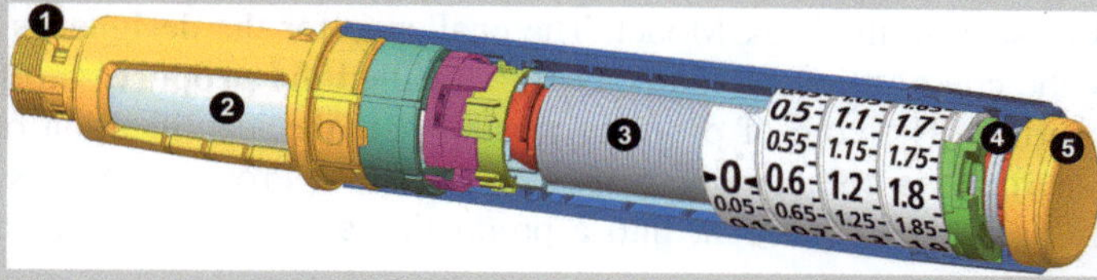

Fig. 12.1 The growth hormone pen, courtesy Novo Nordisk A/S

A key issue for this type of device is the users' trust in the precision of the pen. This is not just a mechanical issue but is also affected by embedded properties, such as how easy the device is to use, the perception of safety, how intuitive the learning process is, and ergonomics. These directly relate to the product's features including the mechanism for setting the dose, the push button, and the 'click' feedback telling the user the injection is complete. The actual mechanical precision of the device is not listed above because the user does not perceive it. However, this is still a critical design element—particularly in the medical context—and must be addressed in order to achieve statutory approval. Precision and control of production variability are key drivers for cost in the manufacturing process. Thus the designer must bring together the user's perception of the product's properties without neglecting the other design elements. This is particularly important as management's overall concern typically focuses on the risks of the development project, both in terms of unforeseen production problems and risks to the patient. Thus, although the user only perceives a small number of properties as important these are embedded in a large number of design elements both at the surface, e.g. the ergonomics of the device, and internally, e.g. the reliability of the spring mechanism.

From the above example, we see how user-oriented properties are embedded in and integral to the design at the surface level and internally. Technical issues and properties hidden from the user internally are still core to the expected functionality. In Chap. 7, we dealt with users' perception of a product's activities and functions, while Chap. 11 discussed behaviour and state changes as the fundamental carriers of functions and properties. Thus, we focus here on two questions that bring these chapters together and form the foundation for property reasoning.

- **What** do users and stakeholders articulate when they talk about properties?
- **How** are these points related to the products and activities that carry the properties?

Users' and stakeholders' articulation of properties is traditionally translated into **goal formulations** describing desirable and necessary functions together with their required properties. These are then modified or corrected as the design evolves and unforeseen problems are encountered (see Chap. 5). The properties link the nature of the product and its related activities, i.e. their structural characteristics as seen in the Link Model. The challenge for the designer is to understand how the composition of characteristics—called the **property model**—leads to the properties. The core of property reasoning is the identification of ways to realize properties via composition of property models. Property models represent a company or designer's insight into a product's nature, its materialization, and production.

Although the idealized conclusion of Product Synthesis is in the production of design documentation, this is obviously not the end of the process. A key characteristic of successful design work is following up the realization phases in order to get real insight into the product's properties. This **property verification** takes two main forms. In the first, the designers' interpretation of the properties is checked by confronting models, prototypes, and test products with real users. In the second, the properties designed into the product are checked against their realization in the actual produced product via experimentation and testing of prototypes. The aim of these is to reduce the risk of product failure by identifying flaws or opportunities prior to product launch. In addressing these various elements, the designer will encounter **trade-off situations** where two (or more) requirements cannot be simultaneously satisfied by the current design. The designer can try to dissolve the trade-off, find a different solution altogether or make a competitive trade-off. In order to help the designer deal with these challenges, Chapter 12 will discuss the following topics with the underlying logic that the secret of good product design lies in effective property reasoning:

- Section 12.2 **talks about properties** where we discuss how to interpret users' articulations about a product's properties and how these can be captured as requirements and proposed characteristic properties.
- Section 12.3 deals with how a product's **properties link characteristics to value** and explores the nature of this link.
- Section 12.4 details how property models can be used to **link issues, properties, and characteristics**.
- Section 12.5 discusses how to deal with **trade-offs** where two properties depend on shared characteristics.
- Section 12.6 deals with **property reasoning during synthesis**, linking the prior sections back to practical design.

12.2 Talking About Properties

When a new product is launched it is not just the main use activity that determines its success. Products also interact with, e.g. services, other products, external systems, recycling, and reuse. Further, products are tied into a wider family of artefacts including the sales material, user manual, spare parts, and utilities. Collectively, these interface with similar elements in competing products, making

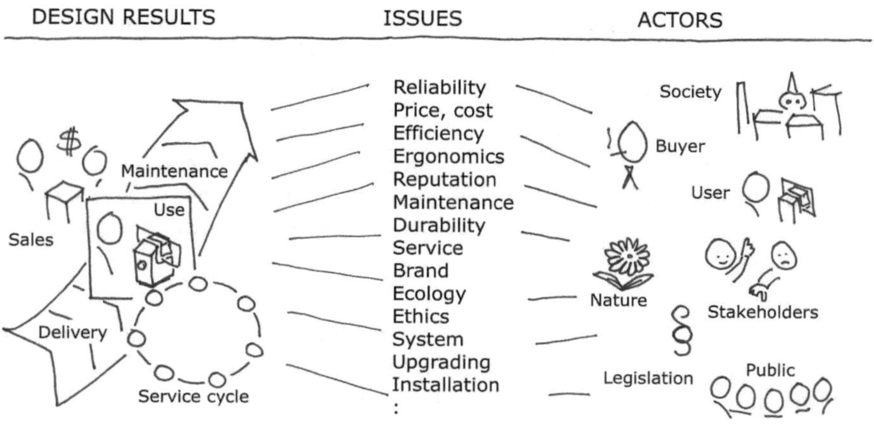

Fig. 12.2 Issues raised by various actors are linked to the different parts of the product and the design progression

the attention of the user a difficult commodity to command effectively. This attention concerns many design issues, such as functionality, properties, and features, as well as larger issues, such as brand and ethics. The Link Model allows for simplification and clarification of these issues. This is illustrated in Fig. 12.2 where some of the design issues raised by various actors are linked to the design entities. This allows the designer to focus on those issues relevant to the issue they are working on without losing sight of the overall aim.

Properties become concrete when we focus on a specific stakeholder, e.g. the manager responsible for product distribution. Here, the property of interest might be the cost of transportation and the associated product properties of weight, volume, and fragility. Other concerns might be the environmental impacts of transport, how the brand is supported by the distribution system, and what ethical issues might be encountered in the target market.

When talking about properties, a certain care needs to be taken in the terminology that is used. Each group of stakeholders use a wide range of articulations and terms to describe their perceptions and wishes. These articulations can, and often do, overlap or have conflicting meanings in different stakeholder groups. Thus the designer must take care when transforming these articulations into technical properties. For example, the **issue** might be efficiency and the **type of property** operational performance. In order to ensure these are properly addressed, they are formulated as a **requirement** with both an **indicator,** e.g. ease of operation, and a **metric,** e.g. operation time less that 5 s. Once the design is complete and the product is on the market, the issue can be assessed in reality: do the users actually find the product easy to use and is it efficient in practice? Overall, these factors feed into the goal formulation.

The goal formulation is developed from the designers' interpretation of need, user statements, and their own perceptions and experience. The goal formulation forms a ‘guiding star’ for the project, being produced in the early stages and evolving with the project. This helps the designers remain focused while allowing new ideas to be incorporated as the project evolves, giving the following heuristic:

An effective **goal formulation** incorporates both the ideas and wishes of the various stakeholders, and those new interpretations of issues that emerge as the project progresses, all of which are transformed into concrete requirements.

Goal formulations are thus composed of multiple individual requirements. In practice, requirements simply describe what is needed, whether it is characteristics (e.g. the size of the product being less than a standard door) or solution focused (e.g. re-use of components from previous projects). As such, requirements can be defined with respect to product properties.

Definition: A **requirement** is a statement about a desirable **property** of a device or activity formulated as a **value statement** with an **indicator** and **metric**.

In industry, it is common to refer to **qualities**, normally defined as the degree of customer satisfaction. Our concept of quality and its relationship with properties is explained in the following example.

Example:
Razor quality. When using a razor, users find weight a key aspect because it influences the shaving activity and also signals something about the razor’s sturdiness. As such, we might define a **requirement** ‘low weight’, composed of the **property** ‘weight’ and the **indicator** ‘low’ (as perceived by the user). However, without insight into customers’ reactions to weight—the **quality**—it is not possible to operationalize this requirement with a **metric,** i.e. why weight is actually ‘low’ and is that positively perceived by the customer? This is illustrated in Fig. 12.3. Based on this assessment, the requirement can be made more concrete via a metric: weight must be between 200 and 250 g or if this is not possible for the company 250 g should be seen as the target or ideal goal. Thus by introducing quality to our indicator, we are able to formulate a much more nuanced understanding of the requirement. In fact, if we only followed the ‘low weight’ aim (as supposed by the original indicator), the requirement becomes unhelpful as customer perception of quality also drops when the weight is below 150 g.

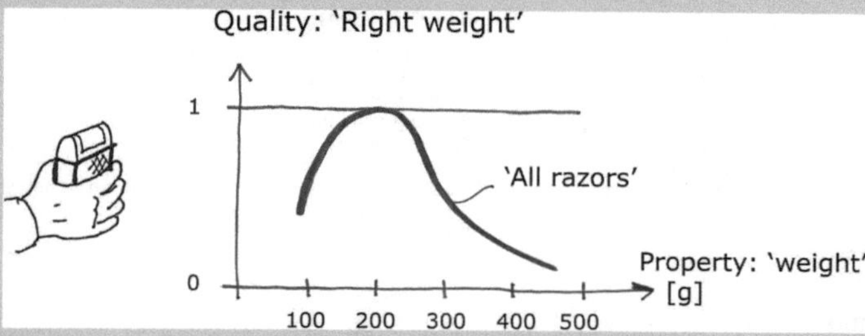

Fig. 12.3 The relationship between a razor's weight and customer reactions visualized as an imaginary curve of customers' reactions to a range of razor weights

In the above example, we use the term quality to describe the range of reactions idealized by the curve shown in Fig. 12.3. For each point on the curve, we can link weight to perceive quality, e.g. 300 g gives 0.5 on the quality axis. This represents the satisfaction of that razor (300 g) for a generic user. Note that this satisfaction value drops in both directions, with both too light and too heavy being undesirable. The designer can use quality assessment to help find the 'sweet spot' which best satisfies the customers.

Although we propose a number of terms and their relationships above, there are many open questions with regards to goal formulation, which are not answered by theories in the literature. In particular, should requirements also describe functions? And should we distinguish between levels of desirability, e.g. demands and wishes (Pahl and Beitz 2007), or need, target, specification and value metric (Ulrich and Eppinger 2004)? Overall, we suggest the comprehensive and helpful discussion of goal formulation in terms of functional specifications, performance specifications, and metrics described by Dym and Little (2000), which we use in our own teaching.

As with goal formulation, property reasoning in general has little theory described in the literature. The fundamental description of property theory is made by Hubka and Eder (1988), while Andreasen (1980) highlights the importance of distinguishing between properties and characteristics. Because of this lack of theory, we propose the 'property model' as a means for understanding how to formulate and satisfy goals. Fundamentally, this builds on an ability to describe issues and properties, which we deal with next.

12.2.1 Issues and Properties

How many issues or properties are there? Kesselring (1951) described more than a thousand types, which is perhaps an unhelpfully large number! One way to gain an overview of these is to consider the factors influencing a product's design. To

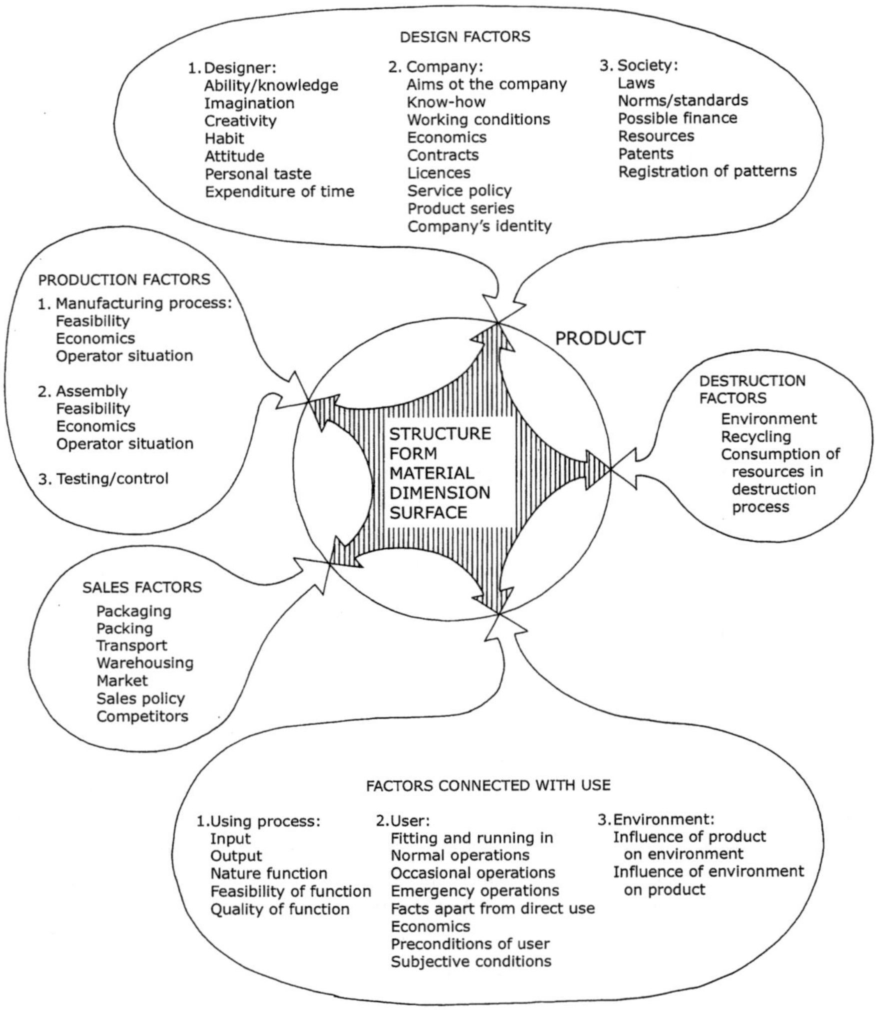

Fig. 12.4 Factors across the product's life to be respected in the product's design (Tjalve 1979)

this end, models related to 'total design' were proposed by both Pugh (1991) and Tjalve (1979). We use Tjalve's model, which is illustrated in Fig. 12.4 showing factors affecting the product design from across the product life. Tjalve (1979) uses a part-oriented terminology with structure, form, material, dimension, and surface as characteristics of the product. We prefer the broader term 'design entities', which includes product, use, service etc. as well as the activities in the product's deployment. Based on this distinction (Fig. 12.4) and that by Hubka and Eder (1988), we propose a classification, which starts by distinguishing between **characteristics** and **properties**, after Domain Theory (Andreasen 1980).

- **Characteristics** define **identity,** i.e. decisions made about mode of use, mode of action, organ structure, part structure, and parts.
- **Properties** define **goodness,** i.e. the impression the user gets of the product's behaviour through its deployment.

We use the term **issue** to link statements to properties, i.e. issues are design-related aspects brought up by users, actors, and designers. Ultimately, some of these issues may result in design work or decisions concerning the properties of the product.

To help clarify the distinction between properties and characteristics, consider the razor example used above. The weight of the razor is a **property**, which depends on the **characteristics**: number of parts and mass of the materials used. Similarly, noise is a property that depends on the characteristics of the motor, structure of the moving parts, and action of the razor.

Although both 'characteristics' and 'properties' are used in everyday language, we refine their definitions with regard to the product's attributes. Here, attribute is a collective term for all the characteristics, properties, and features of an artefact.

Definition: **Characteristics** are a class of **structural attributes** of products and activities determined by the synthesis of the design.

Similarly, we draw inspiration from Smith and Clarkson (2005) and Eder and Hosnedl (2007) to propose the following definition of properties. Note that properties are behavioural, not structural. This is different from everyday usage where, for example, form and dimension would be described as properties. In design, form and dimension are characteristics because they are related to structure.

Definition: Properties are a behavioural class of devices' and activities' attributes, by which they show their appearance in the widest sense and create their relation to the surroundings.

In both definitions, we are concerned with the holistic view of the product including functionality, use activity, distribution and sales, maintenance, and so on. With this in mind, we offer an (rather fanciful) example to help differentiate the two terms—Frankenstein and his monster (Fig. 12.5). Here, Frankenstein has designed his monster, assembled the parts, and is prepared for the lightening! At this point, the monster is still 'dead' and only has the characteristics that Frankenstein has designed via the assembled parts, e.g. the monster is two metres tall. Once the lightening strikes and the monster comes to life, it suddenly carries the properties Frankenstein imagined (hopefully), by being able to interact with the surroundings. Finally, Frankenstein looks upon his creation and assess its 'goodness' based on both its characteristics, e.g. height, and properties, e.g. ability to react to the surroundings (Igor is not required). Put simply, characteristics are determined before and properties are seen after the moment of ignition.

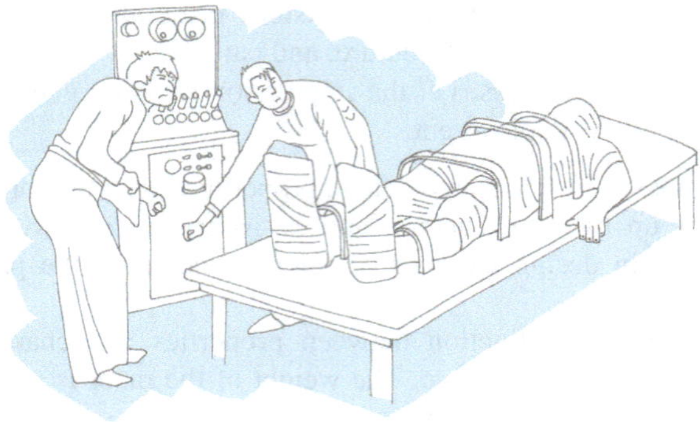

Fig. 12.5 The designer is calling the product to life

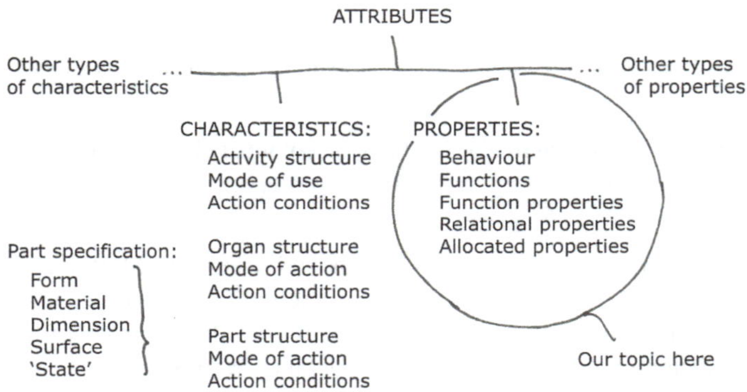

Fig. 12.6 A product's attributes with a focus on properties

Domain Theory allows us to relate the nature of activities, organs, and parts to characteristics and properties as illustrated in Fig. 12.6. Behaviour is a property because although the designer can define mode of action, i.e. how things work, behaviour describes how the product actually acts. In the following sections, we explore the three classes of properties introduced in Fig. 12.6: functional, relational, and allocated.

12.2.2 Function Properties

A product's organs and their composition carry the functions and realize certain properties related to those functions, called **function properties**. Figure 12.7 shows the function properties related to the function *show temperature*, using the example of a traditional liquid/glass thermometer. The figure also shows the function properties related to the function *punch holes*, using the example of a paper punch's punching organ.

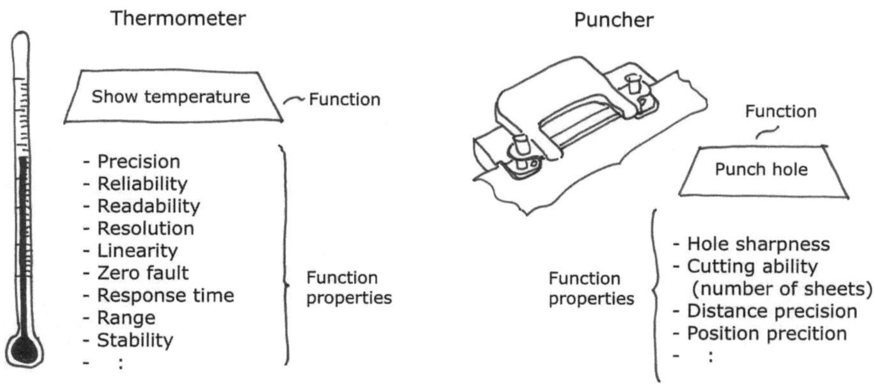

Fig. 12.7 Two examples of functions and their related function properties

Although function properties are closely related to product performance, they are seldom explicitly articulated in a goal formulation. Therefore, we propose the following heuristic.

> In order to understand the **goodness of the functions'** realization and be able to create better solutions it is necessary to describe the relevant **function properties**.

12.2.3 Relational Properties

The properties of interest for users are those that they perceive when they actually use (or misuse) the product. As such, property statements from users cover not only the product's behaviour but also how it behaves in context. It is this behaviour in context that we refer to in **relational properties**.

The relational aspect of a property is considered when we ask: under what conditions can I observe the property? Or at the design stage: what conditions do I imagine will be necessary to create the situation where the property is apparent? In answering these questions, we must consider aspects not just related to the product. To clarify, we offer the following examples:

- Both the product and its surroundings determine a product's temperature sensitivity. This can be simulated based on testing or prototyping different situations.
- Ease of use can only be observed when a user actually tries to use the product, i.e. the ease of use is determined by the interaction between user and product.
- Production cost is related to the manufacturing tasks defined by product and, e.g. their fit with available production equipment, i.e. the interaction between product and production system.

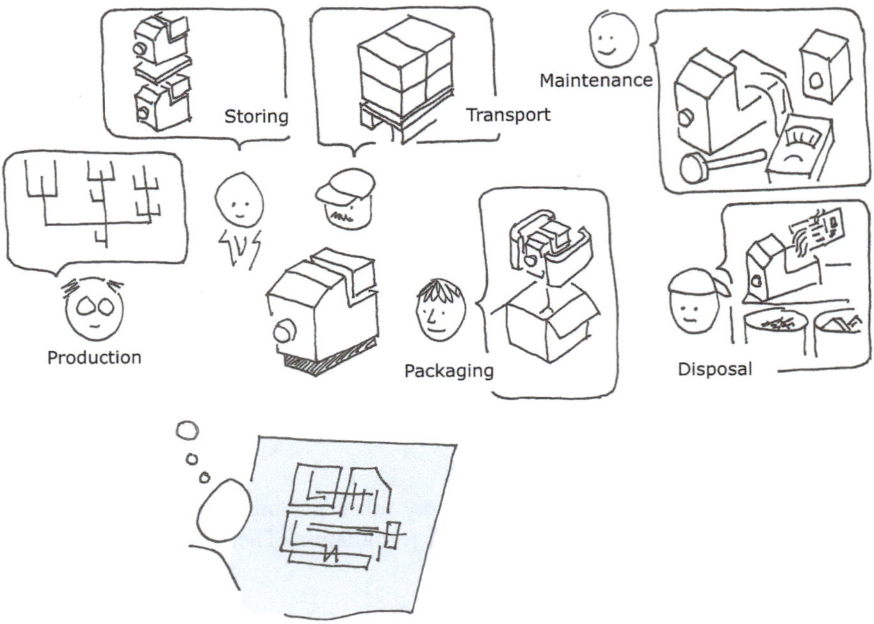

Fig. 12.8 The designer considering the product's fit to its life phases

In Chap. 8, we introduced View Models for obtaining insight into specific product properties. Such view models cover properties related to the product's use activity or to specific lifecycle situations. These situations are referred to as **meetings**, after Olesen (1992). An actor's perception of goodness, related to successfully performing an activity, can be described in terms of seven types of **universal virtues** (Olesen 1992): **cost, quality, flexibility, risk, time, efficiency, and environmental effects**. Looking closer at the issues in Fig. 12.4 many are related to lifecycle activities and can thus be translated into properties related to the product (its fit) and the activity (its performance), Fig. 12.8. An important group of properties related to lifecycle activities are relational:

> It is necessary to get actual, relevant insight into the lifecycle activities and the **meetings** in order to create the product's fit for life and satisfy the product life actors.

12.2.4 Allocated Properties

The final group of properties we will treat here are **allocated properties,** i.e. properties that customers, users, stakeholders, and society relate to products in a symbolic or devotional way. These properties have to do with ideas, such as **excitement** (e.g. attractiveness, 'need to own', au vogue, bad taste, group pressure), **style** (e.g. Scandinavian, neo classical, retro), **trend** (e.g. 'this year's colour is blue', seen on TV, trend magazines), **hobbies** (e.g. gardening, extreme sport, sailing), **origin** (e.g. fair trade, ethnical, 'Made in Germany'), and **brand**. Buyers, companies, and individuals have an influential position that Papanek (1971) visualized in the 1970s. Today, there are many other elements influencing us but the basic premise remains true. In this context, socio-technical design theories deal with allocated properties.

There is a tendency to see properties as being objectively and decisively created by the designer. However, properties seen as representative of the product are really determined by the interaction between people and the wider system. Put simply, without an observer, utilizer, or customer, the product has no properties. A researcher, Mørup (1993), visited the Harley-Davidson Company and found that its designers were scared to make changes that might disturb the customer's perception of a 'real' Harley: the sound? the leaking oil? As such, we see the role of allocated properties in the following heuristic.

> The designer should be conscious of potential allocated properties related to the customer and/or user. In particular, these can often be far from the main focus of the design work.

This section aimed to help answer: how to understand products' properties? This leads to the following heuristic.

> Things get their properties when deployed in context, in a stakeholder's eyes. Functions are necessarily built in but their appreciation depends on the user.

This means that when the used context changes, a new set of properties will be perceived!

12.3 Linking Characteristics to Value

This section uses the Link Model to explore characteristics and value. From the user's perspective, need satisfaction and value are the results of their perception of function, activity result, and properties. How these aspects contribute to overall value perception and thus the buy-decision is very individual and context dependant. From the designer's perspective, the Link Model states that the designer should articulate the mentioned aspects as goals for the design activity. Using property reasoning can give a competitive advantage by satisfying multiple users' perception of value. In order to do this, we need to answer two questions: what in the activity and product is carrying a specific property and how are properties composed?

12.3.1 Properties' Dependency on Behaviour

Most properties, and typically the important ones, are related to behaviour, both of the use activity and the product itself. Thus we can link state changes to properties.

Example:
The Moka Pot. The Moka pot brews espresso type coffee as the result of the use activity shown in Fig. 12.9. Let us try to trace some of the user-oriented properties in this example. First, overall value can be decomposed to reflect ease of operation, quality of coffee, and pride of ownership.
Ease of operation is related to the activity structure, i.e. the basic tasks, dose, assemble, disassemble, clean, and the user's tasks, stop heating or move the pot off the stove. The ease of operation property can be decomposed into properties associated with each of the operations listed above. For example, consider the pot's assembly characteristics (Fig. 12.10), where the *ease of operation* property is decomposed into (*ease of*) *gripping*, (*ease of*) *catching*

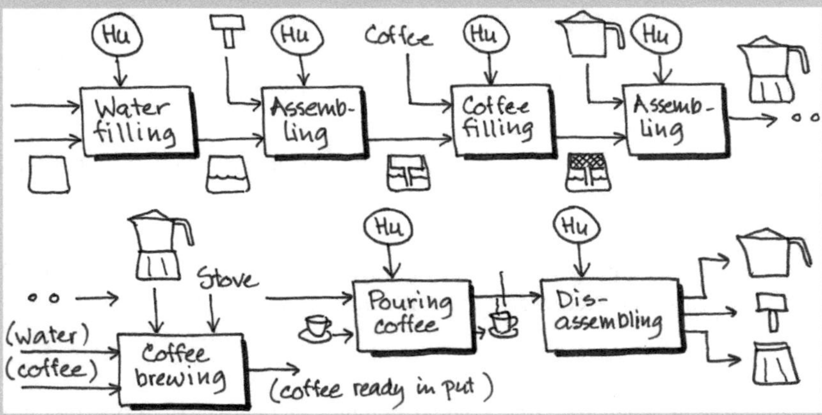

Fig. 12.9 The use activity of a Moka pot

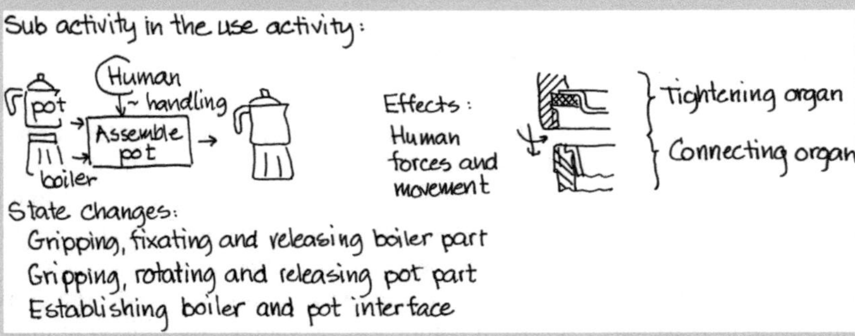

Fig. 12.10 Operations associated with pot assembly

the threads, and *(ease of) tightening* the screw. This property is determined by the design of the connection element (form of tread, number of turns, material pairing), as well as the way the tightening is designed (elasticity of the gasket). The *quality of the coffee* is closely related to the chosen technology and the quality of the ground coffee beans (which is beyond our concern). The technology and the quality relate to the brewing organ's mode of action (proper dosing of water and coffee, temperature, and the build up of pressure in the brewing chamber). However, we see that these conditions do not actually come from the brewing organ but are affects from the boiler organ's mode of action (heat transfer from stove via thermal conductivity). Thus the brewing chamber's performance is dependent on a combination of temperature, pressure, and duration of contact between beans and water. Of particular interest is the build up of pressure, which depends on the dimensions of the holes in the filter and the granularity of the coffee. These are both characteristics but are related, respectively to the product and the operand (the coffee).

Pride of ownership may be related to owing a product in the original design created by Bialetti in 1933, i.e. aluminium pots with orthogonal forms. These parameters are characteristics of the pot.

To sum up, we have linked three properties (*ease of operation, coffee quality, and pride of ownership*) to specific characteristics of the pot. These three user properties give various patterns of composition as shown in Fig. 12.11.

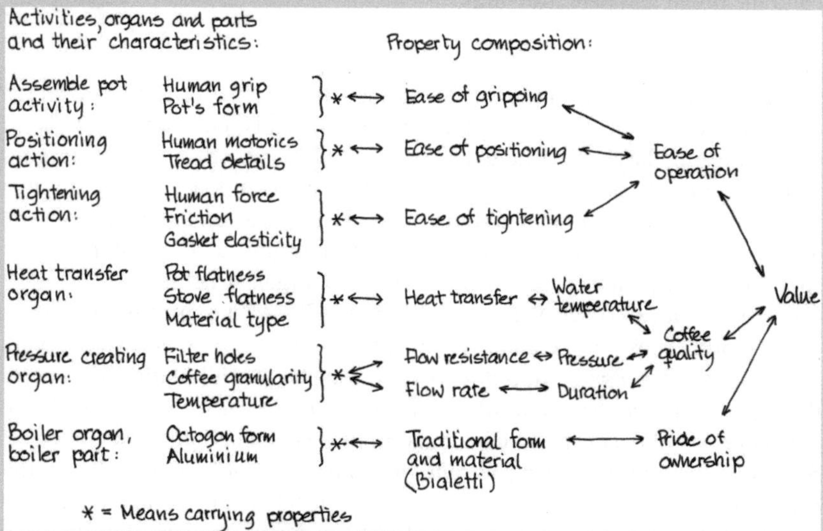

Fig. 12.11 Partial composition pattern for three of the Moka Pot's properties, leading to overall value. The pattern is not complete as more cross relations are to be expected

This example serves to explain the origin of a product's properties as articulated in the Link Model:

- **Product structure** deals with organs and parts and thus the mode of action and specific characteristics of organs and parts.
- **Activity structure** deals with the mode of use and specific characteristics of operands, and delivers the use result, e.g. the finished coffee in the Moka pot.
- Certain properties relate to the **user as operator** and the relationship between the user, product, and use activity.

Each property has its own explanation in, for example, an organ's mode of action or the user's interaction with an assembly operation. We call such an explanation a property model. However, before we can discuss this, we need to understand properties' composition.

12.3.2 Properties' Composition

Here, we focus on how a property can be composed of contributions from multiple entities. In the literature, this is called a mapping (Suh 1990) or transformation, when requirements or properties are linked to characteristics. However, this linking is so complex that these works tend to over-simplify the real situation. We use a detailed example to illustrate the identification of relationships between properties and the organ and part structures.

Example:
Design of a platform. Imagine an indoor lifting platform for raising a craftsman to ceiling level. The platform is manually positioned but raised electrically to the desired height as in Fig. 12.12. A chain of effects goes from the electrical motor, via a transmission, to a bar mechanism that supports the platform. In addition, there is a brake and locking system. The organs for movement, transmission, positioning, and operator support function together in a causal chain to create the organ structure.
Each organ carries a set of function properties contributing to the product's overall properties. For example, the transmission is a wire system with *mechanical efficiency, play, and elasticity*. For the overall platform, some of these properties contribute to the platform's *stiffness, positioning speed,* and *smoothness of movement*, which all add to its mechanical properties.
Similarly, the platform's *size, weight, and manoeuvrability* impact its overall *usability*. The wheel arrangement and turning mechanism create a composed organism. The property *reliability* is linked to many elements but can also be determined by a single weak point. Other properties like *weight or cost* are easily summed for the finished product but difficult to distribute to parts before completion.

Fig. 12.12 Lifting platform used for painting a ceiling

In this example, we imagined a synthesis situation and identified various properties and their relationships to the organs. We also saw how the interaction between organs can carry certain properties. The properties identified above are generally desirable, in the example, and will thus form part of the goal statement as requirements with indicators and metrics. If we switch to an analytical approach, where the platform is known, this interrelation between organs, properties, requirements, and issues can be modelled as in Fig. 12.13 (based on the Link Model). For clarity, we do not distinguish between properties and requirements in this illustration.

Example:
Design of a platform continued. Figure 12.13 illustrates some of the properties we discussed above. The issue utility is composed of the properties: ease of access, ease of moving/re-positioning, and speed of positioning the operator in the work position. These properties relate to the operator support organ, platform mechanism organ, and movement/steering organ. Further, the issue reliability is denoted by a question mark indicating that the entities that critically influence reliability are non-specific, i.e. for a particular design solution, we will be able to point to a specific part, e.g. motor, that is the cause of reliability concern, but any one of many parts could fulfil this role in a general sense.

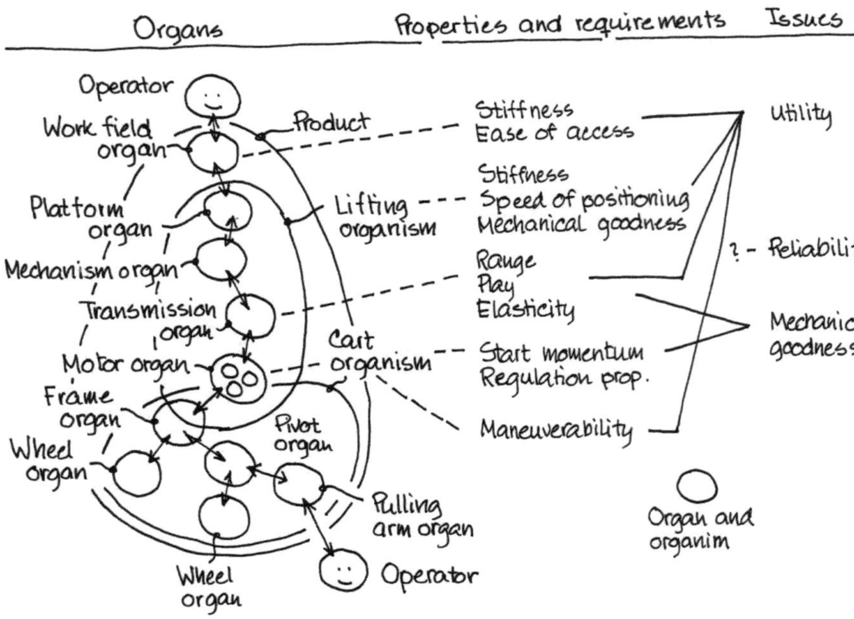

Fig. 12.13 Links between organs, properties, requirements, and issues in the platform example

Our tracing of properties to organs in Fig. 12.13 is qualitative, being based on experience and expectations. However, in Chap. 8 we explained how parts formed building blocks of the organs. Thus, when we need full quantitative explanations for the causes of properties we have to investigate the part structure.

Example:
Design of a platform continued. Certain properties do not appear before the part structure is determined. For example, the issue of *overall size* is determined by parts and their spatial arrangement, which can be derived from a layout drawing. In contrast, *reliability* can depend of a single critical organ or part and thus needs closer study of the structure's different failure modes. Similarly, *appearance* is determined by major parts and assemblies to give a total form (Tjalve 1979). This total form can be modelled in a layout drawing and compared to the brief.
The example in Fig. 12.14 relates *utility* to the *stiffness* of the platform mechanism. Inherent stiffness is primarily determined by the two scissor-like mechanisms carrying the platform. An alternative mechanism principle, e.g. a hydraulic cylinder, would give a different inherent stiffness profile. Add to this the stiffness of the bars and their joints—the rotational and sliding bearings. Together, these two details of the part structure determine stiffness. Of particular importance in this design are the joints, where *play* is damaging, and the traversal beams, which determine lateral *stiffness*.

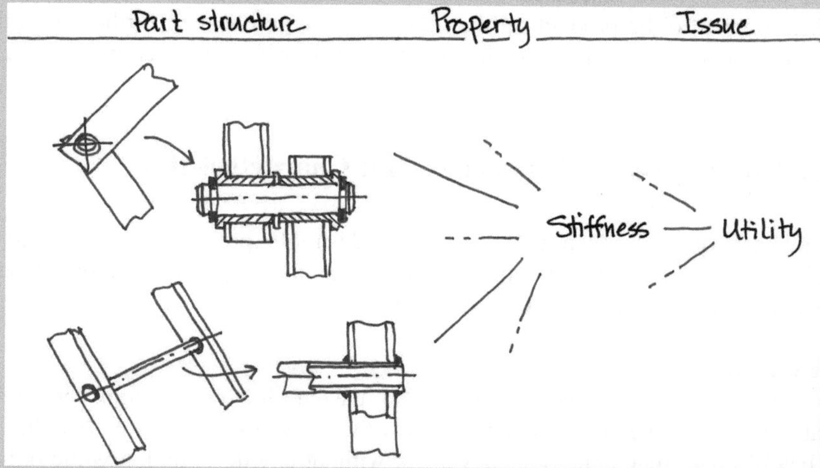

Fig. 12.14 Linking the property *stiffness* to the part structure: bearings and traversal beams

From this example, we can see that tracking certain properties can require some detective work because the sources and propagation is dependant on the specific product composition. For example, if we try and trace the main sources of noise in the lifting platform it is far from simple. We find multiple sources of vibration in the electric motors and lift mechanisms, as well as possible amplification from the platform's structure. A model can be established using a noise simulation program or via a prototype but clearly experimentation and refinement are needed.

Based on the elements discussed in this section, we propose the following heuristic for articulating properties' composition.

When tracing or synthesizing properties be aware that properties can relate to entities' joint contribution (organisms, assemblies) and internal conditions can influence these properties.

Finally, we can answer what in the activity and product is carrying a specific property and how are properties composed?

- Properties relate to use activity, product, and user as shown in the Link Model.
- This relation originates in the characteristics of activities and organs.
- Specific properties, e.g. coffee quality, can be decomposed in a breakdown pattern.
- The breakdown and origin depend on the actual design solutions in the product.
- Certain properties can be understood from just the organ structure while others require materialization in the part structure.

Until now we have discussed qualitative relationships between properties and entities related to the product. In the following, we go one step further in finding the parameters of the entities that influence the properties via the property model.

12.4 Linking Issues, Properties, and Characteristics

In this section, we illustrate how certain desirable properties are created as synthesis progresses and in doing so illustrate the key elements of property reasoning, i.e. thinking properties into the product.

Example:
Creating a bow. When children decide to make a toy bow they look for suitable branches and string. When they test the branches' strength by bending, they discover that some branches break without any deformation while others bend to easily. In this range they find the interesting branches, which store energy from bending and do not break. These branches then form the foundation for the child's tests, which give insight into the design, pointing to the desirability of high stiffness and low weight. The codification of this insight is what we call a **property model**.

Reasoning about properties in practice is normally related to requirements. This leaves the question: how to model the relationship between properties and characteristics quantitatively? Let us start with an example property breakdown.

Example:
Projector. The principle of a projector is shown in Fig. 12.15 where: (a) is a lamp, (b) a reflector, (c) a condenser lens, (d) a filter to reduce heat transfer, (e) and (g) are Fresnel lenses, and (f) is a LCD panel on which the image is created and projected through the lens system (h) to the screen (j).
The core property of the projector is the *contrast* of the image on the screen. This has a number of influences. First, it is dependant on the quality of the screen it is projected on. Second, technical elements such as the LCD panel, filter, and the precision of the lenses affect colour dispersion as well as contrast. The projector also contains a cooling fan that creates vibrations seen on the screen as blur. Finally, operational elements, such as the precision with which the operator sets the focus (or choses auto-focus) also affect performance. As such, there is a wide range of elements that feed into the property of contrast shown in Fig. 12.15. This also points towards possible causes, e.g. the characteristics of product and operation.

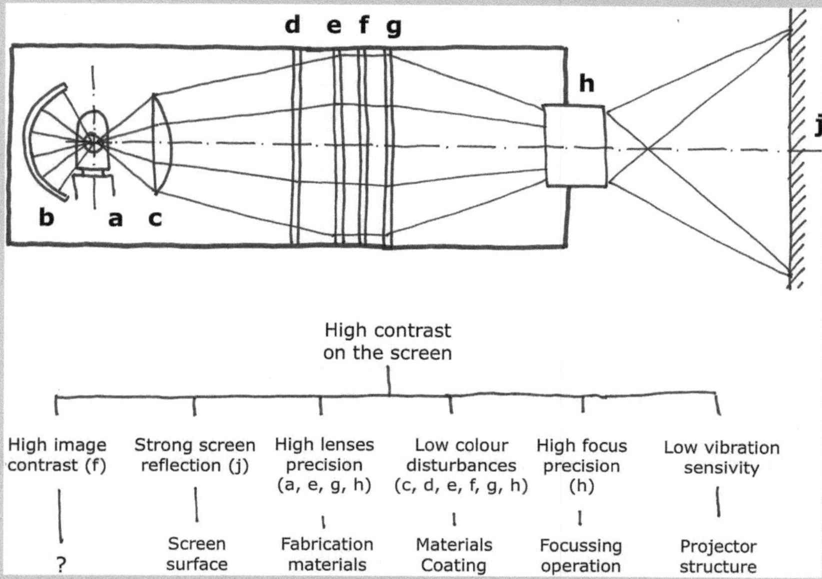

Fig. 12.15 The composition of a projector with a breakdown for the property *contrast*

A key feature of this example is that the various decomposed properties contribute to different aspects of the overall contrast property via different natural phenomena. For example, the quality of the image produced by the lenses is largely determined by their material properties and finishing. In contrast, the vibration effects from the cooling fan are more closely related to production quality and design tolerances. Thus, we need to be careful in assessing the various characteristics' influences and their combination. This carries us to the property model where we bring together the decomposition pattern and link it to the characteristics of the design entities.

Definition: A **property model** describes the insight into the realization of a certain **property** via certain design entities' **characteristics**.

A general illustration of the elements in a property model is given in Fig. 12.16. The composition of a product's deployment (activities, organs, and parts) on the left is linked to certain issues of interest on the right. Requirements (where indicators are known) are linked to characteristics via the properties. A distinction is also made between activity (A), organ (O), and part (P) characteristics.

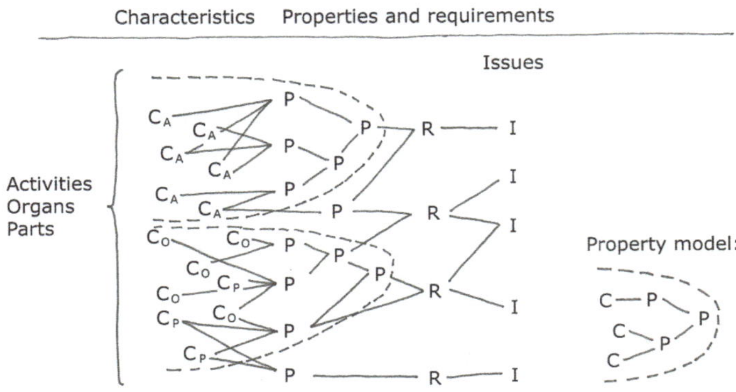

Fig. 12.16 The link model used to create a product model linking design entities, characteristics ($C_{Activity}$, C_{Organ}, C_{Part}), properties *P*, requirements *R*, and issues *I*

With respect to the product, model customers and users tend to focus on certain issues and properties related to their perception of what the company has 'promised' in terms of, e.g. performance or durability. There is a risk that the company is not able to (or never planned to) actually realize these properties. As such, the company needs to have insight into both its product and also the environment into which it plans to launch. The company's property insight typically comes from partial simulation models, testing, data collected from past or running projects, and tacit experience of staff. This type of insight is illustrated in the following example.

Example:
Safe car. A critical issue in car design is passenger safety. This issue is complex, being related to road properties, tire performance, the car's drive and steering systems, braking properties, collision safety, and so on. Some of these properties are distributed across subsystems, e.g. anti-lock brakes, while other properties are achieved through combination, e.g. safety belts, airbags, and role cages work together to protect the passenger (Fig. 12.17).

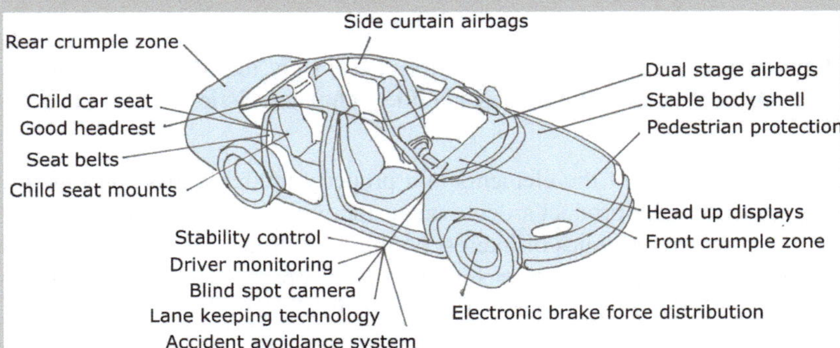

Fig. 12.17 Entities related to a car's safety

For the manufacturer, this means that multiple property models are needed to capture all of the knowledge related to these properties, e.g. ergonomic information, computer simulations of chassis energy absorption design, and the results of crash tests. All of these elements make it necessary to couple the design activity with careful knowledge management.

Many of a product's properties like dynamic response, temperature sensitivity, ease of operation, and robustness can be derived from specific theories, models, or calculation methods. It is the designer's task to handle such property dimensions, combining characteristics from the actual design with the conditions of the products use context. In this respect, these are relational properties, as illustrated in Fig. 12.18. The focus here is on the important role engineering knowledge and scientific insight play in building properties into the product.

The designer should seek out scientifically grounded **best possible knowledge** when building in functions and properties.

A property model is best supported when models of the design are described for all three domains (Mortensen 1999), e.g. following the Chromosome Model. Creating a view model points to the conditions necessary for an organ's

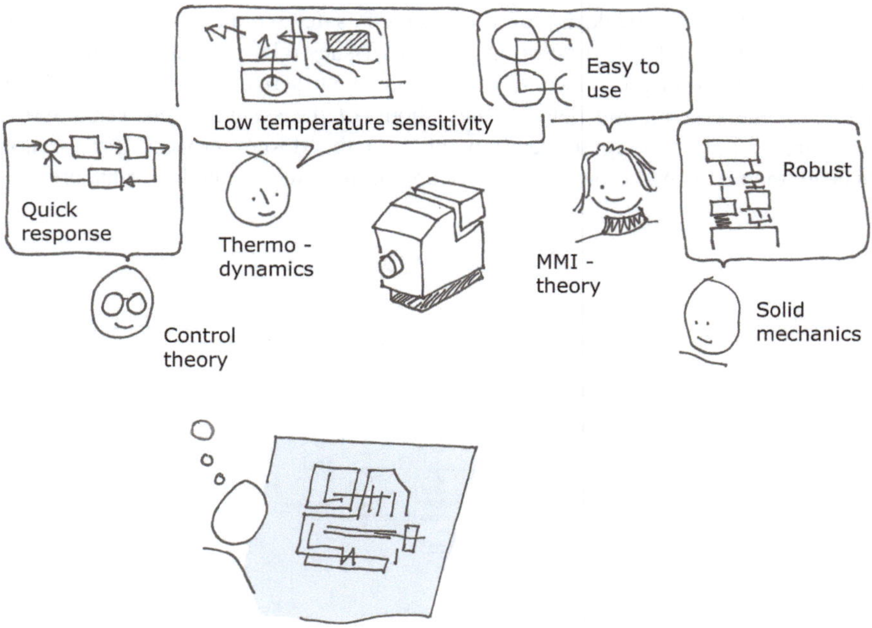

Fig. 12.18 Properties determined from engineering theories (property models) where the product characteristics are determined

functionality to be described and its properties determined. Weber (2014) describes a similar approach in his property driven-development theory where he introduces Characteristics-Property Modelling (CPM) to articulate models of properties' realization. In both approaches the question remains: how are properties decomposed?

12.4.1 Property Decomposition Patterns

Product synthesis can be seen as a progressive concretization and detailing following a composition/decomposition pattern (Svendsen 1994). Organs and parts are added to the structure and gradually the product's behaviour emerges, first functions, then properties. This growth of the structure is followed by a gradual decomposition of the goal structure, which allows the designer to choose between alternatives at increasing levels of detail. This links to the functions/means tree (Chap. 11) where lower levels of functions/means are a consequence of choices that do not satisfy complete solutions. In the same way, fulfilling requirements leads to adjustments in the product structure. This progression gives insight into functions and properties, as well as highlighting where new properties are needed in response to new problem insights.

Many authors recommend an illustrated objective tree, e.g. Cross (2008) and Pahl and Beitz (2007). An example of such a tree is found in Fig. 12.19, which shows objectives for the design of a 'convenient, safe, and attractive new transport system'. Interestingly, the tree brings together issues related to the journey, safety situations, and transport. Two transport-related attractiveness issues are: being a user of the system, and being a 'neighbour' of the system. However, this tree structure is not growing in detail parallel to the design process' progression. Instead, the tree is based on prior knowledge of the issues related to transport.

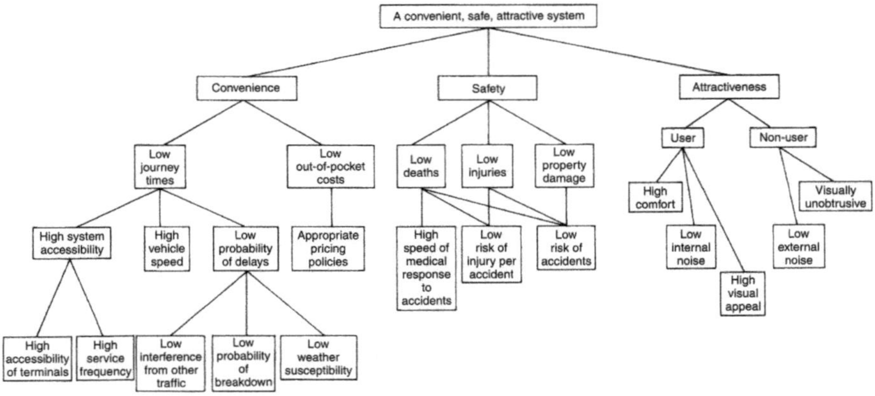

Fig. 12.19 Objectives tree for a new transport system (Cross 2008)

Based on these discussions, we can take two main approaches to making a property decomposition pattern. First, we can create a pattern governed by gradual synthesis and thus merge the properties of sub-systems into the overall property decomposition. Second, we can establish a logical pattern of issues based on our understanding of how the deployed product will interact with actors and its surroundings.

The proper composition of objectives or requirements related to a product depends on the users' concerns and insights into the nature of the solutions.

Here the requirements can change depending on what level of the decomposition we are looking at and with respect to the sub-solution being considered (Fig. 12.19) (Svendsen 1994).

- The requirement becomes one **piece** of the overall satisfaction, e.g. a weight or cost contribution.
- The requirement **changes nature** compared to the higher levels, e.g. time changes to probability or accessibility changes to service.
- The requirement is **unchanged** in the decomposition, e.g. reliability or lifetime requirements are necessarily similar across levels.
- The requirement is **decomposed** into sub-requirements, e.g. the delay probability in Fig. 12.19 having three independent contributions all increasing the overall probability.
- The requirement may lead to **additional functions,** e.g. the car safety example above.

Design can be easily understood as creating ideas and composing a product from many sub-solutions. However, what really matters is the product's ultimate functionality and properties. These depend on the choices made by the designer and thus require a deep understanding of the sub-solutions, the requirements, and their relation to the users' interpretations. This complex web of factors needs suitable visualization models in order to properly stage this design activity. It is this role that is filled by property decomposition patterns. Without them there is little basis for making trade-off decisions in anything other than an ad hoc fashion.

12.5 Trade-Offs

When we analyze a product and the origin of its properties we find that certain properties share characteristics. Here, the properties involved might all be satisfied or one (or more) may be sub-optimal because of the need to give priority to other properties. Finding a balanced solution to such a conflict is called a **trade-off**. Designers often describe this in terms of conflicting requirements. However, in reality the root cause is likely that they have simply not found solutions where this negative dependency is absent.

In engineering, trade-offs are common phenomena in complex decision situations, e.g. balancing development time or product quality against rising costs in a project. We can thus draw on decision-support concepts, such as, sensitivity analysis (Ulrich and Eppinger 2004). Sensitivity analysis seeks to tell us what effect changing one parameter will have on the other parameters, e.g. the effect of delaying product launch on product finish and market confidence.

In designing, trade-offs are typically formulated in terms of pairs of requirements that cannot be simultaneously satisfied (with the current proposed solutions). This situation can be illustrated as in Fig. 12.20, where two properties are compared and the positions for alternative solutions and competing products are shown. In this example, both properties should be minimized for the best product outcome. We can use parameter variation for the property models (A and B) to create the trade-off curves shown. From this analysis, we see that a specific value for concept A gives the best trade-off. Here, the best solutions are situated in the window of ideal values, although solutions in the marginal area are also acceptable.

Example:
Hand dryer. In many public toilets, electric hand dryers are provided as an alternative to, e.g. paper towels. The most common types combine heated air and high airflow rates to dry the hands. These were characterized by a trade-off between airflow and heating elements. This usually resulted in a poor hand drying experience due to the fact that this combination takes a relatively long time to actually dry hands. The Dyson hand dryer radically changed this by using only high-speed airflow. The trade-off is now reduced to: energy consumption v. drying speed. This is generally perceived to be superior because elimination of the heating trade-off means that all the energy can be directed into airflow and thus dry the users' hands significantly faster than the traditional approach.

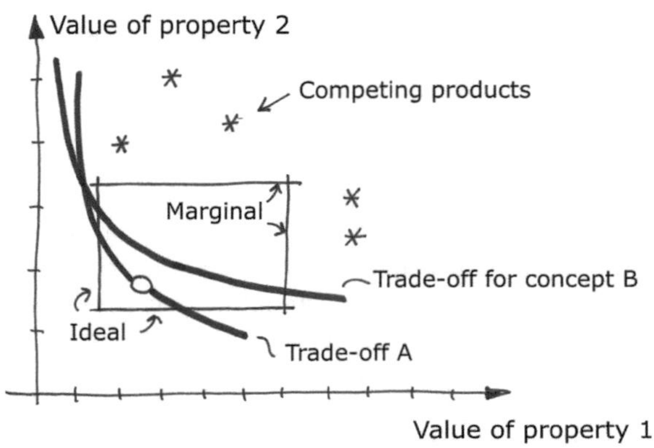

Fig. 12.20 Trade-off space for two properties mapped against existing products and concepts

Ulrich and Eppinger (2004) write about trade-offs: "*An airplane can be made lighter, but this action will probably increase manufacturing costs. One of the most difficult aspects of product development is recognizing, understanding, and managing such trade-offs in a way that maximizes the success of the product*". The key to successfully resolving trade-offs is maintaining focus on those properties of the product that give a competitive edge.

12.5.1 Function Trade-off

Properties in a trade-off situation can belong to different sets of function properties, meaning that one or more functions cannot achieve the desired properties. We call this a function trade-off. For example, consider the combination pocketknife or multi-tool. There is a trade-off between functionality, space, and usability.

In Axiomatic Design Theory (Suh 1990), it is advised that products should be designed such that the functions are independent, i.e. not depending on common characteristics. This allows for optimal conditions for each function to achieve its ideal solution. In reality, this is rarely possible for all functions, and as such, we offer a more pragmatic version: **core** functions should be independent. Many products are designed with some independent core functions and some integrated functions. For example, the core function of the springform (from the egg sausage example in Chap. 11) is to loosen the sausage from the form but also works as integrated part of the transport.

12.5.2 Institutionalized Properties

In an effort to eliminate certain issues important to users, many companies 'institutionalize properties' by arranging persons, teams, or departments to be solely responsible for certain issues, becoming issue specialists. Andreasen and Hein (1987) reported this type of change in Volvo (a car manufacturer). Volvo transitioned from a component-structured organization to a properties-structured organization with groups responsible for comfort, safety, operation, and ergonomics. Today, new institutionalized properties include sustainability and user-oriented issues.

Property models show that issues that are broadly distributed in the product's characteristics structure, e.g. cost, safety, and sustainability, are highly interdependent because they share many common characteristics. For example, design for sustainability highlights how some considerations can also lead to cost and reliability benefits. In contrast, many companies struggle to manage quality and design because the quality group is not involved in the main product design.

The interwoven pattern of property models does not allow the separation of single-issue responsibilities. The project leader has the responsibility for ensuring integration.

12.6 Property Reasoning During Synthesis

When designing, there are a number of critical situations where we encounter problems linked to property reasoning:

- Identifying and understanding the relevant properties: the necessary, the important, and the decisive.
- Creating new solutions whose properties and property models lead to new, competitive products.
- Contributing to incremental development of known solutions' properties so that they dynamically fit with the new context and demands.
- Handling trade-off situations as introduced above.
- Treating properties in the development of product families.

In this section, we offer some advice on how to manage these situations with respect to new product design, incremental design, and platform-based design. In all three types of design goal formulation, especially as a list of requirements, is fundamental.

12.6.1 Designing a New Product: Into the Unknown

Many innovative designs start with new technology, i.e. 'the idea in'. Therefore, their properties are often a consequence of the technology and the subsequent design solutions. These properties can sometimes be unexpected or even harmful. This is due to the fact that the design is based on the designers' ability to predict a product's properties via, e.g. imagination or computational modelling. However, when technologies are unknown, this prediction can be extremely difficult.

History is full of products that were predicted to be successful but never found any utilization or market. Conversely, there are many good products that have been developed through a chain of trial and error, where insights from failures became the driver for the next step in development e.g. steam engines or aeroplanes. Innovation is based on imagination, thus new solutions should be received with scepticism and reactions based on their nature and properties. For example, when it was clear that trains would exceed horse-riding speed, some people believed that the 'extreme' speed might kill the passengers.

Example:
Two-person bicycle. In their teaching, one of the book's authors has for years shown the principal sketch of the bike in Fig. 12.21 and asked: Are you able to balance on a bike with a person on each side? In each year, about half the students believe it is impossible, forgetting their tacit knowledge of balancing the bicycle via the front wheel. In fact, the illustrated bike was very popular between lovers in the 1890s. The moral is that innovation not only comes from the ability to visualize a concept but also the ability to interpret and predict.

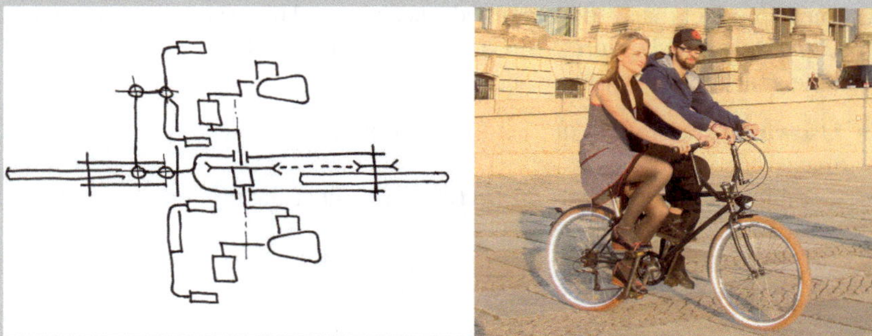

Fig. 12.21 Two-seat bicycle sketch and modern prototype courtesy Folker Silge, Budy Bike

In Chap. 7, we identified four cognitive strategies: problem driven, solution driven, information driven, and knowledge driven (Kruger and Cross 2006). We also quoted the C-K Theory of Hachuel and Weil (2003), linking concepts (new things we do not know about) and the related knowledge to be developed in realizing the concepts. Both of these point to situations where the design and the insight are born together, i.e. the concept and its property models co-evolve.

Innovative solutions may appear as strokes of insight but they typically emerge from extensive information gathering, experimentation, and study of a phenomenon's nature. A challenge of innovation is to what degree we need to know explicitly 'what we are doing'. For example, the invention of the steam engine was the cause for study of machine elements, kinematics, thermodynamics, and design, not the other way round. Thus, property reasoning, i.e. knowledge of certain properties' realization and enhancement (e.g. longer, faster, higher) is a precondition for innovative solutions in many cases and a consequence of the design's characteristics in others. However, as new products are typically based on, as yet, poorly understood phenomenon:

Making things work is a virtue. Supplying the explanation for why they work is a consolidation.

What is the actual role of **goal statements** in innovative design? We see two extremes. The first is where the attempt to formulate goals harms creative activity. The second is where the creative discussion of goals forms the kernel of creative work.

Goal formulations based on requirements originate in knowledge of existing products, predictions, and imagination, thus goals often mirror the past. Where concepts are substantially new most requirements are ultimately superfluous, invalid, or misleading. We have observed that companies often claim to support innovation and make goal formulation open (as prescribed by, e.g. Andreasen and Hein 1987) but gradually restrict the goal to the 'safe side' resulting in extremely traditional products and disappointed designers. This sad situation was also found by Hansen and Andreasen (2007), which raises the question: how can we formulate goals to support innovation?

- Use **value statements** in goal formulation. Describe a good product from the actor's perspective.
- Formulate goals as **desired functions,** i.e. what the product has to do and what it will improve.
- Do not take **requirements** for granted. Make interpretations, explore them, and see what happens if they are neglected.

The mechanisms of creative work are only partially understood. Dorst and Hansen (Dorst and Hansen 2011) explored how we create and dissolve paradoxes using discussion to enhance the creative process. Such paradoxes can be conflicting statements or trade-offs regarding the properties of a concept, leading to new ambitions and innovative ideas.

In innovative design, property reasoning can take the form of creative discussion or constructive conflict based on challenging property statements.

12.6.2 Incremental Design: Learning from History

Incremental design is based on known patterns of how the product works, how it is used, how it is produced etc. Relevant issues or desired properties are known and the design characteristics defined. Incremental design describes both combining existing solutions into new products, and the enhancement, adaption, variation or scaling of established products. For example, the design of the RMS Titanic was created by scaling existing designs. For today's designers, incremental design is referred to as 'drawing on dirty chalkboards', using the old lines and knowledge.

Incremental design is driven by cost reduction and customer demands for better performance and is often bound by severe constraints on the solution space (McMahon 1994). The design is defined by known property models and thus has established guideposts. McMahon identified five modes of incremental design, here articulated in our terminology:

- **Exploring** the **characteristics** of the organs and parts, i.e. searching for better solutions by focusing on key parameters.
- **Improving understanding** of property models via new theory, analysis, or experimental insight. This can give improved properties or the ability to survive new, more demanding, conditions.
- **Changing requirements,** e.g. new market or legal requirements, changes in utility perception, or the addition of function properties.
- **Modifying** the feasible **design space** by searching for new conditions for the design, e.g. related to its manufacture, choice of materials, assembly mode, or new functions like self-adjustment.
- **Changing** to new **design principles**, at least in sub-solutions, to enhance specific functions, e.g. cooling, lubrication, or wear reduction in a motor.

In incremental design, a new product can be described using goal formulations adapted from previous products. Further, the goal formulation only needs to address part of the product when substantial parts of the design are reused. However, it is critical that if a goal formulation is re-used, the new product still satisfies the actual requirements. It is necessary to check that new requirements are satisfied and that the product decomposition is still valid.

Example:
Car safety. In an earlier example, we discussed the issue of safety in car design. Over time, there has been a changing focus on car safety. In 1965, the American Ralph Nader published the book 'unsafe at any speed' (1965) pointing to the horrifying calculations of production costs versus human injuries for optimal profit. The book dramatically changed public and legislative perception of safety and its regulation. Since then, car safety legislation has focused on reducing the risks involved in car accidents, making the safety issue one of the main factors in car design.

12.6.3 Platform-Based Innovation: Distributed Properties

The basic idea of platform thinking is to best utilize a given sum of company assets to manufacture and sell a variety of customized products, sharing common elements. Customization is typically based on combining standard modules and/or parts. We have already discussed how a product's properties are not just based on the organs and parts but also on their structure. Thus it is not always feasible to create a product configuration where the custom properties depend only on the modules and/or parts—their interaction also plays a role.

In Chap. 11, we introduced modularity based on function reasoning, because the combinatory nature of modular product families has the precondition that each module has a distinct function and standardized interaction interface with other modules. However, transforming a range of products into a true modular family (which takes substantial re-design effort) can encounter issues in managing product properties. Out of several module drivers highlighted by Erixon (1998), the following relate to the modules' properties:

- **Carry over**: key modules are used identically in all variants carrying the main functions and properties of the product.
- **Technology evolution**: certain modules are kept under development to give the best possible properties (see incremental design approaches).
- **Planned product changes**: modularity is used to support a partial development strategy focusing on certain modules and enhancement of their properties.
- **Different specifications**: module variants carry specific ranges or values of properties for customization of products.
- **Styling**: The modules are arranged so that styling/identity is preserved across a range of product variants.

Modularization demands and supports a tight focus on module's function properties and property models. This is because properties are key to the creation of a modular family architecture and because configuration (selecting and combining modules) relies on both functions and properties as criteria. Configuration is rule based, with rules relating functions, properties, and module characteristics to ensure compatibility and the desired properties.

Example:
Modular bicycles. Based on 'radical simplification by design' by Mortensen et al. (2012), we can see how modular architecture can yield a product program with a wide range of features and properties mirroring market demands. Several issues need to be defined, e.g. *market range* high-end, mid, and low-end, and the key features that determine price and performance perception. The customers divide by gender and each group has a specific size and weight range, styles, and features. A detailed map of these dimensions is then confronted with business questions: what variants are economically viable? What market segments should be addressed? What performance and features lead to success?
A modular architecture creates a bicycle from function-determined modules, each carrying certain performances and features. Combinations of module variants can be used to satisfy the programs needs but at the same time it is important to retain high commonality for economic reasons. In particular, the elements of the part structure that interface between modules are crucial. Figure 12.22 shows the modules and four female-type bicycles from the program. Here, modularization led to significant changes, respectively, before and after: frame components: 680 $\gg$ 36; possible frame variants: 72 $\gg$ 504; possible bike variants: 72 $\gg$ 2000+.

Fig. 12.22 Modular architecture for a bicycle program (Mortensen et al. 2012)

Platform and modular approaches have, to date, been primarily used by companies with large product ranges and sales volume or with expensive products where reduction of development costs via reuse is critical. However, there is no doubt that many companies can benefit from substituting their wildly growing product portfolio with more manageable modular designs where customization, upgrade, and specification are brought together in a structured way.

12.7 Conclusion

This chapter deals with one aspect of creating a good product based on the Link Model's postulate that users' and designers' interpretation of need satisfaction and product value should be identical. Property reasoning is based on clarifying the basic terms and the nature of properties. These have their origin in the behaviour of the use activity and product but are also related users' perceptions. In particular, developing a competitive advantage relies on strong, original insight into the properties.

Property reasoning applies throughout the Encapsulation Design Model. In the early stages, the reasoning is projection, expectation, and guesswork; later it becomes a matter of synthesis and analysis, and after product realization it becomes verification from sales and user feedback.

This chapter adds to the reasoning about what makes a good product, opened by function reasoning in Chap. 11. However, goodness is not only related to the product and the users; the final element to consider is the many stakeholders who interact with the product throughout its lifecycle. This can be dealt with via dispositional reasoning, which we will explore in the next chapter.

References

Andreasen MM (1980) Syntesemetoder på systemgrundlag (Synthesis methods based upon system theory). Ph.D. Dissertation, Lund University, Sweden

Andreasen MM, Hein, L (1987) Integrated product development. IFS (Publications)/Springer, Berlin (Facsimile edition (2000) Institute of product development. Technical University of Denmark Copenhagen)

Cross N (2008) Engineering design methods. Wiley, Chichester

Dorst K, Hansen CT (2011) Modelling paradoxes in novice and expert design. In: Proceedings of ICED'11, Copenhagen

Dym CL, Little, P (2000) Engineering design—a project based introduction. Wiley, New York

Eder W, Hosnedl S (2007) A manual for enhanced creativity. Taylor and Francis Group Boca Raton

Erixon G (1998) Modular function deployment—a method for product modularization. Ph.D.-Dissertation Royal Institute of Technology KTH Stockholm Sweden

Hachuel A, Weil B (2003) A new approach of innovative design: and introduction to C-K theory. In Proceedings of international conference on engineering design, ICED'03, Stockholm

Hansen CT, Andreasen MM (2007) Specifications in early conceptual design work. Proceedings of 16'th international conference ICED'07

Hubka V, Eder WE (1988) Theory of technical systems. Springer, Berlin

Kesselring F (1951) Bewertung von Konstruktionen. VDI-Verlag, Düsseldorf

Kruger C, Cross N (2006) Solution driven versus problem driven design: strategies and outcomes. Des Stud 27(5):527–548

McMahon CA (1994) Observations of modes of incremental design. J Eng Des 5(3):195–209. http://www.tandfonline.com

Mortensen NH, Gamillscheg B, Bruun HPL, Hansen CL, Cleeman KK, Hendrich K (2012) Radikal forenkling via design (Radical simplification by design). Technical University of Denmark Mekanik, Copenhagen

Mortensen NH (1999) Design modeling in a designer's Workbench. Contribution to a Design Language. PhD-Dissertation, Technical University of Denmark

Mørup M (1993) Design for quality. Ph.D.-Dissertation Technical University of Denmark, Copenhagen

Nader R (1965) Unsafe at any speed. Grossman Publishers, New York

Olesen J (1992) Concurrent development in manufacturing—based upon dispositional mechanisms. Ph.D.-thesis, Technical University of Denmark

Pahl G, Beitz W (2007) Engineering design. A systematic approach, 3rd edn. Springer, London (1st edn. 1977)

Papanek V (1971) Design for the real world. Pantheon Books, New York

Pugh S (1991) Total design—integrated methods for successful product engineering. Addison Wesley, Wokingham

Smith J, Clarkson PJ (2005) Design concept modelling to improve reliability. J Eng Des 16(5): 473–492. http://www.tandfonline.com

Suh NP (1990) The principles of design. Oxford University Press, Oxford

Svendsen K-H (1994) Diskretoptimering af sammensatte maskinsystemer (Discrete optimization of composed machine systems). Ph.D.-Dissertation, Technical University of Denmark

Tjalve E (1979) A short course in industrial design. Newnes-Butterworths, London (Faximile edition (2003) Systematic design of industrial products. Institute of product development Technical University of Denmark, Copenhagen)

Ulrich KT, Eppinger SD (2004) Product design and development, 3rd edn. McGraw-Hill/Irwin

Weber C (2014) Modelling products and product development based on characteristics and properties. In: Chakrabarti A, Blessing LTM (eds) An anthology of theories and models in design. Springer, Berlin

Chapter 13
Dispositional Reasoning

Dispositions are the effects of the designer's decisions about the product's characteristics, which are positively or negatively influential 'downstream', i.e. on the product's realization and life cycle. Through reasoning on and utilizing information and methods related to dispositions the designer can exert an influence that is highly rewarded in later activities.

We introduce a theory of dispositions and show how this theory governs the areas called 'design for X'. Arranging dispositional reasoning in the design process is important because this is normally outside designers' responsibility and knowledge.

M.M. Andreasen et al., *Conceptual Design*, DOI 10.1007/978-3-319-19839-2_13

13.1 Designers' Dispositions

In all life phases we find actors who are responsible for, or involved in activities, related to the product, e.g. supply, assembly, distribution, or disposal. These actors see the effects of the product first hand, e.g. problematic maintenance. We call these effects **dispositions,** i.e. the designer disposes these activities' conditions through the product's design. When designers are able to reason about the relationship between their product designs and the actual effects in the various life phases they are better able to mitigate negative effects and accentuate positive. We call this type of thinking **dispositional reasoning**.

Dispositional reasoning complements the previously discussed function and property reasoning approaches (Chaps. 11 and 12). As such, reasoning in design has two aims:

- To identify those actors influenced by the product and how this relates to the product's design.
- To fit the product's design to the life activities such that the positive dispositions are emphasized in the eyes of the actors.

Traditionally, functions and properties form the basis for formulating a project's goals. However, dispositional requirements are often difficult to define because of questions regarding who the actors might be and how they might interact with the product. Thus, it is highly likely that not all aspects or consequences can be foreseen. As such, dispositional reasoning provides a key tool in setting goals for dispositions and addressing the challenges outlined above.

The idea of dispositional reasoning, as a specific topic, was developed by Olesen (Olesen 1992) in his Theory of Dispositions. However, elements of this type of reasoning have been a part of certain DFX approaches for many years. Ultimately, dispositional reasoning underpins all DFX approaches. As such, we focus on the theory behind dispositions and the nature of the relationship between products and activities. In doing this we also implicitly deal with methods and their staging.

> A disposition describes how a product's characteristics influence the activities expected to be performed during the product's lifecycle, e.g. assembly, supply, or disposal.

13.1.1 Industrial Practice

Organizational structure can easily lead to an 'over the wall' culture where designs and decisions are disconnected from insight into subsequent tasks and ultimately the design outcome itself, illustrated in Fig. 13.1a. Another important feature is the power of influences (Fig. 13.1b), i.e. the actual organizational functions' cost

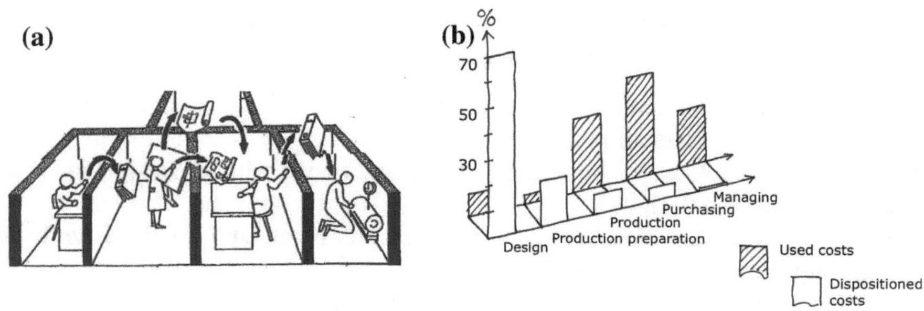

Fig. 13.1 Traditional cost discussions: **a** blind dispositions in the 'over the wall' culture (Ehrlenspiel et al. 2013). **b** "Who throws the biggest shadow?" the cost of functional areas and the associated portion of the total cost they influence

influences and dispositions compared to their costs, shown as shadows. These features point to the need for integration across the design process and the importance of creating effective dispositions.

Although effective organization and management can solve many of the issues highlighted in Fig. 13.1, dispositional reasoning and certain DFX tools are still necessary for successful design. In particular, these are needed to support the design of dispositions when the designer is not familiar with the user. Thus, we can define a number of key concepts that underpin dispositional reasoning.

When a design **characteristic** influences a lifecycle activity we face a **disposition**. This influence is described as a **principle** linking product characteristics to activity characteristics. Together a set of principles related to either a specific lifecycle activity or a general virtue forms a **DFX methodology**. Here, 'X' can stand, for e.g., manufacture, assembly (i.e. lifecycle activities) or cost, time, quality (i.e. general virtues). These DFX principles are **conditionally true** with no immutable 'correct answer'. In each case the designer must verify that the applied rule leads to actual improvement in the specific design situation. For example, if a standard assembly principle, e.g. one-directional assembly, does not lead to cost or time reductions then there is little point investing design effort in this and a more appropriate principle should be sought.

The breadth of DFX principles comes from the fact that many **actors** are linked to the various product life activities and each has their own **values and issues** defining what they see as a good lifecycle activity. Thus it is the designer's task to identify, articulate, and satisfy the requirements that are most important to product success. Ultimately, this comes from an effective **goal formulation** that not only considers the user's perception of a good product but also includes the company's business goals, the necessary product **functions**, and **properties**. All of these elements should then be traceably linked to goodness in the **product life activities**. As such, handling dispositions is a question of combing specific knowledge and underlying principles. Because of this we first outline the basic theory before exploring a specific set of important DFX areas. Finally, we bring these together to explore how dispositions can be used throughout the design activity. As such, we address the following topics in this chapter:

- Section 13.2 explains the fundamentals of dispositional effects and how **dispositions link products and life activities**.
- Section 13.3 explores the various areas of insight dealt with by **Design for X as actor orientated design**.
- Section 13.4 deals with five key DFX areas in terms of **DFX methods and the links between characteristics**: manufacture, cost, quality, and environment.
- Section 13.5 deals with the different roles of dispositional reasoning depending on the type of development project undertaken in **dispositional reasoning during synthesis**.

13.2 Dispositions Link Products and Life Activities

Each of the main lifecycle activities were introduced in Chap. 10 as interactions between the product and the life system, actors, and actor network. As such, the Link Model can be applied to dispositions as shown in Fig. 13.2. This highlights the connection between a specific life phase activity, the designer's goal formulation, and the user's perception of goodness. We can further link these to the wider actor network and life phase systems.

In this model there are two perspectives on 'goodness': the experience of the actors involved in the activity, and the business in which the designer operates. In particular the business perspective ensures that company aims are not ignored, e.g. competition, efficiency, and ethics. In response to these perspectives the designer has three means for influencing the actual life activity. First, they influence the product's design and therefore its fit to the system and activity, e.g. a product's compatibility with an assembly system's grippers and manual operations. Second, they influence the activity itself through the operations required in product use. Finally, they can sometimes influence the life phase system to be more suited to the product, although this is often a given and thus not accessible to the designer.

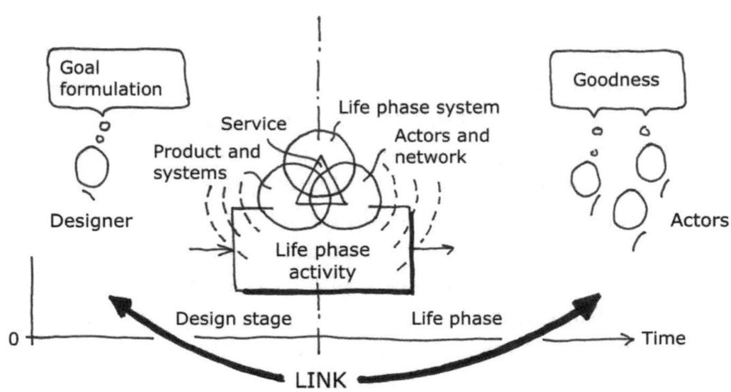

Fig. 13.2 The Link Model applied to dispositional reasoning

13.2.1 Dispositions

Dispositions are conscious or unconscious results of the designer's influence on the lifecycle activities as illustrated in Fig. 13.3. For example, decisions regarding the part structure and materials used for specific parts impact the production machinery required in manufacture. Thus we can define a disposition as follows:

> **Definition**: A **disposition** is a decision concerning a product's **design** that influences the type, content, and efficiency of activities in the product's **life phase activities**.

Based on the model in Fig. 13.3 we can see that dispositions are not a separate entity but are implicitly embedded in the links between the design and life phases activities. For example, a design layout might be optimized for manual assembly but perform poorly on a robotic assembly line—in this case we face a negative disposition. A key question here is what is the extent of the designer's responsibility? Is the designers' task limited to the product or should they also design the production/distribution systems, the users' interaction with the product, and the disposal and reuse systems? Simplistically, the answer is yes because it is necessary for the designer to show due diligence in a general sense. A more sophisticated answer is that task delimitation is a matter of management decisions and project staging.

The dispositions resulting from design decisions are usually easy to identify in hindsight. However, they are much less clear in the design situation and are easily overlooked or misinterpreted. Figure 13.4 shows an idealized design process where potential influences are related to multiple issues. If the actors and designers cannot imagine these dispositions then the risk of later problems is high. For example, in the handover from design to production it is common for problems to occur because the design documents are wrong, not understandable or poorly fitted to the equipment, see the chapter vignette. Therefore, we suggest the following.

> If you cannot see the dispositions caused by your design then ask your 'victims'.

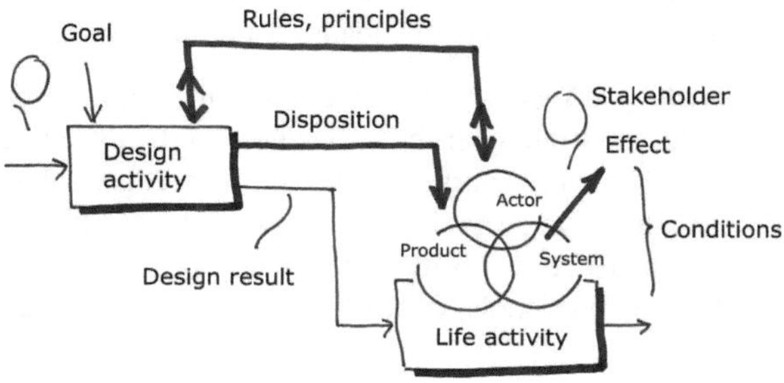

Fig. 13.3 A general model of dispositions, linking the design activity to the life phase activity

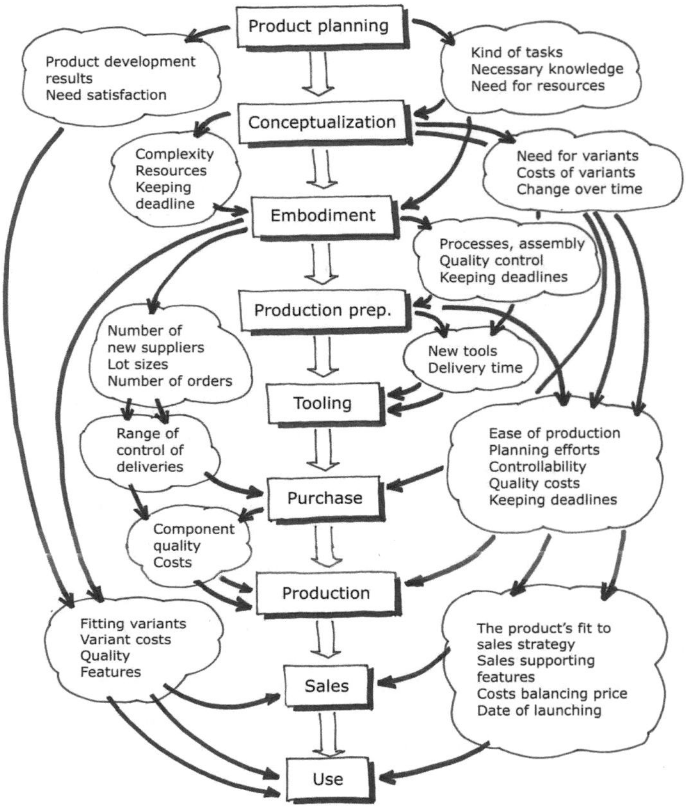

Fig. 13.4 Important issues and dispositional areas in the interaction between the main activities in a product's development, after (Andreasen et al. 1989)

13.2.2 Dispositional Relationships

If we imagine dissecting a company's activities we find that a surprisingly large number are related to the results of the latest product development: purchase, manufacture, assembly, quality control, distribution, sales, installation, and use. In the context of dispositions this leads to a number of effects (Andreasen 1988).

- **Strategy effect**: are we developing the right things? Good dispositions support corporate strategy in both business and innovation. This can take many forms including, machinery, processes, knowledge, sales networks or specific dispositions that give attractive conditions linked to future strategic goals.
- **Group effect**: dispositions unify new products and existing activities via 'groups' like methods, tools, control, service, sales channels, etc. This effect also relates

to the reuse of experiences and supporting learning. Any unnecessary deviation from what 'the machinery' sees as normal is a dispositional disturbance.

- **Standardization effect**: reuse and reduction of variants via standard, e.g. modules, components, materials, suppliers or methods. Standardization aims to reduce cost and increase quality by following rules in disposition creation.
- **Optimization effect**: the analysis and optimization of selected operations, particularly when these are influenced by the product's characteristics. For example, refining the modularization of a product so that it optimally supports assembly line testing.
- **Resource effect**: the optimal utilization or minimization of key resources, e.g. staff or machinery. This effect is illustrated in the computer-supported composition of modular products, which leaves staff free for other tasks.
- **Correctness effect**: the above-mentioned effects disappear if there are faults in the data or information transferred, i.e. it is not correct. For example, when erroneous or insufficient information is transferred from conceptual design to embodiment design flaws can occur.

The examples above focus on effects within a company, however, these types of effects are also found in activities external to the company.

Example:
Strategic materials. Bang and Olufsen decided to adopt aluminium as strategic material to give greater freedom to its designers and to reduce a number of negative environmental effects from previous materials. Design and production possibilities were subsequently clarified and proposed by production staff and then taken up in the product development teams to create a new range of designs (Fig. 13.5).

Fig. 13.5 Bang and Olufsen products with aluminium as strategic material, Beocentre 2 and Beolab 8000, *courtesy* Bang and Olufsen A/S

13.3 'Design for X': Actor-Oriented Design

One way to simplify the various actors' interests is to focus on 'universal virtues', i.e. fundamental dimensions of an activity's goodness (Olesen 1992): quality, cost, time, efficiency, flexibility, risk, and environmental effects (Fig. 13.6). In specific situations we might also add criteria related to the operator's wellbeing or work situation.

'Design for X' or DFX methods aim to answer the question: how can the designer, at the design stage, best fit the product to its life activities? DFX constitutes a large number of methods and guidelines for mastering the issues related to a product's realization and its utilization in the widest sense. Most DFX areas are governed by principles that guide proposals for beneficial courses of action. These principles are often mistaken for criteria, i.e. the belief that simply following the principle will result in a good outcome. Instead, the activities that occur as a result of applying the principles must be assessed with respect to the universal virtues (Fig. 13.7). For example, Design for Assembly efforts might be measured by cost and time of the assembly activity. Linking the DFX areas to the universal virtues two types of DFX approach emerge:

- **Product life phase** DFX topics. These link to activities undertaken during the product lifecycle, e.g. Design for Assembly. Here the universal virtues provide criteria assessing the execution of the activity.
- **Universal virtue** DFX topics. These see the virtues as lifecycle concerns with the virtues specific criteria providing the associated measure of goodness. For example, Design for Cost is linked to life cost optimization.

Although the initial contributions to Design for Manufacture were made in the 1950s the true birth of DFX methodologies came in the 1990s. DFX approaches are traditionally viewed as confronting the principles of an X-area with an actual design proposal. The issues dealt with by DFX can take both product and life activity foci. Thus a broader definition of DFX is:

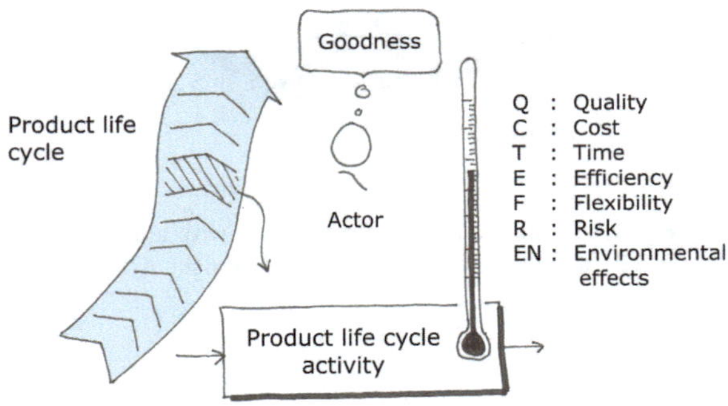

Fig. 13.6 Universal virtues composing the goodness of an activity

Fig. 13.7 The DFX matrix linking product life phases and universal virtues. The marked squares show example areas of interest highlighted in a real project

	Quality	Cost	Time	Efficiency	Flexibility	Risk	Env. effects
Planning			V		V		
Production	V	V					V
Assembly				V		V	
Distribution		V				V	V
Sales	V						
Use	V						V
Maintenance		V		V			
Recycling		V					V
Depose							V

Definition: **Design for X** is a set of product synthesis methods and guidelines that serve to enhance the product life activities by addressing key issues related to the product and its activities.

It can be confusing to distinguish if a DFX effort is focused on the product or the activities. For example, Design for Cost aims to reduce costs associated with the activities, e.g. production or distribution, however, cost is normally seen as a product property. It is even possible that DFX efforts do not make this focus explicit. Figure 13.8 formulates this question as a functional area. For example, should

Fig. 13.8 The design manager **a** measured on his results versus the production manager, **b** measured on his contribution to good production

production be measured on something disposed by others or should those people making the dispositions be responsible for measurement?

13.3.1 Dispositions: A Theory Behind DFX

Dispositions and 'Design for X' methods are most clearly explained with respect to Tjalve's 'design degrees of freedom' illustrated in Fig. 13.9a. This gives a pragmatic hierarchy of product characteristics, each representing a design degree of freedom. Here the hierarchy follows the general rule that each level respects the characteristics defined by the levels above it. However, this becomes more flexible in the lower levels because the products complexity, and the range of alternatives available, makes the pyramid extremely broad at the lowest levels where we define the detailed characteristics.

This pyramid can be used for any life phase system and thus it is possible to examine the links between two pyramids as shown in Fig. 13.9b (Andreasen and Mortensen 1997). In this case the fact that the design is fit for production means that these cross-links have been arranged successfully. These cross-links can be seen as beneficial rules (Fig. 13.9c). For example, a Design for Assembly rule states that "stacked *product structure* allows for 'pick and place' *assembly*, reducing assembly costs". This links the product and assembly pyramids via the 'stacking' rule.

DFX rules can be articulated in a number of ways depending on what is most beneficial for the designer in a given situation, shown in Fig. 13.10. For example, rules be can communicated as tacit experiences from within the team or via more formal documents like a "lessons learned" database, textbooks, archive software, etc. This is represented by the 'shared box' in Fig. 13.10, i.e. by their nature methods are carried in the human mind.

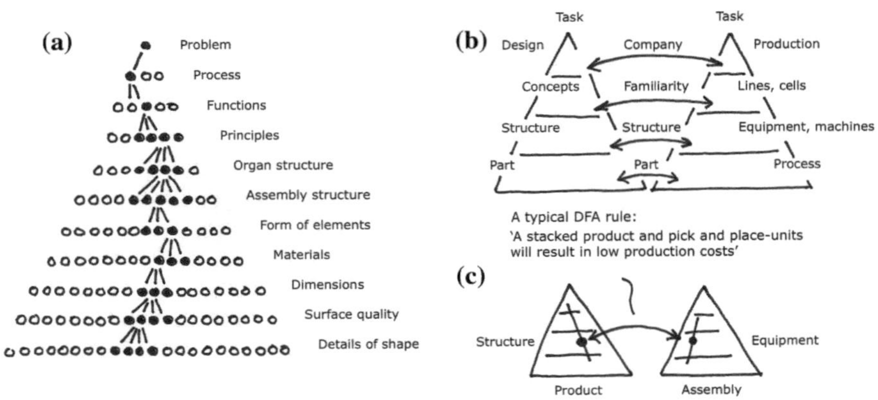

Fig. 13.9 Tjalve's design degrees of freedom **a** general links between the design and life phase system, **b** an example of links with production, **c** a general articulation of a DFA-rule

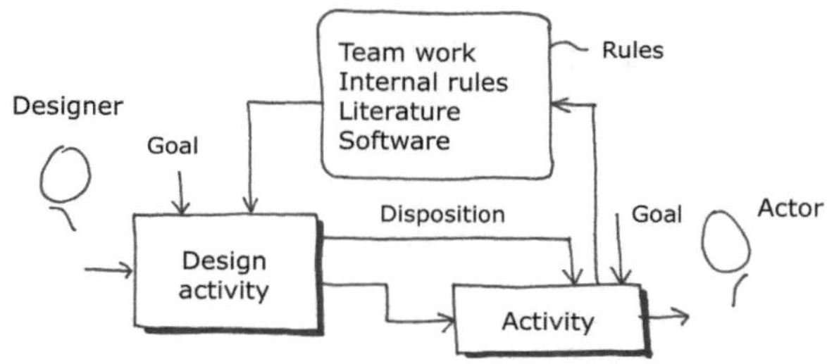

Fig. 13.10 DFX visualized as an interaction managed by a set of rules, which can be realized through, e.g. teamwork, internal rules, general literature, and software

As we mentioned earlier in the chapter the rules governing DFX methods are conditionally valid, i.e. they are dependant on the designer and the overall situation in order to actually lead to the promised effects. For example, the rule 'stacked assembly' requires interpretation by the designer in the context of the real product in order to deliver assembly line cost savings. As discussed by Fabricius (1994), and shown in Fig. 13.9, there are rules at different levels. These not only relate to parts or structure but also to more composed characteristics of products and systems. In this context, matching higher-level rules typically has a greater impact than matching lower-level rules (Fig. 13.9).

Bringing together these features we propose that there is fundamental theory underpinning all DFX types: A disposition (i.e. a design decision influencing an issue 'belonging to the X area') can be articulated as a rule that links certain *design characteristics* to certain *X-area characteristics*, which influence certain *properties* related to the universal virtues. In some DFX areas, e.g. Design for Assembly, these rules are found explicitly in textbooks. However, most areas are so context dependant that pure rule following is not applicable and thus they require effective staging.

13.3.2 Staging DFX

When we compare today's designers with the artificers of the past, e.g. traditional wooden shoemakers, the many stakeholders and DFX areas are striking. The shoemaker created the design, obtained the materials for manufacture, carved the shoes, and sold them. Today, all of these activities are allocated to different specialists or organizational units. As such, dispositional reasoning has become crucial in creating integration across the design process. To illustrate how these DFX efforts are typically distributed Fig. 13.11 shows how cost and resources are committed over time (Andreasen and Hein 1987).

Substantial dispositional influences are established early in the conceptual stages of product development. Thus this must be accounted for when planning the project.

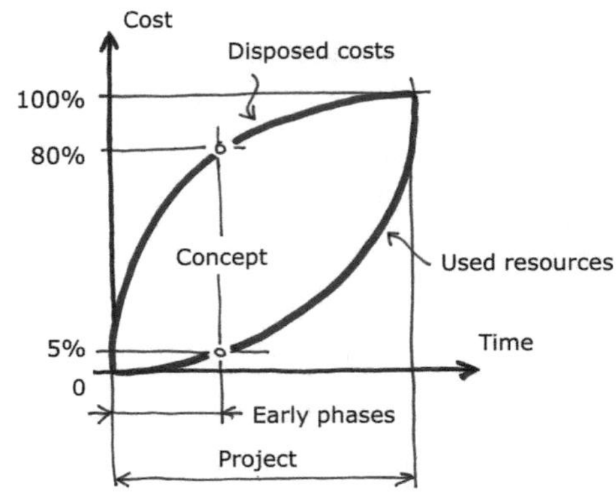

Fig. 13.11 Cost versus resource dispositions in a development project

Olesen (1992) created the Score Model, shown in Fig. 13.12, to support dispositional reasoning. Here the product development progression (top) drives the growing dispositional influence on the DFX areas: planning, production, assembly, etc. This then relates to their positive or negative effects on the product lifecycle activities (right). For example, a project has identified key lifecycle activities where a DFX effort might improve the product's attractiveness. This has lead to the definition of metrics linking the universal virtues to give a measure of 'goodness'. Finally this has resulted in the selection of DFX methods, noted in the matrix on the left.

Olesen's idea is that the project leader acts as a conductor using the DFX matrix as a score, giving the following heuristic.

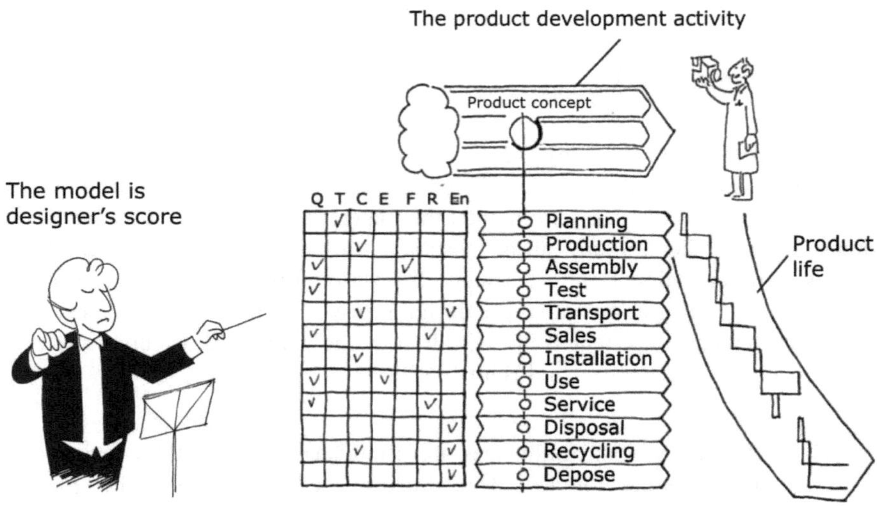

Fig. 13.12 The Score Model directs the designer's attention during a development project

If a DFX effort is to be really influential it must be considered conceptually, i.e. key concepts should be explicitly articulated early in the project and considered throughout in order to have an impact.

In Chap. 3 we highlighted the key role of the team in effective use of design methods—DFX is no different. DFX methods are essential for Integrated Product Development (Chap. 9) and must thus be staged in an integrating manner. For example, they might involve key stakeholders in the team or use gallery techniques to direct discussion of dispositions and the use of knowledge from past projects. Thus DFX methods bring the effects of the design work directly into the design team. This is illustrated by two examples. In the first a consultant counted the number of different screws used in a product, leading to discussions on the cost of purchasing and storage. In the second a company surveyed the various electrical motors used in its product range over time. This resulted in the realization that the variety was much larger than necessary for the product—type suppliers, and specifications—leading to striking price differences between products. How had this happened? On investigation a project leader found that none of the design team had ever visited the production site. They were subsequently amazed by the huge investments in machinery resulting from 'small changes' in their design. Based on this we might ask in what order DFX methods should be applied to avoid these issues? Although some books suggest preferred orderings for practicality or impact, DFX methods are really corrective in nature. This perspective of DFX methods as a corrective feedback loop is illustrated in Fig. 13.13. Using this conceptualization we can better understand why DFX methods are normally a combination of analytical and improvement approaches. This DFX loop links to the iterative nature of design activity where focus is continually shifting from issue to issue (Hubka and Eder 1988; Pugh 1991; Chap. 4).

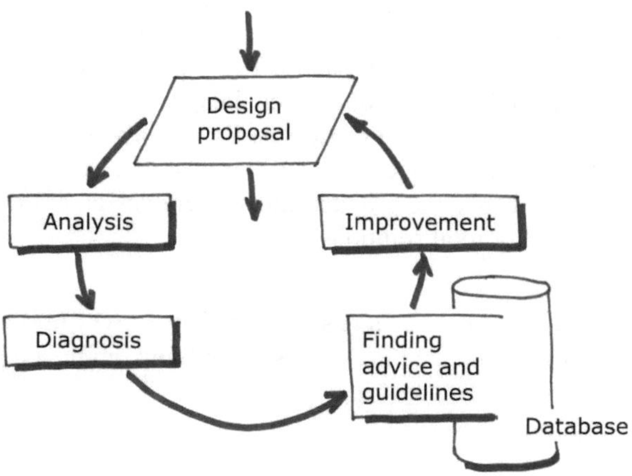

Fig. 13.13 The DFX loop: analysis, diagnosis, and advice, leading to design improvement

13.4 DFX Methods and Links Between Characteristics

In this section we will look at a range of DFX methods in order to distil out the underlying mechanisms linking the dispositions and the rules related to the actual X area.

13.4.1 Design for Manufacture and Assembly

Old engineering design textbooks give rules for 'right and wrong' design based a manufacturing perspective (Matousek 1957). From this DFMA (Design for Manufacture and Assembly) emerged in the 1970s and 1980s as the most influential DFX area. Applications in the automotive industry resulted in part counts being reduced by 50 % and substantially reduced costs. However, the diversity of production methods has changed. What started out as a limited number of manufacturing and design rules exploded into a huge number of special production methods. Unfortunately, it is easy to identify good solutions but difficult to find a way of articulating the problems and actually guiding the designer. Thus we must consider the fundamental dispositional mechanisms.

In the assembly area the assembly activities, available equipment, and key principles (e.g. Fig. 13.9b) can be characterised in two ways (Andreasen et al. 1988):

- To guide the designer to principles and solutions for designing and assembly management based on a set of criteria for optimal assembly (high product quality, high productivity, high profitability, and good working conditions).
- To guide the designer to an 'assembly friendly' design by highlighting principles related to the product's structure and its connections, as well as the individual part's designs.

Even with such guides the effect of a DFA method is fundamentally dependant of the designer's ability to see possibilities. Re-design of existing products follows the analysis > diagnosis > advice sequence (Fig. 13.13) (Boothroyd et al. 2002). A more structured approach is to create an overview of a product's cost structure, 'big Q' and 'little q' quality challenges (see Design for Quality below), its functions, and production processes. Based on this overview (called '*Know your product!*' (Andreasen and Støren 1993) postulates can be formulated for possible better ways to realize the product. Subsequent design efforts focusing on these can result in partial or total solutions.

> Do not expect as a designer that 'they' will find out how to produce it. You must understand the processes and find out how to design to fit them.

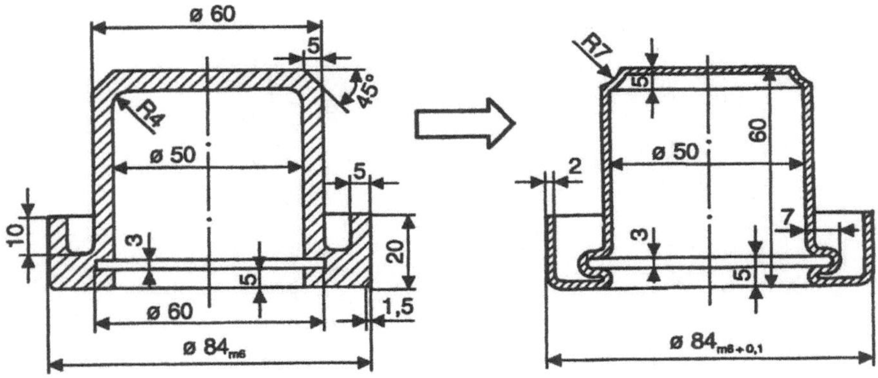

Fig. 13.14 Rationalization of a part: machining (*left*) versus deep drawing (*right*) (Ehrlenspiel et al. 2013)

In contrast to DFA production processes do not have structured DFM methods. Here, the aim is to establish a design that is detailed enough to support analytical approaches for fitting it to the processes, equipment, and tooling, measured via costs, time, quality, and productivity. A typical example is shown in Fig. 13.14. Here deep drawing saves 50 % weight and 50 % cost (Ehrlenspiel et al. 2013).

Example:
Re-design of a welding machine. Here we draw on the well-known consultant work of Fabricius (1994) due its excellent fit with our argument. In the 1990s Migatronic was the first company in Denmark to launch portable welding machines. These products were in demand but suffered significant quality problems. In particular these stemmed from the fact that, in the original design, functionality could not be tested before final assembly. At the same time Finnish and German companies were developing similar products with 40–50 % lower production costs. Thus a re-design was initiated by an analysis of the type illustrated in Fig. 13.15.
Based on this the following postulates for better approaches were produced: using a stacked structure, using 'pick and place' units for assembly, and reducing wiring complexity via a PCB (poly carbonate boards). This last element would also involve heavy coverings and direct connections between the electrical components to enhance reliability. Based on these ideas the new machine was modularised as shown in Fig. 13.16: low current board, high current board, and an extruded aluminium body for housing and cooling the components.
These changes collectively reduced assembly time by 18 %, part count from 1179 to 520, wires from 52 to 0, and the number of machined parts from

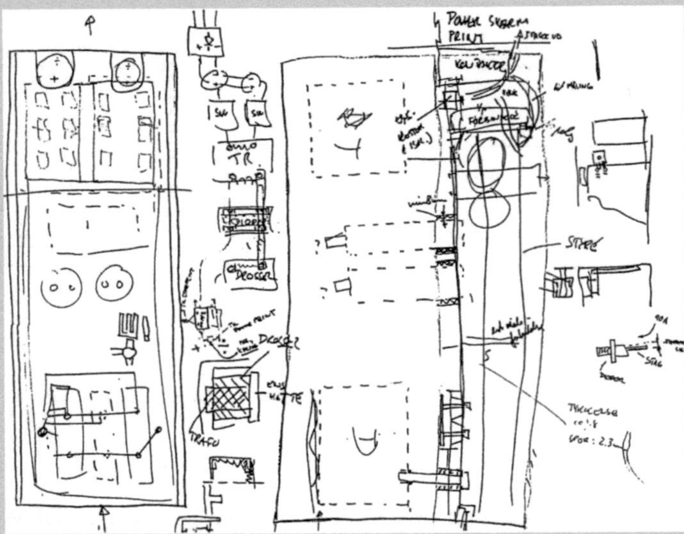

Fig. 13.15 Fabricius' sketch of the new product structure

102 to 10. This resulted in individual parts that were more expensive but gave an overall significant price reduction. In particular this suggests that the Design for Cost approach might not have been as useful here due to the need to increase the cost of some parts. Further, an additional effect was noticed once production of the new design had started: the individual welding machines could be grouped to make larger machines consisting of between two and six combined modules. Thus, products could be better fitted to the customers needs and provided a basis for new platform orientated designs.

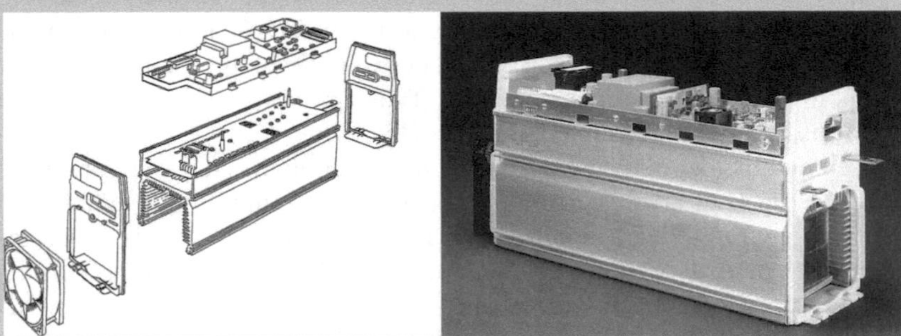

Fig. 13.16 The modular design of the new welding machine (Fabricius 1994)

DFM covers a wide range of situations from simply fitting the product and manufacturing process to specialized production technologies that the designer needs to understand deeply. Typically, design students are advised to create a 'production neutral' design at first, allowing them to exploit different manufacturing processes. However, we argue that a better approach is to start with a plausible 'way of building' in mind (German: *Bauweise*), i.e. a commonly used way of realizing certain functions (Andreasen and Mortensen 1997). These approaches (and DFM efforts in general) are typically measure by cost. As such, Design for Cost naturally links to the DFMA areas.

13.4.2 Design for Cost

Cost reduction and value creation are key factors in competition. Ehrlenspiel et al. (2013) states that cost reduction are a societal task necessary for ensuring trade and welfare. Although many actors influence cost three main elements can be identified in a company.

- **Manufacturing costs**: the manufacturing process, and purchase of materials and components. These costs are variable and related to sales volume.
- **Fixed costs**: the production means, staff, and organizational activities related to the product.
- **Product life costs**: the cost of, e.g. installation, application, maintenance, and disposal. Both the buyer and producer carry these costs.

The designer's influence on cost is relatively easy to trace when considering manufacturing because the costs originate in the definition of the parts and processes needed to create the product. On the other hand fixed costs are related to the operation and utilization of equipment, routines, and practice, which are less directly influenced by the product. Examples of these more complex effects are purchase routines, spare part routines, product modularization, distribution equipment, quality tests, repair routines, etc. Finally, the product life costs are split between the producer and the buyer. Here the designer must decide how this should be distributed, e.g. should the producer spend more on a longer lasting part, or shift the cost to the buyer by forcing them to carry out regular maintenance (this is explored further in Chap. 10).

DFC methods do not have an agreed scope. Some authors see a life cost focus as a virtue while others focus only on *Design to Cost,* i.e. designing to hit the cost goal based on the market price. Pahl and Beitz (2007) recommend interdisciplinary dialogue to identify a broad spectrum of cost influences, as well as Value Analysis to ensure costs are used appropriate. Different cost structures can be established. For example, one might distribute costs to functions/organs based on the importance of these to the user, with 'unbalanced' costly organs being replaced by cheaper alternatives. Conversely, cost distribution based on the parts might focus on materials costs, wages, and machining cost, with the aim to redesign of 'unbalanced' operations. These structures are often associated with **cost drivers,** i.e. modes of action, functionality, or materials that directly cause higher costs and thus form areas of specific focus.

As we noted at the start of this section DFX areas are not independent. It is well known that efforts to create environmentally friendly products often lead to reduced costs. Further, in the welding machine example above the DFM effort lead to substantial cost reductions. Similarly a DFC effort in the welding machine example would likely lead to a different solution and thus influence manufacture.

Many of the cost dispositions made by the designer are routine because they concern commonly occurring, well understood, topics, e.g. connections, welding or choice of electrical motor. As such, few designers expend substantial analytical effort on such common areas.

Designers often claim that 'Design for Cost is what we do every day' but it is important to question the effects of these efforts in the wider design outcome.

13.4.3 Design for Quality

Design for Quality is closely related property reasoning (Chap. 12) due to the fact that DFQ aims to understand how users appreciate properties, and to thus ensure these properties are enhanced in the product. Quality is the customers' reaction to products' properties and can be divided into five main elements (Mørup 1993). Company-orientated properties are:

- **Q-properties** ("big Q"): properties satisfying external stakeholders, the user being the most important. The challenge here is to identify the most important qualities, leading to successful, competitive products.
- **q-properties** ("little q"): properties related to internal efforts to reach appropriate 'big-Q' values. These can be seen as quality efficiency. The challenge here is to minimize quality efforts (e.g. measured in cost) whilst ensuring full parameter control: all products identical.

Customer-oriented properties are:

- **Position properties**: properties linked to the product's raison d'être, by positioning the product between the best on the market and satisfying customers. These properties can be related to what buyers see as innovative or when they have specialist's knowledge that allows them to appreciate the product's peak performance. Specific properties or features are often enhanced to give the product its individual character.

- **Expectation properties:** properties promised by the company's reputation or advertising. These might not be explicitly articulated by the customers but can have a major impact on how products are received. These properties relate to buyers' basic expectations, such as safety, reliability, eco-responsibility, etc. Superior performance in this category does not necessarily increase customer satisfaction (Mørup 1993).
- **Obligatory properties**: properties following the state of the art in a class of products. These qualities are not promised but are instead conditions for being able to enter the market. The customer expects these properties based on, e.g. experience, expectation or legal regulations. Omission of obligatory properties results in negative customer reactions.

All these properties dynamically follow a product's evolution, i.e. over time a property might gradually become common across all products, moving from expected to obligatory. As such, qualities are not only considered at the point of sale but are also experienced throughout a product's service life. Here, **Robust Design** (Matthiassen 1997) aims to make the product inherently impervious to the many types of variation that occur in the product's realization, e.g. changing material properties, poorly toleranced dimensions, or new assembly conditions. This focuses on ensuring undisturbed functionality despite variation. Similarly **Design for Reliability**, (Crowe and Feinberg 2014; Blanchard and Fabrycky 1990) aims to maintain the Q-properties throughout the products life despite disturbances from use. The mindset model in Fig. 13.17 illustrates this reliability challenge. Here, the product's quality composition is described as layers of big-Q qualities supported by technical qualities, such as robustness and reliability. Further, product life qualities, q-qualities, occur in distinct life phases, e.g. production and maintenance. As such, these underpin the pyramid of big Q qualities.

The dispositional mechanisms behind the quality properties link them to the product's characteristics, behaviours, and state changes (Chap. 12). Mørup (1993) identifies three types of methods related to DFQ: goal formulation, synthesis (manipulating the characteristics to obtain good properties), and verification. A unique aspect of quality is that users are required for verification through, e.g. customer dialogues, workshops or field tests.

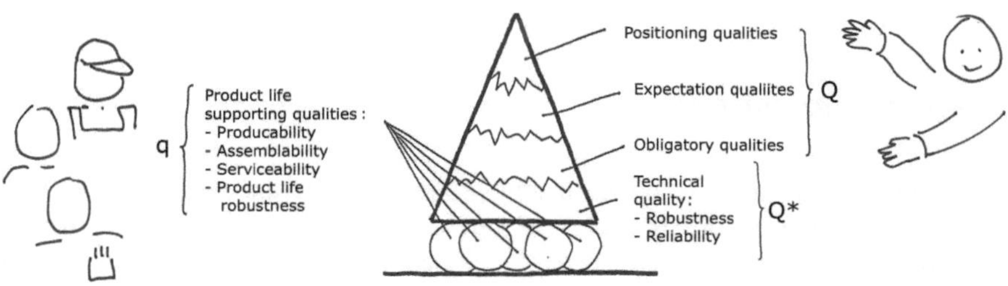

Fig. 13.17 Mindset model for the composed quality pattern (Andreasen and Hein 1998)

Example:
Design for Q and q. The Bang and Olufsen CD player illustrated below has two sliding doors that open when you approach the device to get access for CD loading and the control panel (see Fig. 13.18). In particular customers like the quick, soundless movement, while the producer knows that this Q property should be maintained for 17 years. The supporting technical properties 'Q#' include: low wear, robustness in the man/machine interface (for instance when one of this books authors 'just want to see what happens when you obstruct the doors with a pencil'), robust to environmental input (e.g. different cleaning sprays), and robust to changing parameters (e.g. when the plastic guides shrink over time). Little q efforts relate to easy alignment of the guides and doors, adjustment of individual products, and successfully debugging early generations of the software.

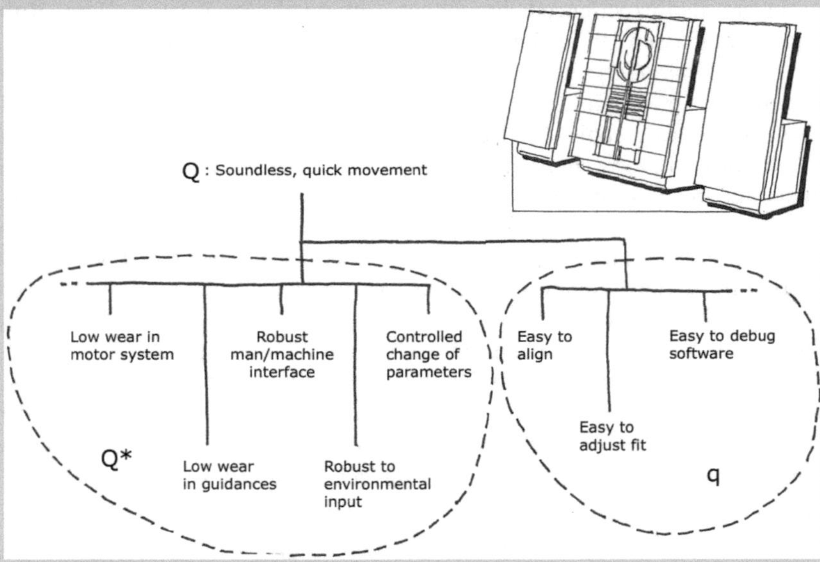

Fig. 13.18 Quality aspects in a Bang and Olufsen CD player

13.4.4 Design for Use

At its core design aims to create useful products, however, the use activity has its own particular nature in terms of interaction between human and device. Therefore, **Design for Use** has been developed with the aim to create products that are problem free, flexible, socially acceptable, practically acceptable, and ultimately useful for the user.

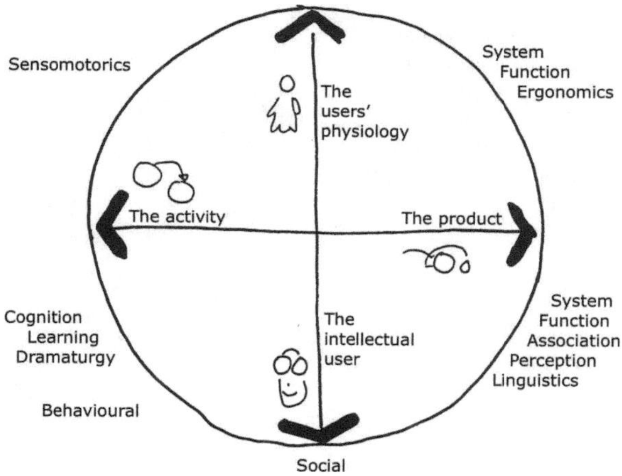

Fig. 13.19 Four dimensions of interaction

The dispositional mechanisms in DFU link the product characteristics to the intended/executed use activity and the user's characteristics. For example, Hede Markussen (1995) used interaction design to identify the key operational characteristics or design degrees of freedom that determined the quality of a product's use. Hede Markussen aimed to balance engineering understanding (product's causal effects on the use activity) and experience-based understanding (the user's experience of and reaction to the design and its use). To do this he described the four dimensions shown in Fig. 13.19.

Interaction is a living, dynamic process. It is invisible and complex, and it claims the use of a special language in the design and articulation of the user interface.

Examples of advice from interaction design not surprisingly focus on behaviour: 'Give the user control over the product', 'It must be easy to give up an operation procedure and escape the action', 'Make shortcuts for experienced users', 'All interaction operations should give an information response'.

Example:
Manual control of a bed. User involvement in the design of a controllable bed resulted in the interface shown in Fig. 13.20. Here the main challenge was to link the user's own position and intentions to the bed's response when a button was pressed. In order to achieve this, the bed's four motors were required to activate safely, in unison, and without disturbing the user.

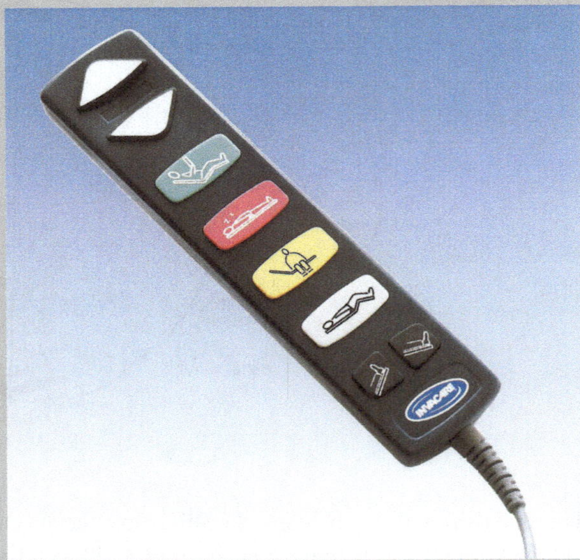

Fig. 13.20 Visual instructions for use scenarios controlling the bed's four motors, courtesy Institute for Product Development, DTU

Users' personality, motivation, intellect, experience, and cultural background all influence the success of product interactions. Capturing the core identity of potential users and effectively mapping user groups that need special attention is thus central to DFU. Closely related to this is **Inclusive Design** (Keates and Clarkson 2003). Here design interaction efforts specifically aim to include those users who are excluded for whatever reason, e.g. too young/old, less able or having functional impairments. The effects of such a focus can lead to greater social inclusion, higher user satisfaction, and growth in sales volume.

13.4.5 Design for Environment

The question of **sustainability,** i.e. maintaining the Earth in a functioning condition for future generations, poses two significant questions to industry. First, industry is one of the main sources of pollution and users of natural resources. Second, the flood of new products produced by industry leads users to consume even more resources and pollute in the form of product waste. Efforts to combat this can be found on many levels, from international initiatives down to policies that effect individual designers. This has resulted in a huge number of environmental methods and buzzwords. Bringing some order to this profusion a distinction can be made between (Dusch et al. 2010):

- **Green Design**: individual product features are optimized for environmental performance.

- **Eco-Design** (DFEn): the entire product lifecycle is optimized for environmental performance (McAloone and Bey 2009).
- **Sustainable Product Design**: social aspects are also considered.
- **Design for Sustainability**: complex societal systems are also considered.
- **Transformative Design**: totally new scenarios are envisioned.

One of the main approaches in this area is to quantify the 'environmental goodness' of a product by creating an ideal reference model. This uses Life Cycle Analysis (LCA) to model the resources used, the emissions caused, and the end of life deposition related to the product's composition. These quantitative models are complex—meaning that it is difficult to assess the veracity of their predictions in the real product context. In retrospect it is not difficult to create an LCA and to map the use of resources and harmful effects on the surroundings, i.e. nature and humans. However, from the designer's perspective it is difficult to integrate these complexities when also pressured by the more tangible aspects of product design, e.g. DFMA. As such, it is beneficial to design based on real situations or scenarios that allow for more direct reasoning.

The dispositional mechanisms in Design for Environment link the compositional elements of the product to the activities in its life phases, including its use (Olesen et al. 1996). Environmental effects are created in each of the product's life phases. These effects can thus be identified by confronting the product and its elements with the life phase activities; what we call **meetings**. Figure 13.21 shows an example of such a mapping, linking a product's composition to its life phases to identify key meetings and their environmental effects. This leads to the following heuristic, which is illustrated in Fig. 13.22.

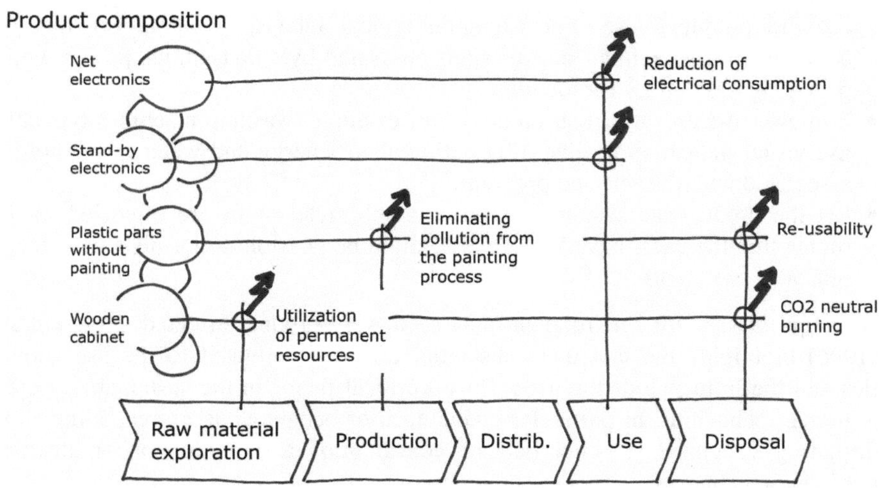

Fig. 13.21 Environmental effects stemming from interactions between the product, actor and a product life system, after (Olesen et al. 1996)

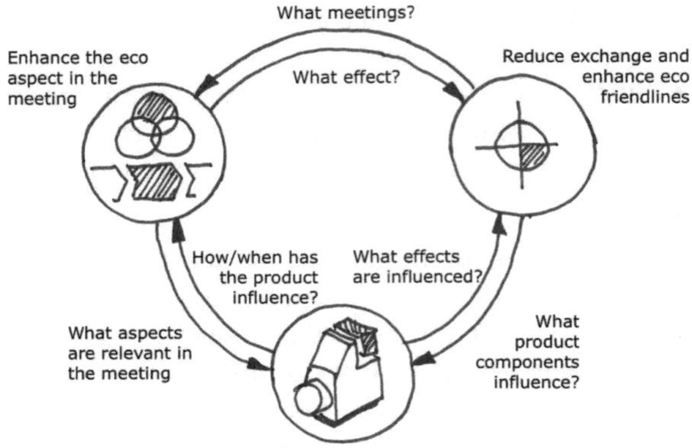

Fig. 13.22 Mindset model for Design for Environment (Olesen et al. 1996)

Environmental effects, the product, and the meeting should all be confronted; these are the key dimensions to be manipulated.

The questions that drive the search for alternatives are shown in the illustration. These are based on the idea that a design result is available, at least as a concept, or that past products and their life phases can be used for finding viable design directions. This leads to the identification of several basic design principles that link the product characteristics to the meetings, i.e. the interaction of product, operator, and lifecycle system in a life activity. Such principles point to positive or negative environmental effects. Some examples are highlighted below.

- Reduce energy consumption by avoiding functions that require energy throughout the product's use. For example, typical televisions contain numerous components that remain active when on 'stand-by', consuming power while providing little utility for the user.
- Empower the user to act sustainably. For example, washing machines typically use visual indicators to give information about energy and water consumption associated with the selected program.
- Let the product act sustainably. For example, televisions are often left on in rooms that the users have left. Thus it might be possible to install a movement sensor to switch off the TV.

Although Design for Environment may seem very product oriented Olesen et al. (1996) highlights the fact that substantial effects are related to the use activities and the influence of the user. Thus a critical factor in the sustainability area is human behaviour. In particular changing user behaviour is a precondition for designing systems that better balance consumption and production or creating totally new scenarios.

13.5 Dispositional Reasoning During Synthesis

Although DFX methods are both corrective and generally rule or principle based it is still important to consider how best to integrate them in product synthesis. For example, the general roles of DFX methodology are in making the product 'fit for life', enhancing competitive aspects, and supporting the 'idea with' the product. However, there are many more roles to be respected in product development. Thus, we take a closer look at dispositional reasoning in the three design situations: designing a new product, incremental design, and platform based design.

13.5.1 Designing a New Product

When a company decides to create a new product the focus is on innovation in the design. In this situation "*The designer and the team may see the project as aiming at an innovative product and may be too focused upon the product, forgetting its use, users and product life stakeholders*" (Dorst 2006). The fact that a product exists based on its interaction with the context is easily missed in such a situation. To repeat Chap. 10: *the life cycle should be designed before the product*!, i.e. the designer should not blindly create a new product without knowing its effects, especially its environmental effects.

> Efforts are needed to bring product life realities into designing a new product.

These efforts include a greater focus on finding product life information in the research phase, describing the actual product types, their users and lifecycle. One strategy is to launch the product as early as possible, in order to learn from real-world experience. Another is to arrange design reviews combined with DFX loops to give the product the best possible start conditions.

In new product design DFX can be mirrored as XFD: efforts in the X area support the product design. In an earlier example we described the selection of aluminium as a strategic material, based on feedback from the production (Sect. 13.3). Other examples include the development of service efforts supporting the product, radical cost efforts to make the product in 'its own class', creating scoping projects to find new products utilizing a unique production technology, or utilizing new reuse technologies. Just as designing can start from user's needs it can also begin with stakeholders' needs in a specific life phase.

13.5.2 Incremental Design

In incremental design the product's main function, its application, and a major part of the product are unchanged. Thus the new design aims to enhanced performance and add features or new forms. We are able to use insights from past products and their dispositional links as starting points for enhancement or innovation in the new product's fit for life. Here, the Score Model (Fig. 13.12) can be used as a framework for describing concrete insights, supported by stakeholders' perceptions of fit and criteria for goodness.

Another practical approach is to introduce a formal concept design report, where all relevant life phases are treated and proposals made for re-design—allowing clarification of possible new designs. Unfortunately current industrial practice has a weak definition of what requires innovation and what should be reused, resulting in extremely wasteful redesign efforts where new components, new processes, and new spare parts are created with no value for the business.

Example:
Scania heavy vehicles: The Swedish company Scania follows an interesting incremental design strategy. Here, the basic idea is not to launch distinct lorry models but to create continuous enhancements, and offer the customers highly individualized products. The preconditions for this are dynamic and precise reactions to the market, respect for key requirements, and tight control over quality and cost. The designer's priority in executing a customers order is safety, environment, quality, and cost. As such, the strategy demands high levels of dispositional insight to avoid decreasing volumes by, e.g. over production of new components.

13.5.3 Platform-Based Design

In platform based design products have many variants with high commonality, with the aim to satisfying individual customers' wishes whilst also reducing design and manufacture effort. This requires alignment across the whole product lifecycle. Alignment is established when the different structures in the product life systems are coordinated and fitted for a total optimization, i.e. positive dispositional effects are successfully utilized and managed. This is illustrated in Fig. 13.23, which shows the alignment challenges for a modular product family. Alignment requires the optimization of commonality, as well as controlled variation to manage the down stream effects on the product life phases. Further, this also builds on optimal fit between the products, their supply, manufacture, quality control, distribution, etc.

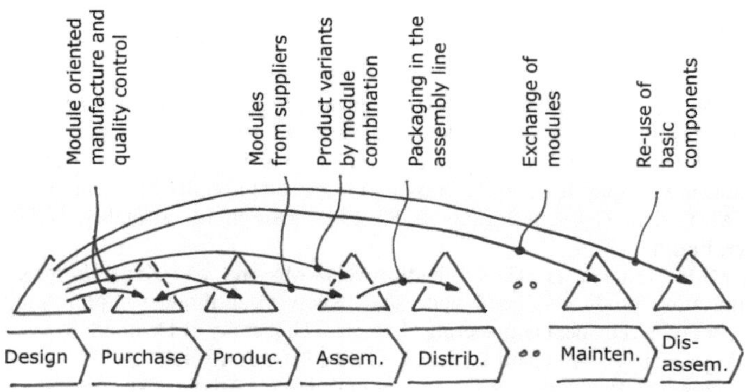

Fig. 13.23 Advantageous alignment between the structures of product life systems

Several types of DFX can be used in alignment. For example, Design for Assembly has rules regarding modularization and improving manufacture, supply, and distribution; and Design for Manufacture has rules for creating commonality between variants. As such, platform based design generally demands a modular focus.

13.6 Conclusion

Dispositions provide an underlying pattern in design supporting the identification of flaws and weaknesses but also innovative possibilities related to the product life and the complex systems in which the product exists. Modularization brings these dispositions to the surface by seeking to answer: who should benefit from the modular structure? To answer this we must understand the dependencies and possibilities linking the modules in order to create the best conditions, for e.g., quality control or recycling.

Dispositions point to important questions about the organization of tasks and responsibilities in a company. Should the design manager be measured on their design for manufacture effort? How should design for environment be weighed against cost reduction or production deadlines? Answering these with reasonably valid statements demands experience from previous designs, competitor's products, and best practice because not every issue can be considered for each new project.

The three topics function-, property-, and dispositional reasoning together deal with the difficult matter of creating good products. They collectively highlight how, if we wish to succeed, we must understand the mechanisms underpinning design, their interaction, and how they can be arranged. This leaves one final question: *What is a good product?*

References

Andreasen MM (1988) Udviklingsfunktionens disponeringer (The development function's dispositions). Institute of Product Development, Technical University of Denmark, Copenhagen

Andreasen MM, Hein L (1998) Quality-oriented efforts in IPD—a framework. Integrated product development workshop IPD, vol 1. Otto-von-Guericke University Magdeburg

Andreasen MM, Kähler S, Lund T, Swift K (1988) Design for assembly. IFS (Publishing)/ Springer, Berlin

Andreasen MM, Støren S (1993) Produktfornyelse—baseret på "Kend dit produkt"-filosofien (Innovating products—based upon the "Know your product"—philosophy). Lyngby, Department of Mechanical engineering, Technical University of Denmark

Andreasen MM, Hein L, Kirkegård L, Sant K (1989) Udviklingsfunktionen—basis for fornyelse (The development function—seed of innovation). Jernets Arbejdsgiverforening, København

Andreasen MM, Mortensen NH (1997) Basic thinking patterns and working methods for multiple DFX. In: Meerkamm H (ed) Report on Beiträge zum 7. Symposium "Fertigungsgerechtes Konstruieren", Universität Erlangen-Nürnberg

Andreasen MM, Hein L (1987) Integrated product development. IFS (Publications)/Springer, Berlin. Facsimile edition (2000). Institute of Product Development, Technical University of Denmark, Copenhagen

Blanchard BS, Fabrycky WJ (1990) Systems engineering and analysis, vol 4. Prentice Hall, Englewood Cliffs

Boothroyd G, Dewhurst P, Knight W (2002) Product design for manufacture and assembly, 2nd edn. Marcel Dekker, New York

Crowe D, Feinberg A (eds) (2014) Design for reliability. CRC Press, Boca Raton

Dorst CH (2006) Design problems and design paradoxes. Des Issues 22(3):4–17

Dusch B, Crilly N, Moultric J (2010) Developing a framework for mapping sustainable design activities. In: Design research society conference 2010, Montreal Canada (unpublished)

Ehrlenspiel K, Kiewert A, Lindemann U (2013) Kostengünstig Entwickeln und Konstruieren, (Cost beneficial development and engineering design), 3rd edn. Springer, Berlin

Fabricius F (1994) Design for manufacture (DFM). Institute for product Development, Technical University of Denmark, Copenhagen

Hede Markussen T (1995) A theoretical basis for creating interaction design (in Danish). Ph.D thesis, Technical University of Denmark

Hubka V, Eder WE (1988) Theory of technical systems. Springer, Berlin

Keates S, Clarkson J (2003) Countering design exclusiveness—an introduction to inclusive design. Springer, London

Matousek R (1957) Konstruktionslehre des allgemeinen Maschinenbaues (Study of design in general mechanical engineering). Springer, Berlin

Matthiassen B (1997) Design for robustness and reliability. Ph.D dissertation, Technical University of Denmark

McAloone TC, Bey N (2009) Environmental improvement through product development—a guide. Danish Environmental Protection Agency. ISBN 978-87-7052-950-1

Mørup M (1993) Design for quality. Ph.D thesis, Technical University of Denmark

Olesen J (1992) Concurrent development in manufacturing—based on dispositional mechanisms. Ph.D dissertation, Technical University of Denmark

Olesen J, Wenzel J, Hein L, Andreasen MM (1996) Miljørigtig konstruktion (Design for environment). Technical University of Denmark, Institute for Product Development

Pahl G, Beitz W (2007) Engineering design. A systematic approach, 3rd edn. Springer, London

Pugh S (1991) Total design—integrated methods for successful product engineering. Addison Wesley, Wokingham

Chapter 14
Good Design

This chapter closes this book by bringing together our exploration of conceptualization. This needs one final element to meet our ultimate goal of empowering the creation of influential, sustainable, and useful products. We must answer: what is a good product; and what composes good practice, research, and education? This last chapter knits together all the threads into a comprehensive worldview of design that supports the readers' activities as practitioners, researchers, and students.

14.1 Effects of Design

The Industrial Design Engineering (IDE) department at Delft University of Technology uses the mission statement "*Creating successful products people love to use*". They highlight products' role in both satisfying needs and creating successful business, all in a sustainable way. We started this book by setting the scene for conceptualization before exploring the product development process and lifecycle. Seen in this broad perspective, *what is a good design*?

As we noted in Chap. 1 good design practice also fundamentally links to research and education. We therefore see 'good design' (at the highest level) as an interaction between four elements:

1. **Design results** as satisfiers of human and societal needs, and as statements of what matters in society and how problems are solved.
2. **Design practice** as a driver for innovation and creator of new products.

M.M. Andreasen et al., *Conceptual Design*, DOI 10.1007/978-3-319-19839-2_14

3. **Design research** as a supplier of design knowledge for practice and education.
4. **Design education** as a supplier of innovators and skilled designers able to utilize new technologies and stage groundbreaking new companies and social initiatives.

In the following sections we examine each of these four dimensions of good design, before combining them in Sect. 14.4 to discuss **the designer's challenge**: the necessary, the important and the decisive. Finally, Sect. 14.5 discusses **goodness in practice, research, and education**.

14.2 Good Products: The *Idea with* and *Idea in*

A good product is created as the "**the strong concept**", i.e. it successfully connects market opportunities, new technology, a company's resources (development, production, distribution, and sales), and sales success. It provides an outstanding answer to the need, either through quality or by creating a new way of articulating the question.

Regardless of the market's demand for a product we should always strive for the "**good product**"; one with the best possible design, functionality, and properties. This can only be realized by staging conceptualization so that need, business, market, and value dimensions are balanced against the technical elements related to the physical product, shown in Fig. 14.1. On the one hand we deal with the product's utility. Here, conceptualization helps us create the **idea with the product,** i.e. new, important conditions for the product's use, utility, social importance, need satisfaction, and marketing. On the other hand we deal with the creation of the product. Here conceptualization helps us create the **idea in the product,** i.e. the product's mode of action, structure, and form in accordance with the desired functions. As such, conceptualization leads to the design of concepts with strong ideas both *with* and *in* the product.

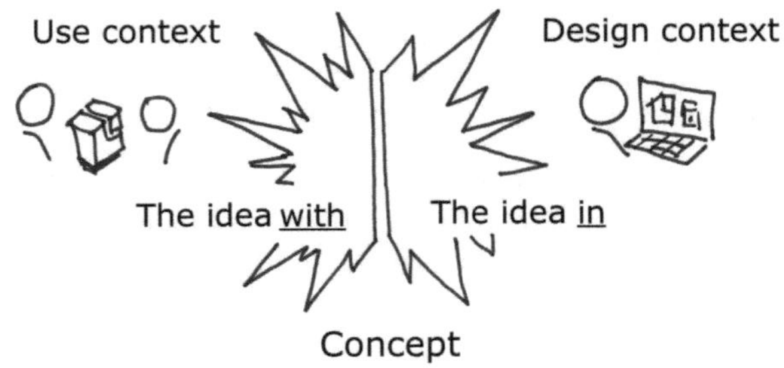

Fig. 14.1 Two sides of a product concept (Hansen and Andreasen 2002)

Example:
The Danish train system. The Danish intercity rail network is characterized by a large number of lines with short distances between stations. Trains are expected to cover longer distances as well as smaller lines. The established solution shown in Fig. 14.2 is the IC3, which has a lightweight aluminium chassis and standard diesel engines on each bogie. This gives, together with automatic couplings, a modular train. The *idea with* the trains is to serve the traffic need, i.e. they are flexible, have short travel times between stations, and support frequent service. The *idea in* the trains is to support this flexibility through low weight, high acceleration, modular design, and easy coupling.

Fig. 14.2 The IC3 train and the net of lines it covers

In our Encapsulation Design Model conceptualization captures the multiple issues identified during Exploration and gradually brings these into the design activity. Historically, conceptualization has started from the technical problem and goal formulations, where the ideal product and technical considerations are key. In reality new products' success is also dependant on launch strategy, market share, financial result, and technical performance (Hultink et al. 2000). However, we might still ask: what makes customers buy a product? The user appreciates utility and performance, particularly when buying technical 'rational products', e.g. washing machines.

For less 'rational' products users and buyers' value perception is derived from the product's utility, yield, user's excitement, visual appearance, brand, reputation, and technical features. The buyer often focuses on secondary features in contrast to the concentration of the design effort. In particular, price can outweigh a buyer's quality evaluation. Often, price perception is governed by brand and a naïve belief in a relationship between price and quality. Thus we must ask: how to bring the *idea with* into design? Chapter 6's five feed chains give us a basis for interpreting need and balancing task and idea elements against insights from the use situation. This gives the *idea with* the design.

The mindset 'idea with/idea in' allows us to fundamentally question the product's 'right to exist', no discussions of goodness can exist without it.

A concept proposal should always contain, and argue for, both the 'idea in' and the 'idea with'. Here the 'idea in' deals with the technical aspects and the 'idea with' utility and market.

14.3 Stakeholders for Goodness

The good product is one with positive effects in terms of sales success, high market share, good fit with the company's functions (e.g. manufacture, distribution), and setting a new innovation paradigm (e.g. the transition from normal to 'smart' phones). Obviously this presents us with a multidimensional problem; the designer faces multiple actors, must balance multiple wishes for properties and functions, and must deal with multiple design entities. Further, all this is underpinned by the product's lifecycle and the subsequent actor interactions. From this we can identify the four key stakeholder groups shown in Fig. 14.3, inspired by Tan and McAloone (2006).

1. **Users** whose application of the product leads to the need satisfaction and value perception. Successful products balance these effectively against their price.
2. **Manufacturers** who see the product in a business light and appraise its contribution to the company's profits. Successful products drive profits and effectively utilize company resources.
3. **Society** where the product (and company's activities) are observed as resource consumption and value creation. Successful products create value while respecting ethics and sustainability.
4. **Lifecycle actors** who interact with the product in a line of different roles, from suppliers to disposal. Successful products balance lifecycle needs with immediate demands for profit.

When searching for a unifying view on 'good design' these four groups are key. Thus we explore each in turn.

14.3.1 The Users' Perception

Products only truly show their utility and value when someone actually uses them. Thus it is the user who primarily determines if a product is a good one, but we

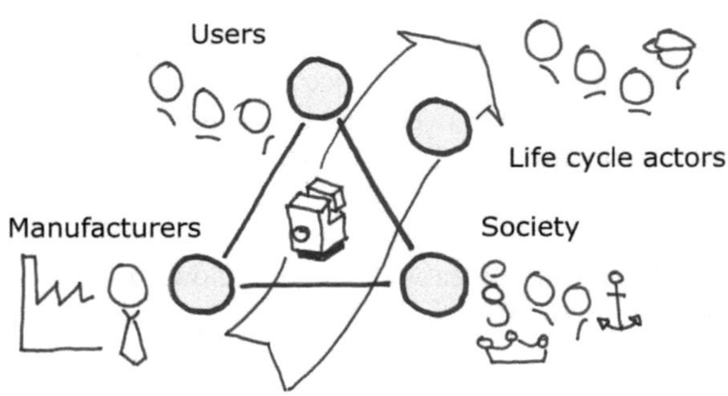

Fig. 14.3 Key stakeholder groups related to a product, its realization, and its lifecycle

should remember buyer and user are not always the same person. When users buy a product they bring it into a distinct context that is more or less related to the intentions of the designer and manufacturer. Each utilization is uniquely determined by product, use, time, and place. This individualization of the product is a problem for the designer at the conceptual stage: what will the product encounter in the hands of the user; and what will the user experience and value in this context?

A traditional, if slightly bizarre, example for illustrating this context dependency is the *Penny-farthing* (a type of bicycle with an extremely large front wheel), which was popular in the 1870s. It was mainly bought by rich young men and used for Sunday performances in parks observed by young ladies. It took high elegance to mount and part of the sport was to fall off gracefully. Thus, the manufacturers promised in their advertising that the bicycles could withstand a certain number of falls. In this context the Penny-farthing reminds us of *skateboards* today. Neither product is really intended as a means of transport in the traditional sense.

This brings us to the idea that a concept is not something inherent to a product, instead it is composed of a product's functionality, form, and performance, as well as the product's use, meaning, and utility in relation to existing solutions, tradition, reputation, value, etc.

A concept description should identify users and context as two sides of the same coin.

In the Link Model we find that the user's appreciation of a product is composed of the use result, the user's experience of owing and using the product, and the product's functions and properties. These are further supplemented by properties such as esteem as discussed in Chap. 12. The value concept describes the ideal product and use based on the user's situation and the context in which the product is used.

A product's **use** and (use related) **value** is determined by the user's experience. This value is determined by the product's properties, use process, output, and need satisfaction, but is also dependant on the user's situation and context.

Throughout this book we have used 'value' as a general denominator for goodness and perceived utility in a broad sense. It is well known that the human perception of value is a 'gestalt', i.e. it is perceived as a totality that cannot be explained by its composition. However, this is not very helpful for the designer. As such, we have discussed a number of value dimensions that can be addressed through design. Here, other terms like usability and applicability merge.

Usability describes how well a product supports the user in achieving their tasks or goals, quickly and easily. Nielsen (1993) states that: "*To some extent usability is a narrow concern compared to the larger issues of systems' acceptability, which basically is the question of whether the system is good enough to satisfy all the needs and requirements of the users and other potential stakeholders such as the users' clients and managers*".

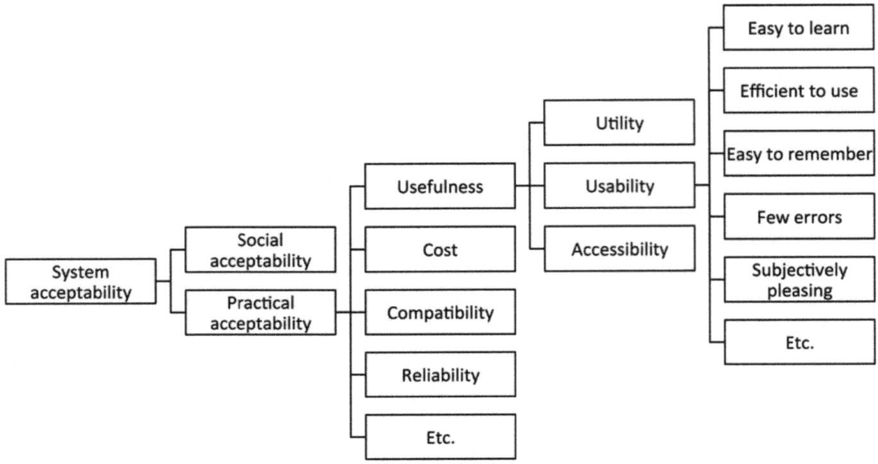

Fig. 14.4 Usability as part of system acceptability

Nielsen's breakdown of acceptability is shown in Fig. 14.4, to which we have added accessibility (Keates and Clarkson 2003). It is interesting to read the illustration from left to right and focus on barriers. For example, acceptability is judged with respect to the users' preferences, as well as cultural or social norms and economic or technical elements. Accessibility describes barriers to product use related to the user. For example, if users are excluded from using the product because its specification does not respect their abilities or deviation from 'normal': sight, hearing, mobility, strength, etc., or the chosen technologies exclude users unfamiliar with them.

The Link Model highlights the challenge of obtaining reliable insight into users' perception of goodness, and the Encapsulation Design Model highlights the importance of early need and problem exploration. However, true insight can only come from users' interactions with models, prototypes, and test products. This leads us to a summary checklist for exploring users' perception.

- The product should be **acceptable**. This should consider the total system including, suppliers of accessories, components or knowledge, services and upgrade, disposal, and economy. Further, there should be no social, cost, or technical barriers to accessibility.
- The product should be **useful** (and show it). This includes quality, an exciting use activity and result, good need satisfaction, and high usability, i.e. ease, efficiency, reliability, and robustness (quality).
- The product should bring **esteem** and **pleasure** to the user. This includes pride of ownership, pleasure in the appearance, social esteem, and the enhancement of a user's identity.
- The product should allow the user to show their **responsibility**. This includes ethical and ecological responsibility (all the way down to supplier), as well as supporting the user in ecological behaviour.

14.3.2 The Manufacturers' Perception

In the eyes of the manufacturer product goodness looks radically different from the user's perception; the user can choose between many products with only the price at stake. In contrast the manufacturer is reliant on their need interpretation being accurate and the product being able to carry the investment and risk of development, i.e. actually sell! Although we focus on the product as physical artefact the manufacturer sees the product and business as closely linked, as illustrated in Fig. 14.5. This highlights the socio-economic function and need satisfaction as the basic value interpretations of the user, and the business economy and company income as the core value interpretations of the manufacturer.

In the context of Fig. 14.5 it is worth distinguishing several product types: *commodities*, normally natural bulk products offered through market trade; *goods*, which are produced, standardized, tangible, and offered to multiple users; *services*, which are intangible, delivered on demand, and offered to specific clients; and *experiences*, which are staged, memorable and personal (Tan 2010). Our discussions of conceptualization are generally applicable but should always be adapted to the specific project context.

Satisfying user need in a new or better way is the core of all new product development (Andreasen and Hein 1987). Thus, the manufacturer's challenge is to be aware of and identify new interpretations of need, to find new means of satisfaction, and to convey these virtues to potential buyers. Typically value chains are established, where competing products are offered but each manufacturer utilizes many degrees of freedom to create its own approach to need satisfaction and sales. Examples include offering trucks that 'help reduce the lifecycle transportation costs', offering fasteners that 'reduce operational costs', and offering lubricants that 'increase machine performance and uptime' (Tan 2010).

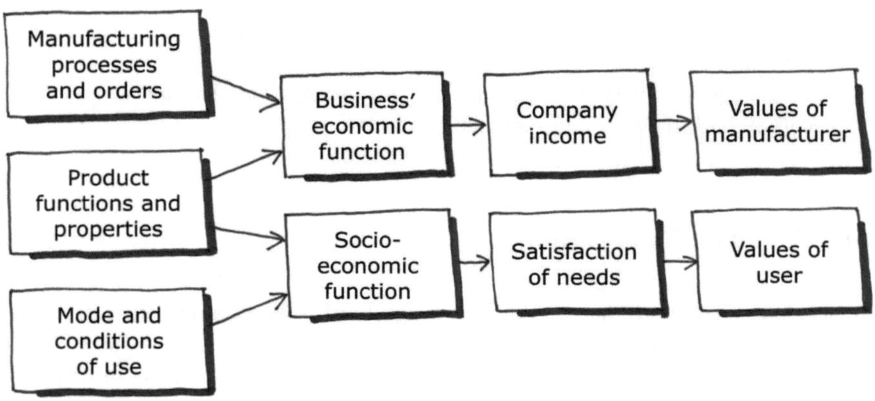

Fig. 14.5 The dual function of a product, after Roozenburg and Eekels (Roozenburg and Eekels 1996)

For the manufacturer need articulation, need satisfaction, value proposition, and holistic product performance are key to the 'good product'.

Our advice is to follow the Encapsulation Design Model and invest in Exploration. In particular, it is important to continue the clarification of user and market aspects throughout the design process. It is critical to understand what actually delivers value and competitive advantage. At stake are the resources invested in the project. It is not enough to simply record costs during the project; the designer must balance cost and income in real time. Figure 14.6 illustrates this with respect to a project's decisions on technology, investment, resources, and time, as well as the cost of designing itself. Here, dispositional effects are crucial to a sound economy (Chap. 13). Ultimately, investment should lead to profit mediated by market share, sales, and product price. The role of design is to identify and justify the need, articulate the design, clarify the situation, and build in added value directed at increasing price and sales.

In a company new products are a means for expanding business and market. Therefore, the 'good product' is often described in terms of both a product goal statement and a business goal statement (Andreasen and Hein 1987). These set goals for the ideal product and the ideal business. Again we summarize these factors in a checklist:

- The product should create **new markets** and increase **market share** via strong sales arguments, high added value for the customer, precise need satisfaction, and high, controlled quality.
- The product should support **alignment** between it, the business, and the company's efforts. This includes supporting company vision and strategy, manufacture, sales, quality, sustainability, etc., as well as maximizing the company's network relations.
- The product should minimize **development costs** and maximize **resource utilization**.
- The product should enhance the company **brand**, identity, and reputation with regard to ethics and sustainability, etc.

14.3.3 Society's Perception

Society and design are interwoven in the supply of food, communication, schools and hospitals, energy, and so on. New products can be introduced to this web with positive and/or negative effects. Thus, we can put society's interests on the same footing as citizens' interests, i.e. maintaining and developing welfare in a balance between creation of value and consumption. For the designer this societal web is not only where products end up but also where their responsibility and ethics will

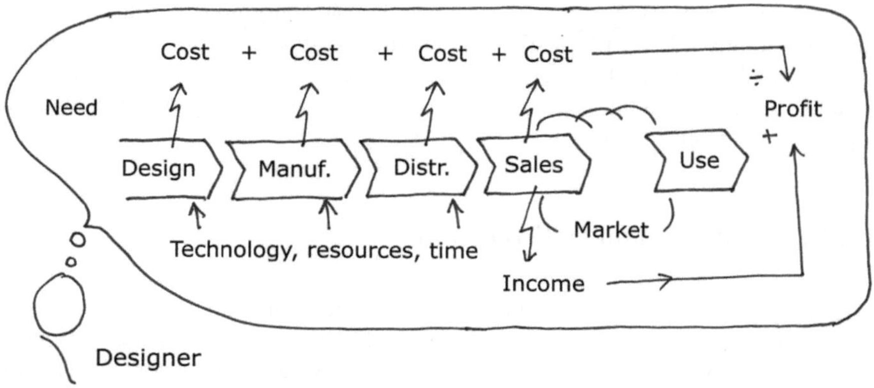

Fig. 14.6 A simplified model of project economy and the designers' responsibilities

be realized in terms of sustainable effects. It is also important to remember that the designer can take advantage of this web by creating concepts based on establishing new links between existing activities. In the societal context the 'good product' has several dimensions distinct from the user or manufacturer:

- Products and services **create value** in a society via trade and are thus vital economic elements.
- Products and services support societal structures and act as a means of **taxation**.
- Products and services lead to employment and thus support **individual welfare**.
- Society **frames** industrial activity economically, socially, and environmentally, representing the public's interests.
- Society sets the **game rules** by protecting intellectual property and at the same time giving designers responsibility for the safety and consequences of their products. In particular ethical, environmental, and societal standards must be respected.

Sustainability is influenced by the need and task formulation, as well as the nature of the concept and the final product's context. Our philosophy is that serious attention should be given to ecology and sustainability up-front in the goal formulation. This is discussed with respect to dispositional reasoning in Chap. 13, which highlights the need to consider environmental consequences.

When a new product is created a spider web of influences is established across environmental, economic, social, and cultural dimensions. Here, effects can be difficult to trace and assess. In particular, the users' and other actors' utilization of the product can create unexpected positive and negative effects. In societal terms it is the designer's responsibility to map this spider web.

Society sees a good product as one where environmental, economic, social, and cultural effects are understood, controllable, and sustainable.

Society's expectations concerning sustainability are enormously complex and it can seem impossible for the designer to see the possibilities and influences emerging from their product. However, this does not mean efforts should not be made to establish an understanding, for example through modelling the product lifecycle to capture sustainability issues.

14.3.4 Lifecycle Actor's Perception

In Chap. 10 we introduced the product's lifecycle as a design entity, i.e. something that can be designed or at least influenced by the designer. In each life phase 'meetings' happen between the product, life phase systems, and actors. The actors' roles can be, e.g. operator, supplier of accessories, or consultant. Here, the company is interested in the goodness of the activities performed in these meetings. This is linked to the actors' satisfaction, affects business and reputation, and can expose issues that the company is responsible for, e.g. concerning environmental effects or recovery of materials. In Chap. 13 we explored how products interact with the lifecycle and how designers' dispositions can understood and managed.

> The product should be **'fit for life'**. This includes, good task execution, i.e. the activities' properties and effects on lifecycle actors, and the activities' impact on the product.

14.4 The Designer's Challenge

These four types of actor collectively create an apparently unmanageably complex web of issues and properties linked to the 'good product'. However, we can make this mess manageable by asking: *what is important*? Our understanding is that the product's primary function is king: a knife must be able to cut; a car must be able to move. Products should be stable, safe to use, function every time, and support environmentally responsible deployment. But in the end it is in the **need** and the real **situation** where the few, primary properties are decisive, leading to product sales. We bring together these multiple issues and actors through a focused understanding of value, illustrated by the mindset in Fig. 14.7. This highlights three types of property: **necessary**, **important**, and **decisive**.

14.4.1 What Is Necessary?

Synthesis starts on the inside, i.e. to create something functioning, useful, and tractable. These properties (functionality, usefulness, and tractability) all belong to the necessary. They are critical and require gradual refinement throughout concept

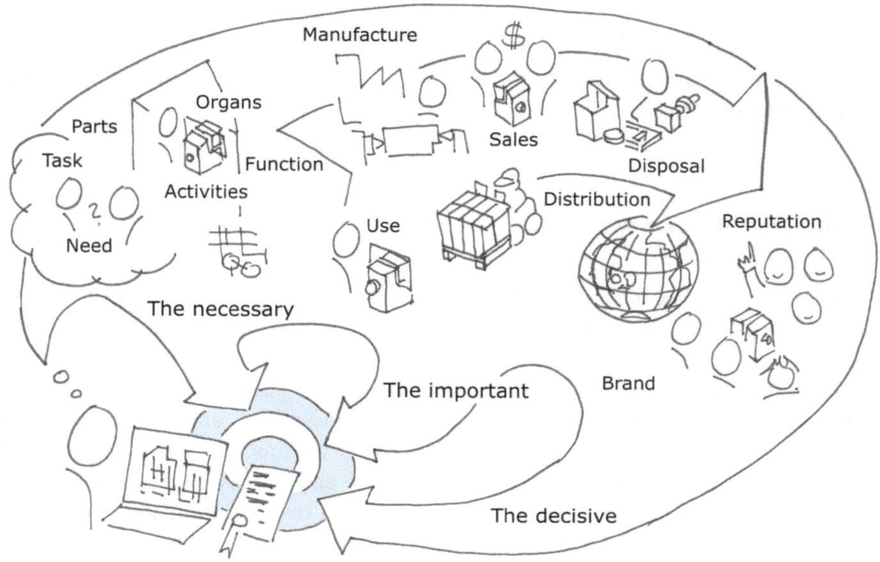

Fig. 14.7 The goal formulation exploring what is necessary, important, and decisive

synthesis to ensure their alignment with the goal formulation. If 'the necessary' cannot be addressed then solutions are rejected or projects abandoned. This links to Pugh's (1991) warning to avoid conceptual vulnerability, which "*usually manifests itself in two ways: Either the chosen concept is weak due to lack of thoroughness in the conceptual approach, or the chosen concept is strong, but due to lack of thoroughness in the conceptual approach the reasons for its strength are not known or understood*".

An important task in concept selection is thus choosing appropriate criteria to ensure a competitive product. In the Link Model we emphasize how the technical criteria must be confronted with the business and user criteria to predict the competitive power of a concept. However, an explicit statement of 'the idea with' the product is seldom asked for.

In order to identify necessary elements we must consider the different aspects of design reasoning, in particular, functionality: the product must actually function! The principles 'simplicity, clarity, and totality' proposed by Aguirre (Pahl and Beitz 2007) help keep synthesis on track, at all levels of detail. 'The necessary' is the foundation of the conceptualization.

The necessary are the core of a concept, the design's principal function. We should strive for simplicity, clarity, and totality, as the necessary are key to projects' tractability and results.

The necessary can also be found in a goal formulation's requirements, e.g. the desire to reach a new level of performance. If no solution is found with the required level of performance, this necessary requirement is not satisfied and the project will be stopped.

14.4.2 What Is Important?

Independent of what function a product offers it has no meaning if the function is not followed by "yield, reliability, and economy", which Wallace (2014) describes as a key set of properties. This means that the product should perform as promised, reliably, and economically throughout its service period. 'The important' are the properties that the product reveals in the hands of an 'unprepared user'. As such, they are at the root of success or failure, as well as personal safety (not be confused with reliable function). Lifecycle actors and society should also be considered with respect to 'the important'. Specifically, major negative effects should be avoided at a personal level (noise, pollution, waste, etc.) and at a societal level in the form of ethics and sustainability (unethical production conditions, neglecting environmental and resource restrictions, etc.).

The key question when considering 'the important' is: does it really give the user utility, in a sustainable way?

> **The important** satisfies the need as promised, reliably, safely, and economically.

It is interesting to observe that users experience 'the important' during deployment and thus these factors are not necessarily decisive arguments when purchasing a product.

14.4.3 What Is Decisive?

The credo of this book is that a concept should create 'a difference that matters'. The idea being that the designer should be conscious of the properties that set their product apart from the mainstream and above the company's past products. This difference should be key to the concept proposal; it will not be created in the design work following conceptualization. Further, the difference can be anywhere in the product, its use, or its related entities. A major problem here is communicating the concept's virtues to the customer because these virtues can be radically new ways of seeing a need, use or utility.

Although some textbooks describe decisive as 'functionality, safety, aesthetic, ergonomic, and timeliness', It is evident that these are a mix of necessities (functionality, safety), means (aesthetic, ergonomic), and what might be decisive, e.g. timeliness. In our view 'the decisive' belong to the use activity: what does the product do for the user, and how do they see the utility of this when buying? This dimension is emphasized in the Link Model and our articulation of the universal virtues. However, it also relates to the product's timeliness with respect to trends, seasons, emerging needs, etc. Therefore, our advice is limited to the following:

The decisive lie in the product's sales, use, and reputation dimensions. It is important to build the decisive into the design work so that potential buyers can identify it at point of purchase.

Bringing these three elements together we might ask in what order these properties should be considered. The simple answer is *necessity* comes first (something must be synthesized before we can judge its properties), second *important* (a concept which is not robust and tractable should be rejected at an early stage), and third *decisive*. However, decisive elements should be identified early and be incorporated during all stages of design. Unfortunately design is not so simple, and in order to clarify these properties we need to progress iteratively because they are not all obvious from the start. This demands prototyping, experiments, and user involvement.

This closes our discussion of 'good design' with respect to the product. Next we turn our attention to 'good design' with respect to the process.

14.5 Goodness in Practice, Research, and Education

Design practice is central to the creation of powerful products and also the subject of design research and education. Although we focus on good design there are many other factors that affect success, e.g. management, which each have their respective research fields and professional practices. In design there is an interaction between practice, research, and education as shown in Fig. 14.8. The figure also highlights the ideal nature of these interactions.

With respect to this interaction we are interested in how these three elements can reach a high level of yield and support societal welfare through good design.

14.5.1 What Is Good Practice?

The most pragmatic answer is that only the results count, i.e. success on the market. Designers are not measured on their use of procedures or methods, but on their productivity. However, this effectiveness dimension is closely related to the

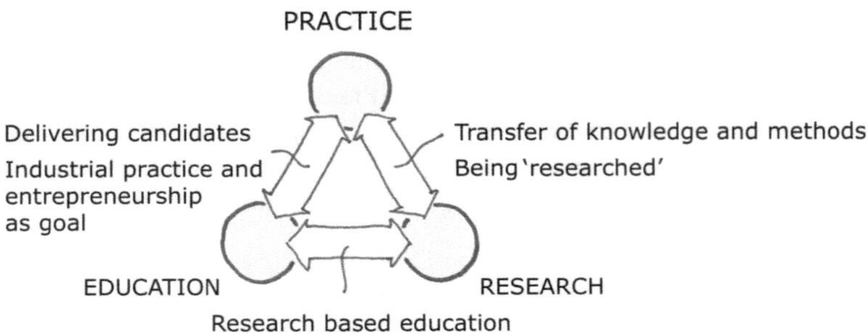

Fig. 14.8 The interaction between practice, research, and education

efficiency dimension, i.e. the designer optimizes their use of company resources by identifying opportunities to improve the business while controlling potential. "*We need to be perfect in all aspects*" said a manager, referring to quality assurance, distribution efficiency, environmental certificates, ergonomics, supply, legislation, and many other issues. In this domain dispositional thinking is essential.

In terms of good design in a team, three dimensions are important: professionalism, skills, and knowledge. A designer must be able to adapt to a wide range of tasks, from explorative and creative to analytical, corrective, optimization. No one can master all of these but a professional can adapt and perform as needed (using, e.g. this book for support). There are many buzzwords for describing best practice in companies', e.g. agility, voice of the customer, set based or lean. However, these all come back to different perspectives on efficiency and design professionalism as discussed in this book.

Industrial practice is rapidly evolving and highly complex, although some fundamental characteristics remain constant. This makes the ability to adapt and build on these fundamentals key attributes of effective design practice—hence why we dedicate so much of this book to design reasoning. This is particularly important because practice often leads research, meaning designers often encounter situations where there is little formal guidance. Wallace (2014) noted, "*Products these days are much better... much cheaper ... Many companies and design teams are doing an excellent job—even if they are not using the 'specific new methods' proposed by design researchers*".

14.5.1.1 What Is Good Design Research?

As noted above many design situations are encountered for the first time in practice, leading to the rather unscientific origin of many models, methods, tools, and principles. Design research seeks to understand the core of these insights and subsequently synthesize fundamental design knowledge to give back to the designer. Blessing (2002) defines design research as:

Definition: '***Engineering design research*** *involves: the formulation and validation of models and theories of the design phenomenon; and the development and validation of knowledge, methods, and tools (based on these models and theories) to improve the design process (i.e. support industry producing successful products)*'.

This definition reflects our aim in this book and highlights the questions fundamental to design research: what is a successful product; how can it be created; and how can we improve the chances of being successful? (Blessing 2002). We have addressed these throughout this book and bring their answers to the fore in this chapter. Throughout, we have built on the fundamentals of design as defined by design research. Thus, this text provides the reader with a foundation for further reading. In particular we recommend the works of Roozenburg and Eekels (1996), Dym and Little (2000), and Andreasen and Hein (1987) to the interested reader.

In Chap. 3 we discussed designers and their knowledge, and pointed to three types of knowledge: design, professional, and branch. Here, design knowledge is the type of insight found throughout this book. This is fundamental and applies to all aspects of design, independent of discipline. In particular, our Encapsulation Design Model, the Link Model, and the three modes of design reasoning fall into this category. When we consider good design research we have three thoughts in mind: **radical**, **relevant**, and **rigorous**. *Radical* denotes importance, i.e. contributing to one aspect of design knowledge. *Relevant* denotes how the research links to industrial practice, problems, and situations, in a timely way. *Rigour* denotes the quality of the research and the researcher's line of reasoning. Here we might focus on clarity of research question, proper use of concepts and theories, and proper handling of methods, data, and argumentation, Blessing and Andreasen (2005). Finger and Dixon (1989) state that: "*An ideal research institution should be based upon a solid basis and should master 'best practice'*". However, design is complex and the designer cannot hope to simply pick up research and apply it without adapting its lessons to the context of their own project—just as with design methods and staging.

Habermas (1998) discusses how research supports practice by providing underlying truths that are explainable via subjective decision making, e.g. concerning industrial experiences and attitudes. As such, design research results support design by providing insight, models, and methods. In practice these contribute to higher productivity and increased probability of success. Wallace (2014) sums this up as "*Any design theory is worth precisely little until it has been applied and validated in practice*". This means that the research taken up by industry is sometimes radically different from what is praised by other researchers. In particular we see that clarity, impact, and ready applicability are winning criteria for industrial adoption, with rigour not necessarily considered. This leads to the situation where industry uses an eclectic mix of research results (Araujo 2001; Grabowsky and Geiger 1997). In particular successful research (from an industrial perspective) must be applicable for new products, with relevant goals, effective use of resources, appropriate results for practice, and contribution to design knowledge and innovation (Andreasen and Wallace 2011). This is illustrated in Fig. 14.9, showing the link between research and practice, as well as their different goals.

Based on this we can see that there is a transfer problem, the missing link between researcher and designer. Although, when researchers and designers work together directly there is often a very profitable synergy—possible because this becomes a joint learning experience.

- Researchers should demonstrate the real world utility and benefits of their results if they wish designers to adopt them.
- Designers can gain substantially from involving researchers as specialist contributors in their design staging.
- There is much to be gained from an up to date understanding of design research findings and methods, although the rules of staging still apply: they must be understood and adapted to the specific situation faced by the designer.

14.5.2 What Is Good Design Education?

Design is taught in engineering schools, architecture schools, and industrial design schools (which focus on aesthetics, use, form, and materials, aimed at non-technical products and sometimes even non-industrial products). The background of the authors is in teaching Bachelor's and Master's degrees in design and innovation, producing design engineers. Their role is to be skilled designers who are also able to stage design, coordinate, and integrate (McAloone et al. 2006; Jørgensen and Boelskifte 2005).

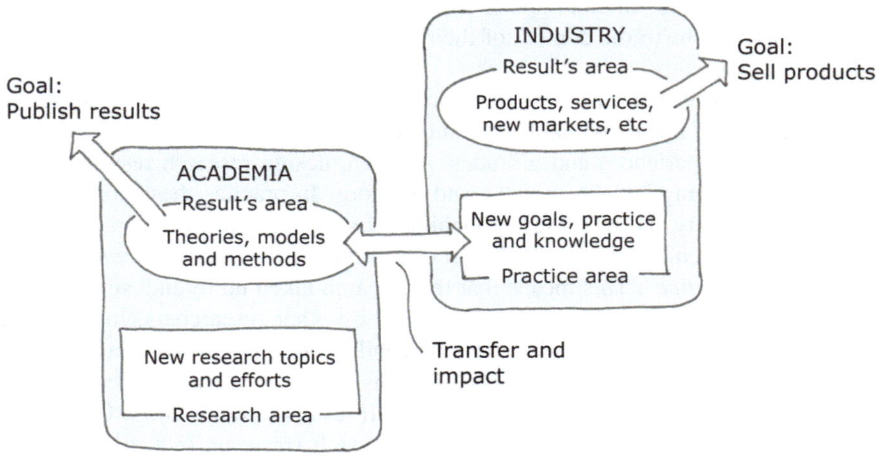

Fig. 14.9 The link between research and practice

Designers' knowledge and competences are realized in practice and moderated by their skills and attitude. Many knowledge areas contribute to understanding the nature of artefacts, context, and social environments, but not necessarily design knowledge. As such, the key characteristic of education is learning to bring these competences together effectively. In particular, awareness, synthesis, innovative thinking, and integrative reasoning. These must be trained as skills through project work tied to real design practice. In part, this balance between fundamental training and practical work helps designers to address the transfer problem we discussed in the 'good research' section. Further, this realization illustrates the importance of designers challenging their own learning by integrating design research in their practice. As such, we can summarize our thoughts on good education as follows.

- Design education should reflect design practice in bringing together both technical and social elements into the broad multidisciplinary endeavour described throughout this book.
- In order to improve in design it is not possible to just read or just practice. Design education builds on expanded knowledge, application in practice, and reflective learning. This is as true for designers in practice as it is for students in more formal education.

These heuristics apply to both formal design education and designers seeking to educate themselves. Design is fundamentally a discipline that is learned through reading, practice, reflection, and improvement—no one operates at 100 % capacity on day one! This comes back to the need to develop the three different aspects of design knowledge, in conjunction with softer skills and attitudes. A strong community of practice supports this, with other designers help facilitate the read, practice, reflect, and improve cycle, reflected in Fig. 14.10.

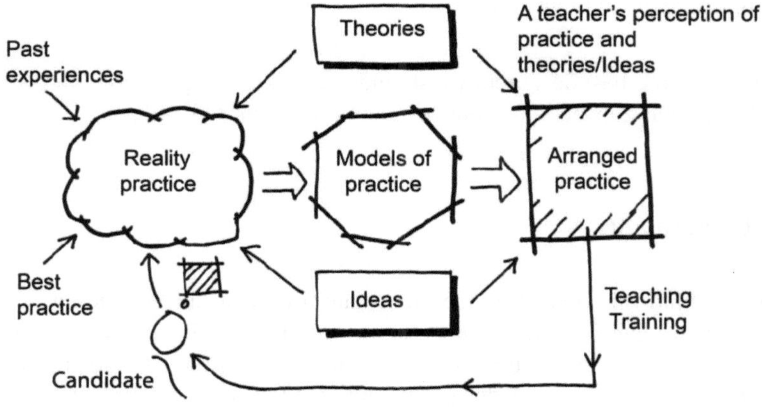

Fig. 14.10 Practice and its relation to theory and education (Andreasen 2009)

Key to building on this learning cycle is encountering different projects and situations. In particular we highlight a number of differentiating dimensions that should be considered.

- **Design focus**: which element of the Encapsulation Design Model is the learning focused on, e.g. exploration or product development. Further, what specific issues are to be addressed in the product domain, e.g. technical (material strength, mechatronics, etc.) or social (user perception, actor interaction, etc.).
- **Competences and skills focus**: reflecting the design focus, what skills are to be developed, e.g. analysis, synthesis, visualization, communication, etc.
- **Project context**: which elements define the current project and how can they be learned from, e.g. different managers, team composition, goals and requirements, etc.

This composition of project variety mirrors our Encapsulation Design Model and its stages, the treatment of issues and design entities, and the reasoning approaches.

14.5.2.1 Bringing Together Practice, Research, and Education

Based on our discussions above we see that design practice, research, and education are fundamentally related. This is reflected in Fig. 14.11 where these worlds collide. Here, design practice is both the source for potential research insight (through empirical study) and the receiver of this refined reflection (in the form of theories, models, and methods).

Based on this holistic perspective we strongly believe in the need for designers to understand the totality of design, envisaged in the Encapsulation Design Model, as well as the social and ethical considerations involved. Risk taking and a belief in one's own abilities is the foundation for daring to create new companies and products. Design is much more than a simple technical challenge, it is also a personal, intellectual, and innovative one.

> If we want the best designers we should recognize design as the intellectual challenge that it is, and promote designer's freedom, reflection, and learning.

14.6 Conclusion

A driver behind this book has been to create a holistic understanding of design from which we can empower conceptualization. In this chapter we have focused on the ultimate goal of design—the creation of good products. To this we have added the creation of good designers and good design practice. As such, we have not discussed methods at length, not only because they are well addressed

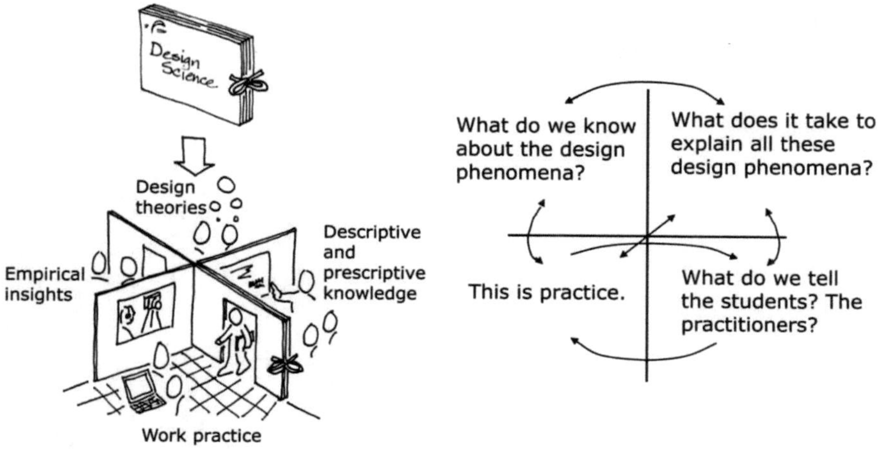

Fig. 14.11 Design science as four interrelated worlds (Andreasen 2008)

elsewhere, but because our focus is an understanding the underlying nature of design. With this understanding we stand ready to tackle design problems in many disciplines or situations, with many different goals or needs, under many different types of constraint, and (almost incidentally) are able to adapt and apply methods as needed for our situation. This is in contrast to the 'method and tool' focus where we might have the equipment but not the understanding or mindset required to adapt and use it. This is jokingly illustrated in Fig. 14.12, and in the common phrase 'all gear, no idea'.

In bringing this book to a close we look to the future.

Outlook

In Chap. 1 of this book we illustrated our stepping stone metaphor for design. In this we can reflect on the composed nature of design, related to artefacts, people, and process. Further, it helps us realize that there are many ways to understand one's own role and work as a designer. Many approaches and compositions can be created that will prove powerful for the next generation's culture and values.

The issues brought together in this chapter challenge our perception of the practitioner, researcher, and student as totally distinct. Rather, each draws on and reciprocally strengths the others. As complexity and project scope increase this relationship can only grow in importance. In particular, practitioners face an ever-growing challenge in maintaining an overview and planning where best to spend their limited resources. Thus, practitioners, researchers, and students share a common future of increasing design complexity. This brings us back the dimensions of change highlighted in Chap. 2: globally distributed design, increasing social responsibility, the need to address both the lifecycle and the product, the shift towards service, and the growing responsibility of the designer.

Fig. 14.12 A 'well prepared' designer with all the methods and tools he might need

Today we are preoccupied with innovation and creativity as the conjuring tricks that will solve all our problems, creating new industry, employment, and welfare. However, these are only possible when coupled with professional designers who understand design and conceptualization as the real drivers of new product development. Design efforts are the seeds of our future; our challenge is to create value, sustainability, and ultimately the future we want!

References

Andreasen MM, Wallace KM (2011) Reflections on design methodology research, keynote speech at International Conference on Engineering Design ICED 2011, Copenhagen. Unpublished

Andreasen MM (2009) Complexity of industrial practice and design research contributions—We need consolidation. In Meerkamm H (ed) Proceedings of 20th symposium design for X, Neukirchen. Erlangen, Lehrstuhl für Konstruktionstechnik

Andreasen MM (2008) Consolidation of design research: symptoms, diagnosis, cures, actions?. Unpublished presentation at DS board meeting, Eltville

Andreasen MM, Hein L (1987) Integrated product development, IFS (Publications)/Springer, Berlin. Facsimile edn. (2000) Institute of product development, Technical University of Denmark Copenhagen

Araujo CS (2001) Acquisition of product development tools in industry: a theoretical contribution. Ph.D. dissertation, Technical University of Denmark

Blessing L (2002) What is this thing called design research? In: Annals of 2002 International CIRP Design Seminar Hong Kong

Blessing L, Andreasen MM (2005) Teaching engineering design research. In: Clarkson J, Huhtala M (eds) Chapter 4 in engineering design, theory and practice. A symposium in honour of Ken Wallace. Engineering Design Centre, University of Cambridge, UK

Dym CL, Little P (2000) Engineering design—a project based introduction. Wiley, New York

Finger S, Dixon JR (1989) A review of research in mechanical engineering design: part 1: descriptive, prescriptive, and computer based models. part ll: representations, analysis, and design for life cycle. Res Eng Des 1: 51–67, 121–137

Grabowsky, H, Geiger K (eds) (1997) Neue Wege zur Produktentwicklung (New directions for product development). Raabe Berlin

Habermas J (1998) On the pragmatics of communication. Cooke M (ed) Blackwell Publishing, Cambridge

Hansen CT, Andreasen MM (2002) The content and nature of a design concept. In: Boelskifte P, Sigurjonsson JB (eds) Proceedings of NordDesign 2002 Trondheim, pp 101–110

Hultink EJ, Hart S, Robben HSJ, Griffin A (2000) Launch decisions and new product success: an empirical comparison of consumer and industrial products. J Prod Innov Manage 17(1):5–23

Jørgensen U, Boelskifte P (2005) Design and Innovation—developing a curriculum for future design engineers at the Technical University of Denmark. Engineering product design education conference. Napier University, Edinburgh

Keates S, Clarkson J (2003) Countering design exclusion—an introduction to inclusive design. Springer, London

McAloone T, Andreasen MM, Boelskifte P (2006) A scandinavian model of innovative product development. In: Krause F-L (ed) Procedings of 17th CIRP design conference 'The Future of Product Development'. Springer, Berlin, pp 169–278

Nielsen J (1993) Usability engineering. Academic Press Inc., Boston

Pahl G, Beitz W (2007) Engineering design. A systematic approach. 3rd edn. Springer, London (First edition 1977)

Pugh S (1991) Total design—integrated methods for successful product engineering. Addison Wesley, Wokingham

Roozenburg NFM, Eekels J (1996) Product design: fundamentals and methods. Wiley, Chichester

Tan AR (2010) Service-oriented product development strategies. Technical University of Denmark, Ph.D. thesis 2010

Tan AR, McAloone TC (2006) Characteristics of strategies in product/service-system development. International Design Conference DESIGN 2006 Dubrovnik, Croatia

Wallace KM (2014) Personal correspondence

Appendix

List of definitions

This list is composed of the definitions used in this book. The concepts related to the nature of products and activities are defined in relation to other definitions, indicated by bold text. The definitions' location in the book is also shown.

Action conditions are the arrangement of external **effects** and **interactions** between bodies, which create the physical conditions for utilizing a natural phenomenon to create **state** changes, and subsequently effects (Sect. 11.3).

Behaviour is the complex of **state changes** that occur in an activity or device based on natural phenomena (Sect. 11.3).

Characteristics are a class of structural attributes of products and activities determined by the synthesis of the design (Sect. 12.2).

Collaborative design is the process through which actors from different disciplines share their knowledge about the design process and the design itself. This creates shared understanding related to both process and artefact, helps integrate their knowledge, and helps them focus on bigger common objectives—the final product to be designed (Sect. 4.4).

Community of practice is a socio-technical pattern that evolves in a team and its space as a result of experience, cooperation, learning, knowledge creation, and sharing (Sect. 4.4).

Competences are the ability to actually realize knowledge as rational actions depending on the goal and the activity (Sect. 3.5).

Concept (1) is a design proposal that is detailed enough to justify if it is a good answer to the task and intention, and show a high probability of realisation and success (Sect. 2.6).

Concept (2) is a proposal for a product's composition and issues that is detailed enough to justify it as a good answer to the task and intention. Further, the task and intention are justified with respect to the conceptual need satisfaction and the knowledge required, i.e. the probability of successful realization, need satisfaction, and success in the widest sense (Sect. 2.6).

M.M. Andreasen et al., *Conceptual Design*, DOI 10.1007/978-3-319-19839-2

Concept Synthesis is the phenomenon of creating a kernel of insight and ideas in the form of concepts. This provides the answer to need and intention, and is a proposal of the probable tractability and success in its development, realization, sales, and use (Sect. 7.1).

Design for X is a set of product synthesis methods and guidelines that serve to enhance the product life activities by addressing key issues related to the product and its activities (Sect. 13.3).

Design practice is a **work pattern** based on the type of product to be designed, the company—its past design activities and future aspirations, and an understanding of which approach leads to the highest probability of success. (Sect. 3.2)

Design procedure is a design process model fitted to a specific context. It is used as the basis for a procedural plan when a design project is executed (Sect. 5.3).

Disposition (1) is a decision concerning a product's **design** that influences the type, content, and efficiency of activities in the product's life phase activities (Sect. 13.2).

Disposition (2) *is the part of a decision taken in one functional area that influences the type, content, efficiency or progress of activities within other functional areas* (Olesen 1992) (Sect. 10.3).

Effect is a state change in a mode of action or mode of use, which leads to interaction with other entities (Sect. 11.3).

Embodiment is the design activity that determines the complete **part structure** of a product. This is based on the organ structure and the satisfaction of manufacturing requirements. This activity results in a full justification of the product's functionality and properties in its final manufactured state (Sect. 11.5).

Engineering design research *involves: the formulation and validation of models and theories of the design phenomenon; and the development and validation of knowledge, methods, and tools (based on these models and theories) to improve the design process (i.e. support industry producing successful products),* (Blessing 2002) (Sect. 14.5).

Exploration is the upfront design activity that leads to the initiation and argumentation for a project. It is also the continuous process that supplies data, information, and knowledge (Sect. 6.1).

Functions are a product or activity's ability to do something actively or be used for something, i.e. deliver an **effect** (Sect. 11.2).

Interactions are the propagation of effects between and inside **organs**, explaining the **behaviour** of the organ structure and each organ's **mode of action** (Sect. 11.4).

Knowledge *is a competence notion that promotes rational action depending on goal(s) to be achieved for a particular activity* (Sim and Duffy 1998) (Sect. 3.5).

Method is a goal-oriented rationalization or simplification of engineering work in the form of a standardized work description (Sect. 3.4).

Mindset is the proper understanding of a method's use in accordance with the designer's reality (interpretation of task, situation, execution, validation, etc.), and the method's background and proper use (Sect. 3.4).

Mode of action and **mode of use** are phenomena where **effects** from the surroundings and interactions between the **action conditions** realise **natural phenomena** resulting in a desired **effect**. One or more effects trigger the activity or organ (Sect. 11.3).

Model is a human creation that carries attributes similar to the modelled phenomenon or object (Sect. 3.3).

Organ (1) is a functional unit of a product where the arrangement of **mode of action** is based on **effects** from other organs, **action conditions**, and **interactions**, by which it creates **functions** (Sect. 11.3).

Organ (2) is a system element of a product (when we see the product as a system from the function perspective). An organ is characterized by its function and mode of action, i.e. what it does and how it works (Sect. 8.6).

Part (1) is a material element of a product. The part materialises the bodies and their interactions and is characterized by its form, material, dimensions, and surface qualities (Sect. 11.5).

Part (2) is a system element of a product (when we see the product from the embodiment perspective). A part is characterized by its physical properties, e.g. form, material, dimensions, and surface qualities (Sect. 8.7).

Product development (**company perspective**) is the use of exploration, design, manufacture, and marketing/sales to launch new, product-based business utilizing the company's resources (Sect. 2.4).

Product development (**user perspective**) is the creation and launch of products with new or different functions and/or properties, which offer new or added value to the customer/user (Sect. 2.4).

Product development is a company's activity associated with creating new business based on developing and launching new products. The activity is initiated by need and market research, as well as ideation, and ends with production, distribution, and sales (Sect. 9.2).

Product (1) is a general denominator for materialised, executable artefacts, i.e. artefacts able to carry behaviours and realize functions and properties through a **use activity** (Sect. 11.3).

Product (2) is any kind of materialized and executable artefactm i.e. able to carry behaviour and properties in order to realize functions and be deployed in a **use activity** (Sect. 8.5).

Product life cycle is the totality of activities related to an individual product's life, from its establishment through its deployment to its disposal (Sect. 10.2).

Product service system is a marketable set of products and services able to jointly fulfil a user's need. One company or a network can provide PSS (Sect. 10.2).

Properties are a behavioural class of devices' and activities' attributes, by which they show their appearance in the widest sense and create their relation to the surroundings (Sect. 12.2).

Property model describes the insight into the realization of a certain **property** via certain design entities' **characteristics** (Sect. 12.4).

Requirement is a statement about a desirable **property** of a device or activity formulated as a value statement with an indicator and metric (Sect. 12.2).

Skills are learned abilities and the capacity to carry out a task (Sect. 3.5).

Staging is the act of establishing and fitting a team's space to a project, to best support the design activity (Sect. 4.2).

State is a description of an entity in terms of parameters (physical quantities), e.g. temperature, pressure, composition, phase, momentum etc. (Sect. 11.3).

System (1) is a model of an object (a real or conceived product or activity) based on a certain viewpoint, which defines the elements of the system and their relations. A system carries **structure** i.e. the elements and their relations (arrangement, architecture), and **behaviour**, i.e. the system's response to a stimulus depending on stimuli, structure, and state (Sect. 8.3).

System (2): From a functional perspective a product is a **system of organs**. As such, a product's **organ structure** is defined by the organs (its elements) and their interaction (its relations) (Sect. 8.6).

System (3): From an embodiment viewpoint a product is a **system of parts**. This **part structure** consists of parts, seen as elements, and their assembly interfacing, seen as relationships (Sect. 8.7).

Team *is a small group of people with complementary skills who are committed to a common purpose, performance, goals, and approach for which they hold themselves mutually accountable* (Katzenbach and Smith 1993) (Sect. 4.4).

Technology is a combination of material devices, procedural prescriptions, and intentions that are interwoven with humans' work and social activities, articulating and structuring humans' behaviour and life in society [after Jørgensen (2008)] (Sect. 2.3).

Use activity is an arrangement of mode of use. This brings **together natural phenomena** and state changes, **effects** from the product, humans, and active surroundings, and **action conditions** (Sect. 11.3).

View model is a model derived from the product model able to articulate a certain product property (Sect. 8.9).

MIX
Papier aus verantwortungsvollen Quellen
Paper from responsible sources
FSC® C105338

If you have any concerns about our products,
you can contact us on
ProductSafety@springernature.com

In case Publisher is established outside the EU,
the EU authorized representative is:
Springer Nature Customer Service Center GmbH
Europaplatz 3, 69115 Heidelberg, Germany

Printed by Libri Plureos GmbH
in Hamburg, Germany